R.R. Vondrak • J.W. Keller • C.T. Russell
Editors

Lunar Reconnaissance Orbiter Mission

Previously published in *Space Science Reviews* Volume 150, Issues 1–4, 2010

Springer

R.R. Vondrak
NASA Goddard Space Flight Center
Greenbelt, MD, USA

J.W. Keller
NASA Goddard Space Flight Center
Greenbelt, MD, USA

C.T. Russell
Institute of Geophysics & Planetary Physics
University of California Los Angeles
Los Angeles, CA, USA

ISBN 978-1-4419-6390-1
Springer New York Dordrecht Heidelberg London

Library of Congress Control Number: 2010929578

Cover illustration: Artist rendering of the Lunar Reconnaissance Orbiter in orbit about the Moon.
Courtesy of Chris Meaney/NASA.
Back cover illustration: Launch of the LRO and LCROSS spacecraft on board an Atlas V on June 18, 2009.
Courtesy of NASA.

Printed on acid-free paper

Springer is part of Springer Science+Business Media (www.springer.com)

Contents

Space Sci Rev (2010) 150: 1–2
DOI 10.1007/s11214-010-9640-4

Preface

R.R. Vondrak · J.W. Keller · C.T. Russell

Received: 17 February 2010 / Accepted: 17 February 2010 / Published online: 2 March 2010

One of the most important celestial bodies, the Moon, is literally at the Earth's doorstep. The Moon is important for what it can tell us about the formation and evolution of the solar system especially the inner solar system. It is important because it can serve as a veritable planetary laboratory enabling us to understand physical processes that take place on the Moon as well as on other similar solar system bodies. It is important because we may use the Moon as a platform for scientific instruments and a place where astronauts can learn to live and work on planetary surfaces. Yet it is insufficiently explored.

This collection of articles describes an attempt to redress this oversight, the Lunar Reconnaissance Orbiter mission, its flight system, instruments and scientific investigations. It begins with a foreword by Wendell Mendell that describes the events that have culminated in the current return to the Moon. The foreword is followed by an introduction to the mission and its objectives by R.R. Vondrak and colleagues. An important aspect of any space mission is the flight system that has to survive the rigors of space and provide the needed support for the investigations. The spacecraft and mission design are described by C. Tooley et al. The Laser Ranging system, a spacecraft subsystem for precision orbit determination, is described by M.T. Zuber and colleagues. These articles are followed by discussions of each of the investigations and their attendant hardware: the LRO Camera by M.S. Robinson and colleagues; the LRO Diviner IR Radiometer by D.A. Paige et al., the LAMP Lyman Alpha Spectrometer by G.R. Gladstone and colleagues, the LEND Neutron Detector by I. Mitrofanov et al., LOLA the Lunar Orbiter Laser Altimeter by D.E. Smith and colleagues; the CRaTER Cosmic Radiation detector by H. Spence and colleagues; and the LRO Miniature Radio Frequency Technology Demonstration by S. Nozette and colleagues.

R.R. Vondrak (✉) · J.W. Keller
Greenbelt, MD, USA
e-mail: richard.vondrak@nasa.gov

C.T. Russell
Los Angeles, CA, USA

The success of this volume is due to many people, the hardest working of whom are the authors who have had to distill a set of highly readable papers from the myriad documents and charts that such missions produce. The editors benefited from an excellent audience, the referees who carefully read each article and provided feedback to the authors that strengthened the volume even more. These referees included O. Aharnson, J.L. Bandfield, B. Bills, G. Born, G.T. Delory, W. Feldman, N.S. Hegde, B. Klecker, W. Mendell, R. Mewaldt, S. Mottola, S. Nakazawa, L. Paxton, R. Phillips, J.E. Pirozzoli, T. Prettyman, B.R. Sandel, R. Simpson, H. Weaver, M. Wieczorek.

Equally important has been the strong support this project received at Springer especially the extra effort expended by Fiona Routley, Randy Cruz and Harry Blom. At UCLA we were skillfully assisted by Marjorie Sowmendran, who acted as the interface between the authors, referees and publisher.

Space Sci Rev (2010) 150: 3–6
DOI 10.1007/s11214-010-9639-x

Foreword: An Exploration-Driven Renaissance in Lunar Science

Wendell Mendell

Received: 17 February 2010 / Accepted: 17 February 2010 / Published online: 2 March 2010

In the 3½ years between the first human lunar landing and the final launch, the Apollo engineering team enormously expanded the range and performance of surface systems and added orbital mapping capabilities. High-resolution cameras and a laser altimeter in the Service Module Bay photographed and profiled the lunar surface, including areas that might be future Apollo landing sites. A gamma ray spectrometer and an X-ray spectrometer provided data for construction of maps indicating the concentration of certain elements. In today's mode of doing business, a similar systems evolution would take a decade or more.

While the gains in surface mobility, astronaut EVA time, and precision landing were impressive and productive, the first geochemical (elemental) maps from the orbital spectrometers startled many planetary scientists. Those maps could be interpreted in terms of the rock types identified in the returned samples. In principle, scientists could begin to infer the geological evolution of a planet had the mapping been global.

To that end, leaders in the lunar science community approached NASA about the addition of a small, unmanned polar orbiter mission. When the Administrator suggested that Apollo 17 could be an orbit-only mission, the conversation ceased.

In 1973, the Manned Spacecraft Center (now the Johnson Space Center) submitted an unsolicited proposal to NASA Headquarters for an inexpensive lunar polar orbiter with only geochemical instruments, i.e., no cameras. The Office of Space Science was taken by surprise and eventually decided to compete the mission. Of the three NASA Centers that proposed, MSC had the most robust lunar science capability; but the evaluation board doubted that a human spaceflight center would be capable of cost containment on a small spacecraft project. The mission was first assigned to the Goddard Space Flight Center and, after a year of little progress, was transferred to the Jet Propulsion Laboratory. There, the mission acquired the designation Terrestrial Bodies Orbiter—Lunar (TBOL) to be consistent with the recently issued report of the Terrestrial Bodies Science Working Group. That advisory com-

W. Mendell (✉)
NASA/JSC, Houston, TX, USA
e-mail: wendell.w.mendell@nasa.gov

mittee had drawn up a comprehensive exploration plan for all "rocky" bodies in the solar system.

The JPL Director had made a famous speech to his employees, urging them to conceive of visually exciting, Congressionally-attractive missions to far flung destinations using the latest technology with lots of imaging (which he called "purple pigeons"), in contrast to mundane science spacecraft sent to nearby destinations (which he called "grey mice"). A T-shirt was sold that year at the Lunar Science Conference depicting a purple pigeon and a grey mouse attempting to coexist on a cratered terrain. JPL nevertheless produced a fine lunar spacecraft design with a capable science team before the project somehow faded away. The reports are still in my file cabinet.

Thus began a 30-year drought on lunar initiatives from the Office of Space Science. However, the lunar science community had not given up on a new lunar mission.

European scientists unsuccessfully proposed an ESA Polar Orbiting Lunar Observatory (POLO). Italians reworked the concept into a Moon Orbiting Observatory (MORO) proposal to their space agency invoking a launch from the Space Shuttle. A Lunar Geoscience Observer appeared in a list of second tier recommendations for a new line of planetary spacecraft in the 1983 report to NASA from the Solar System Exploration Committee. The failure and the high cost of the Mars Observer doomed the new program and any possible lunar mission.

At the Lunar Science Conference in 1987, Dr. Tom McCord chaired an informal meeting of international lunar scientists to formulate a coordinated strategy for proposing lunar missions to three major space agencies. The discussion went so far as to "allocate" different themes to the various national groups represented in the room. Each contingent planned to approach their own agency with a lunar mission concept and point out a complementary concept being considered in another agency. Professor Hitoshi Mizutani was one of the Japanese representatives, and his successful Lunar-A proposal to ISAS was the only tangible outcome of the conspiracy. Unfortunately, his mission development, built around penetrators with geophysical instrumentation, faced formidable technical challenges and was never flown.

Two years later, a U.S. President stood on the steps of the National Air and Space Museum and proposed a broad program of human exploration, within which the Moon would be an important element. During the brief existence of the Space Exploration Initiative, Associate Administrator for Exploration Michael Griffin championed an inexpensive, fast-track project called Lunar Scout for orbital surveys. Its track was not fast enough, and it vanished in the political storm that sank the SEI ship.

In 1994, a U.S. spacecraft arrived in lunar orbit carrying an array of sensors developed within the Department of Defense, with some guidance from NASA and the planetary science community. The mission, described as an in-space technology demonstration, was the brainchild of Pete Worden, then an Air Force colonel and later Director of the Ames Research Center. Of particular importance was the multispectral imaging system and the lidar rangefinder device that could precisely determine the distance between the spacecraft and the physical surface beneath it with 20 m to 30 m resolution. For 71 days, Clementine mapped the Moon. The data from the laser ranging device significantly improved our knowledge of lunar topography, and from the multispectral imagery were constructed the first global maps of lunar mineralogy. Many people, particularly from outside the U.S., erroneously believe that Clementine was a NASA mission.

The Clementine operations team noted that two specific contiguous orbits passing over the lunar south pole were oriented with respect to the Earth so as to allow a bistatic "radar" measurement potentially capable of detecting pure water ice inside permanently shadowed craters. During those specific passes, an Earth-based antenna monitored the spacecraft's telemetry signal as it was reflected from the lunar surface. Similar bistatic measurements had been performed during the Apollo 15 and 16 missions, though not over the polar regions. The team published results claiming to observe characteristics of the expected water ice signal in the coherent backscatter opposition effect associated with the reflected beam. A subsequent paper by different authors failed to find the reported signature in a reexamination of the data. The lively controversy over the bistatic radar result lives on today.

In 1994 NASA instituted a new program, named Discovery, to create a line of low-cost planetary mission spacecraft with associated scientific investigations. Lunar Prospector was the third Discovery mission but was the first to be competitively selected from team proposals led by a Principal Investigator. Its selection was largely attributed to an extraordinarily low cost estimate in an era whose motto was "faster, better, cheaper". Some senior members of the lunar science community publicly complained that such a spartan spacecraft could not possibly yield data of any utility and might "kill" lunar science for another 30 years. However, the unusual spin-stabilized polar orbiter performed admirably and produced major discoveries at its promised low cost. The Lunar Prospector gamma ray spectrometer mapped out a large region rich in radioactive elements, undoubtedly the source of the enigmatic KREEP component found in the Apollo samples. (KREEP is a quasi-acronym for an enigmatic geochemical component of Apollo samples that was rich in potassium, rare earth elements, and phosphorus.) The discovery of the Procellarum KREEP Terrane (PKT) solved a geochemical mystery in the lunar samples but toppled simplistic spherically symmetric scenarios of lunar thermal evolution. Meanwhile, the Lunar Prospector Neutron Spectrometer was returning valuable information on unanticipated concentrations of hydrogen in the lunar surface layer at the poles. Although the spatial resolution of the data is very coarse, oversampling in the polar regions permits investigators to suggest tantalizing relationships of the proton excess with the interiors of large, permanently shadowed craters, adding fuel to the "lunar ice" fire.

In January 2004, another President of the United States reincarnated human exploration of the solar system with a surprisingly detailed charge to NASA that featured lunar exploration as a major near-term theme. On March 1, 2004, senior members of the lunar science community convened a workshop at the Lunar and Planetary Institute (LPI) to discuss what characteristics of the lunar environment should be determined by the first precursor lunar orbital mission for the new initiative. Workshop chair G. Jeffrey Taylor reminded the attendees that policy goals gave priorities to support of human exploration over those of pure science, even though lunar science would clearly benefit. On the other hand, NASA Chief Scientist Jim Garvin (by phone) urged the attendees not to neglect advancement of lunar science in their recommendations because of its increasing relevance to the understanding of the entire Solar System.

Working from these somewhat discordant charges, all attendees agreed that the first priority was to ensure a safe landing and return of human explorers. Much better photography as well as topographic and cartographic mapping was needed for the Moon to pick landing sites and to enable precise navigation. (Photographic coverage of Mars was much better than the lunar data set, much to the astonishment of NASA engineers.) Next on the list was the ability to find and characterize resources as an aid to extended human presence and for future lunar economic development. From the perspective of exploration, any polar volatiles and the nature of the polar environment must be understood. Direct measurement of surface temperature was desirable, particularly in the unseen depths of the permanently shadowed polar craters. The radiation environment must be monitored to model its effects on long-duration human lunar missions. As the lunar surface directly experiences the solar wind plasma, surface electrical fields also were of some concern.

Immediately following the 1½-day workshop, a select group of lunar scientists met with Chief Scientist Garvin to begin the process of defining the robotic mission that would be called Lunar Reconnaissance Orbiter. Of course, moving from a set of data needs to a mission concept with an instrument complement involves a number of considerations such as instrument technology readiness, budget, and schedule. The LRO mission as flown represents certain compromises, but every instrument onboard addresses the needs of future exploration in some way, and most are clearly tied to the priorities articulated by the March meeting attendees. Indeed, many of the attendees of the March 1, 2004 workshop would never have believed that LRO could obtain sub-meter resolution imaging, 10 km scale hydrogen mapping, 5–10 cm vertical accuracy topographic mapping, and temperature mapping down to 20K!

As the LRO project gathered steam, the lunar community witnessed a multinational fleet of spacecraft arriving at the Moon. The first to arrive was Small Missions for Advanced Research in Technology One (SMART-1) in November 2004 from ESA. Using ion engines for a slow but highly fuel-efficient lunar transfer orbit, the small spacecraft carried a suite of instruments intended as proof of technologies. Next to attain orbit was the huge *Kaguya* spacecraft from JAXA, which entered its operational orbit in December 2007. Its impressive suite of instruments mapped the Moon for 18 months. With its two sibling satellites *Okina* and *Ouna*, it performed the only true mapping of the lunar farside gravity field. The Kaguya gravity field model and the topography data has been used to plan the LCROSS impactor mission that was selected to ride along with LRO. *Chang'e* from China arrived shortly after Kaguya and quickly began orbital operations from a 200-km altitude, mapping for 16 months. Approximately a year later, India's *Chandrayaan* arrived and began collecting information at its planned 100-km altitude. After LRO arrived on the scene in June 2009, it coordinated observations with the Indian spacecraft to perform a two-spacecraft version of the Clementine bistatic measurement of the reflected signal from the interior of polar craters. Unfortunately the experiment was unsuccessful.

The missions of Smart-1, Kaguya, and Chang'e ended with controlled impacts into the lunar surface, while Chandrayaan collected data up until a communications failure. Japan, China, and India have announced plans for follow-on lunar missions with landings on the surface.

As of this writing, the Augustine Commission Report is in preparation and will be an important element in a review of current priorities in U.S. civilian space policy, including reassessment of human exploration of the Moon. Irrespective of that decision, we can now say that the polar orbiter mission conceived at the end of the Apollo Program has finally been flown, not once but several times! We have new participants in lunar exploration, both international space agencies and private entities, whose motivations and interests are not directly tied to NASA programs. Data from LRO will be placed in the public domain expeditiously, and we hope that the findings (and particularly the measurements) from other missions will also be accessible to the international scientific community. Privately financed operations are under no such imprimatur for disclosure of results, but such missions can carry scientific instruments as customers. I am optimistic that this panoply of new activity will generate a great deal of new understanding of the history and current state of the Moon and will engender an exploration-driven renaissance of lunar science.

Space Sci Rev (2010) 150: 7–22
DOI 10.1007/s11214-010-9631-5

Lunar Reconnaissance Orbiter (LRO): Observations for Lunar Exploration and Science

Richard Vondrak · John Keller · Gordon Chin · James Garvin

Received: 12 October 2009 / Accepted: 11 January 2010 / Published online: 23 February 2010

Abstract The Lunar Reconnaissance Orbiter (LRO) was implemented to facilitate scientific and engineering-driven mapping of the lunar surface at new spatial scales and with new remote sensing methods, identify safe landing sites, search for in situ resources, and measure the space radiation environment. After its successful launch on June 18, 2009, the LRO spacecraft and instruments were activated and calibrated in an eccentric polar lunar orbit until September 15, when LRO was moved to a circular polar orbit with a mean altitude of 50 km. LRO will operate for at least one year to support the goals of NASA's Exploration Systems Mission Directorate (ESMD), and for at least two years of extended operations for additional lunar science measurements supported by NASA's Science Mission Directorate (SMD). LRO carries six instruments with associated science and exploration investigations, and a telecommunications/radar technology demonstration. The LRO instruments are: Cosmic Ray Telescope for the Effects of Radiation (CRaTER), Diviner Lunar Radiometer Experiment (DLRE), Lyman-Alpha Mapping Project (LAMP), Lunar Exploration Neutron Detector (LEND), Lunar Orbiter Laser Altimeter (LOLA), and Lunar Reconnaissance Orbiter Camera (LROC). The technology demonstration is a compact, dual-frequency, hybrid polarity synthetic aperture radar instrument (Mini-RF). LRO observations also support the Lunar Crater Observation and Sensing Satellite (LCROSS), the lunar impact mission that was co-manifested with LRO on the Atlas V (401) launch vehicle. This paper describes the LRO objectives and measurements that support exploration of the Moon and that address the science objectives outlined by the National Academy of Science's report on the Scientific Context for Exploration of the Moon (SCEM). We also describe data accessibility by the science and exploration community.

Keywords Moon · Lunar · Remote sensing · Spacecraft

R. Vondrak (✉) · J. Keller · G. Chin · J. Garvin
NASA Goddard Space Flight Center, Greenbelt, MD 20771, USA
e-mail: richard.vondrak@nasa.gov

1 Introduction

The Lunar Reconnaissance Orbiter (LRO) was successfully launched on June 18, 2009 and joined an international fleet of satellites (Japan's SELENE/Kaguya, China's Chang'E, and India's Chandrayaan-1) that have recently orbited the Moon for scientific exploration purposes. LRO is the first step to fulfill the US national space goal to return humans to the Moon's surface, which is a primary objective of NASA's Exploration Systems Mission Directorate (ESMD). The initial LRO mission phase has a one-year duration fully funded under ESMD support. LRO is expected to have an extended phase of operations for at least two additional years to undertake further lunar science measurements that are directly linked to objectives outlined in the National Academy of Science's report on the Scientific Context for Exploration of the Moon (SCEM). All data from LRO will be deposited in the Planetary Data System (PDS) archive so as to be usable for both exploration and science by the widest possible community.

A NASA Announcement of Opportunity (AO) solicited proposals for LRO instruments with associated exploration measurement investigations. A rigorous evaluation process involving scientific peer review, in combination with technical, cost and management risk assessments, recommended six instruments for LRO development and deployment. The competitively selected instruments are: Cosmic Ray Telescope for the Effects of Radiation (CRaTER), Diviner Lunar Radiometer Experiment (DLRE), Lyman-Alpha Mapping Project (LAMP), Lunar Exploration Neutron Detector (LEND), Lunar Orbiter Laser Altimeter (LOLA), and Lunar Reconnaissance Orbiter Camera (LROC). In addition, LRO also accommodates a hybrid-polarity synthetic aperture radar system (Mini-RF) as a technology demonstration supported by NASA's Space Operations Mission Directorate.

The Foreword to this volume provides an overview of the history of lunar robotic exploration since the Apollo program. An early review of LRO exploration mission objectives and instrument description can be found in a recent summary paper (Chin et al. 2007). Other articles in this volume provide detailed descriptions of LRO instrument capabilities, performance, and calibration during ground-based integration and testing. A description of the LRO spacecraft and mission design can also be found in this issue (Tooley et al. 2010).

In this paper, Sect. 2 describes the Exploration Mission objectives. Section 3 summarizes the instrument capabilities and LRO mission implementation. Section 4 describes the options and objectives for the extended Science Mission. Section 5 gives the specific Planetary Data System nodes where each instrument's data will be archived for widespread access and the LRO data deposition schedule.

2 LRO Exploration Objectives

The goal of returning humans to the surface of the Moon is a major objective of the US national space program (National Space Exploration Policy Directive, NSPD 31). LRO is the first mission undertaken by NASA's Exploration Mission Directorate (ESMD) and LRO's primary goal is to provide the critical supporting measurements that will enable cost-effective return of human explorers to the surface of the Moon. Some of the required ESMD measurement data include:

- Knowledge of the radiation environment in lunar orbit and the biological effects caused by exposure to this space radiation environment
- Global geodetic lunar topography tied to an accurate center-of-mass grid at suitable spatial scales (relevant to precision navigation and surface mobility)

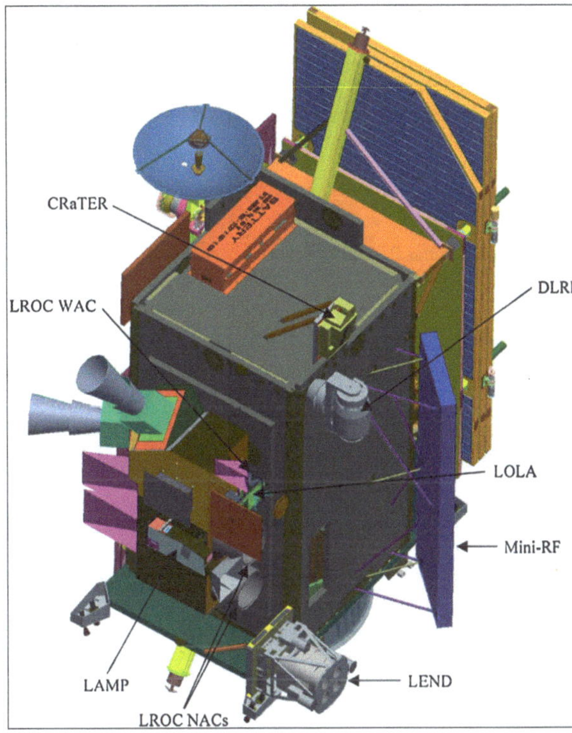

Fig. 1 Placement of the instruments on the LRO spacecraft

- High spatial resolution maps of enhanced hydrogen deposits in the Moon's regolith
- Temperature maps of the Moon's polar permanently shadowed regions (PSR's)
- Landform-scale images of lunar surfaces in permanently shadowed regions
- Identification of putative deposits of appreciable near-surface water ice in the Moon's polar permanently shadowed regions (PSR's) or evidence that they do not exist
- Images at meter and smaller scales to assess the safety of potential lunar landing sites (with definitive meter-scale feature identification)
- Characterization of the illumination in the Moon's polar regions at relevant spatial and temporal scales (i.e., in terms of hours)

NASA released an Announcement of Opportunity (AO) that called for a suite of instruments to provide measurements that will fulfill Exploration requirements on the basis of an independent assessment of measurement objectives essential for human lunar exploration. After a rigorous evaluation process of scientific peer review, as well as technical, cost and management risk assessments, six instruments were selected for LRO. The selected instruments are: Cosmic Ray Telescope for the Effects of Radiation (CRaTER), Diviner Lunar Radiometer Experiment (DLRE), Lyman-Alpha Mapping Project (LAMP), Lunar Exploration Neutron Detector (LEND), Lunar Orbiter Laser Altimeter (LOLA), and Lunar Reconnaissance Orbiter Camera (LROC). In addition, LRO also accommodates a compact, hybrid-polarity, dual-wavelength synthetic aperture radar system (Mini-RF) as a technology demonstration focused on advanced telecommunications and polarimetric SAR imaging, supported by NASA's Space Operations Mission Directorate. Figure 1 shows the mounting of the instruments on the LRO spacecraft flight system.

Table 1 Key LRO Data Products Advance ESMD Exploration Goals

Instrument	Key LRO Data Products	Examples of Exploration Benefits
CRaTER: Cosmic Ray Tele-scope for the Effects of Radiation PI: Harlan Spence, University of New Hampshire	Lunar and deep space radiation environment and tissue equivalent plastic response to radiation	Safe, lighter weight space vehicles. Determination of the radiation environment for human presence at the Moon and journeys to Mars and beyond
DLRE: Diviner Lunar Radiometer Experiment PI: David Paige, UCLA	500 m scale maps of surface temperature, albedo, rock abundance, and ice stability	Measures thermal environment in permanent shadow and permanent light, ice depth map
LAMP: Lyman Alpha Mapping Project PI: Randall Gladstone, SWRI	Maps of frosts and landforms in permanently shadowed regions (PSRs)	Locate potential water-ice on the surface, image shadowed areas, and map potential landing areas in PSRs
LEND: Lunar Exploration Neutron Detector PI: Igor Mitrofanov, IKI	Maps of hydrogen in upper 1 m of Moon at 10-km scales using neutron albedo	Locate potential water-ice in lunar soil or concentrations of implanted hydrogen
LOLA: Lunar Orbiter Laser Altimeter PI: David Smith, GSFC	~25 m scale polar topography at <10 cm vertical, global topography, surface slopes and roughness	Identify safe landing sites, image shadowed regions, map potential surface ice, improve gravity field model
LROC: Lunar Reconnaissance Orbiter Camera PI: Mark Robinson, ASU	Thousands of 50 cm/pixel images, and entire Moon at 100 m in UV, Visible. Illumination conditions of the poles	Surface landing hazards and some resource identification, including locations of near constant solar illumination
Mini-Rf Technology Demonstration PI: Ben Bussey, APL/JHU	X and S-band radar imaging and repeat-pass interferometric SAR	Demonstrate lightweight SAR and communication technologies, locate potential water-ice using radar scattering

Table 1 provides the list of LRO instruments, the Principal Investigator, key data the instrument plans to provide, and the benefits to the Exploration program. The instrument performance parameters are summarized in Table 2, and the wavelength coverage and spectral resolving power of the LRO instruments is shown in Fig. 2.

Except for calibration activities, the LRO spacecraft is generally pointed in a nadir direction to allow mapping of the lunar surface by the remote sensing instruments. The LRO spacecraft capabilities are summarized in the companion paper by Tooley et al. (2010). A unique feature of the LRO spacecraft is the precision orbit determination that is provided by one-way laser tracking of LRO from Earth-based laser ranging facilities, which is enabled by the Lunar Orbiter Laser Altimeter (Zuber et al. 2010, this issue).

Each LRO instrument has strong individual capabilities for providing both scientific and engineering-enabling measurements of the lunar surface and environment. In concert, the LRO instrument suite addresses LRO measurement goals in a robust and resilient manner by using several complementary remote sensing methods that reinforce the discoveries made by an individual instrument. Table 4 summarizes the LRO measurement objectives for ESMD and the complementary nature of their investigations. Figure 3 illustrates the role that each instrument plays to contribute to the success of the LRO ESMD mission, as stated by the mission's Level 1 requirements, and the required performances to reach minimum and full mission success.

Table 2 Instrument Performance Parameters

Instrument	Instrument Classification	Characteristic Range	Characteristic Resolution	Spatial Resolution (from 50 km)	Spatial Coverage	Data Rate Gbits/day
LOLA	Laser Altimeter	Range Window 20–70 km	10 cm vertical	Five 5 m laser spots 25 m spacing	Polar Grid 0.001° latitude 0.04° longitude	1.4
LROC NAC	High Resolution Camera	Broadband Centered at 550 nm	±150 nm	50 cm/pixel	Targeted >10% Lunar Surface 100% >85.5° lat.	515
LROC WAC	MultiSpectral Camera	315–680 nm	Spectral filters centered at 315 nm 360 nm, 415 nm, 560 nm, 600 nm, 640 nm, 680 nm	100 m/pixel visible; 400 m/pixel UV	Full lunar surface at each wave-length and various lighting angles	41
LEND	Neutron Detector	Thermal to 15 MeV	*Four Bands* Thermal <0.4 eV Epithermal 0.4 eV–10 keV Fast 10 keV–1 MeV Energetic 1 MeV–15 MeV	Epithermal 10 km FWHM (see text for other bands)	Full lunar surface and deep space	0.26
DLRE	Radiometer	20 K to 400 K	5 K	400 m	Full Lunar Surface Day/Night Temperatures	3.5
LAMP	UV Imaging Spectrograph	52 to 187 nm	3.5 nm	260 m	Full Lunar Surface	2
CRaTER	Primary and albedo cosmic ray sensor	LET spectra 0.2 keV/μm to 7 MeV/μm	<3%	77 km	Full lunar surface and deep space	7.8 (peak)
Mini-RF	X- and S-band Synthetic Aperture Radar	4 cm (X-band) 12 cm (S-band)	Sensitivity (NES0): −30 dB (S-band) −25 dB (X-band)	150 m 15 × 30 m (zoom)	Limited during the nominal mission	7.7*

*mini-RF 7.7 Gbits for 4 min. data collection interval

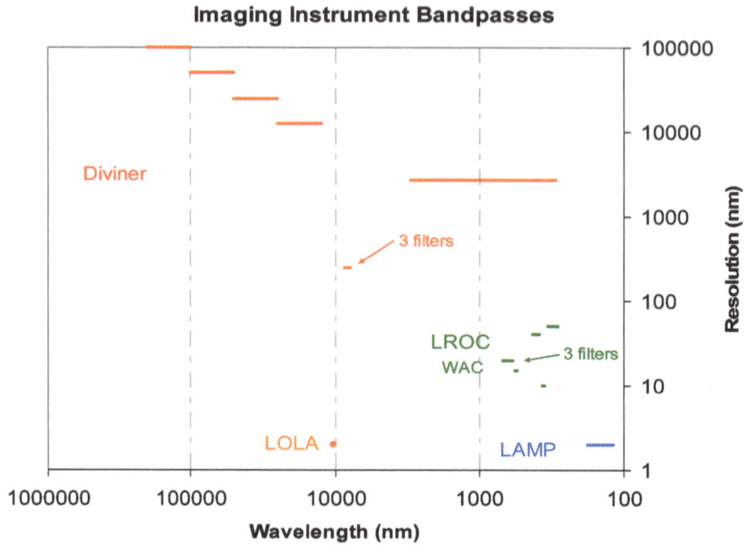

Fig. 2 Instrument bandpasses. LOLA instrument included for completeness, but not scaled to resolution

	Objectives	LRO Requirements	CRaTER	DLRE	LAMP	LEND	LOLA	LROC	Mini-RF
1	Find Safe Landing Sites	M30 M40 – Global geodetic grid 10 cm vertical, and at the poles, 50 m horizontal resolution					▲	▲	
		M80 – Identify surface features & hazards		▲			▲	▲	
2	Locate Potential Resources	M90 – Characterize the polar region illumination environment		▲			▲	▲	
		M50 – Provide lunar temperature map from 40 - 300K, 5 K precision over full diurnal cycle.		▲					
		M60 – Image the permanently shadowed regions.			▲		▲	○	
		M70 - Identify putative deposits of water-ice			▲	▲	▲		○
		M100 - Characterize lunar mineralogy		▲	○	○	○	▲	○
		M110 - Hydrogen mapping				▲		○	
3	Life in the Space Environment	M10 - Characterize the deep space radiation environment at energies in excess of 10 MeV	▲			▲			
		M20 - measure the deposition of deep space radiation on human equivalent tissue.	▲						
4	New Technology	P160 - Technology demo	○				○		▲

Minimum Success (rows M30 M40 – M90)

Full Success (all rows)

▲ Level 1 Requirements

○ Contributions– may require additional analysis to exploit

Fig. 3 LRO level 1 Requirements and Criteria for Minimum and Full Mission Success of the ESMD Mission

3 Exploration Mission Implementation

After LRO payload selection in December, 2004, the LRO development was begun early in 2005. A summary of key dates in the development of the LRO spacecraft flight system is given in Table 3.

As described in Tooley et al. (2010), LRO was a rapid development project. The launch date was moved from the fourth quarter 2008 to 2009 due to delays in availability of the launch vehicle and due to conflicts with the Atlas V launch facilities at Cape Canaveral.

Table 3 Summary of key events in LRO development

Event	Date
Preliminary Design Review	2/7/2006
Confirmation Review	5/17/2006
Comprehensive Design Review	11/10/2006
Pre-Ship Review	2/9/2009
Orbiter delivered to the Cape	2/12/2009
Launch Readiness Review	6/15/2009
Spacecraft Launch	6/18/2009
Lunar Orbit Insertion	6/23/2009
Begin Instrument Activation and Calibration	6/30/2009
Insert into Mission Orbit (50 ± 15 km circular)	9/15/2009

LRO was launched from the Cape Canaveral Air Station in Florida on June 18, 2009 and was inserted directly into lunar orbit on June 23. LRO entered into an initial polar-inclination (89.7) elliptical orbit (30×216 km) with the periselene over the southern hemisphere. After reaching this initial orbit, a commissioning phase began during which the spacecraft and instruments were tested and calibrated. This phase was planned to last for approximately 60 days.

During the commissioning phase the spacecraft systems were activated and tested. Each of the instruments was activated and numerous calibrations were successfully conducted. Data were collected in the nominal operating configuration with the instruments nadir pointing to verify that they were functioning properly and to calibrate against well known lunar features, as well as with deep space. Also, there were specific star calibrations where the LRO orbiter was pointed or rastered across selected stars to verify the pointing accuracy of the instruments and to determine the sensitivity. The LRO spacecraft was reoriented to point at the Earth to calibrate the Laser Ranging instrument and the Mini-RF Technology Demonstration Experiment with Earth based assets, such as the radar at the National Astronomy and Ionosphere Center Arecibo Observatory for the Mini-RF instrument and one-way laser ranging from LOLA to a receiver on the Earth, and from an Earth based laser to the LOLA receivers. Finally a number of roll, yaw and pitch maneuvers were undertaken to assess light scattering in the instruments and relative inter-pixel sensitivity (flat field).

Because of the MiniSAR S-band polarimetric radar on the ISRO Chandrayaan-1, there was a unique opportunity for bistatic radar measurements during the commissioning period when the LRO and Chandrayaan-1 spacecraft were at approximately the same altitude. A dual orbiter spacecraft bistatic experiment was attempted on August 19, 2009, but was not successful due to operational difficulties with the pointing of the ISRO Chandrayaan-1 spacecraft. Another dual-spacecraft bistatic radar scattering experiment attempt was planned for the following week, but was canceled due to the loss of communications with the Chandrayaan-1 spacecraft on August 25, 2009. These coordinated operations resulted in an extension of the LRO commissioning activities for an additional two weeks, relative to the original baseline plan.

On September 15, 2009 the LRO orbiter spacecraft was moved into a polar-inclination (89.7 degrees) circular 50-km mean altitude orbit for a planned one-year duration to execute its baseline ESMD mission phase ("the Exploration Mission"). The orbital period is typically 113 minutes. This polar orbit allows repetitive measurements at high latitudes, producing a dense net of measurements as illustrated in Fig. 4, where the polar groundtrack is shown for a one month period. Since the Moon rotates once each sidereal month, successive

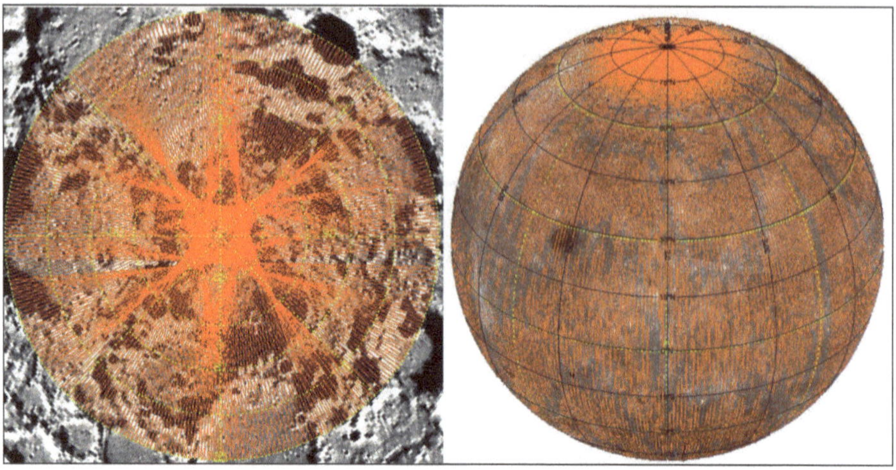

Fig. 4 LRO groundtrack emphasizes remote sensing of the Lunar Poles

groundtracks are separated by about 31 km at the equator. Because most instruments make measurements over a narrow swath, at lower latitudes there will be inter-orbit ground track gaps or gores in the mapping during one lunation. However, measurements accumulated during the one-year Exploration Mission result in complete global coverage.

An important consideration in selecting the launch date was the variation of the LRO orbital plane during the course of the primary mission. Even though the tilt of the lunar spin axis with respect to the ecliptic plane is very small, only 1.557 degrees, there will be some seasonal variation of illumination and resulting surface physical temperature in the lunar polar regions. To measure this effect, it is desirable to have the LRO ground track be very near the noon-midnight meridian at the solstices. As shown in Fig. 5, at the first solstice on October 22, which corresponds to midsummer at the South Pole and midwinter at the North Pole, the LRO ground track will be very close to local noon/midnight at that time.

The LRO mission played a major role in the support of the Oct. 9, 2009 LCROSS impact experiment. When LRO was manifested on an Atlas V (401) rocket, NASA took advantage of the additional payload capacity to select the Lunar Crater Observation and Sensing Satellite (LCROSS), a lunar impactor mission to launch simultaneously with LRO. LCROSS consists of the Centaur stage as a primary impactor, accompanied by a shepherding remote sensing satellite for post-impact observations.

The role of LRO is to make detailed observations before, during, and after the LCROSS impact in a permanently shadowed crater near the lunar South Pole. The impact will occur within the highlands impact crater Cabeus on October 9, 2009. LRO has made detailed observations to support the LCROSS selection of this crater for the impact. The LRO team used laser altimetry data from both the LRO LOLA and JAXA Kaguya laser altimeters to determine regions of permanent shadows that would be the most likely regions to harbor frozen volatiles, if any are preserved in significant concentrations. The topography measurements also identified regions within the polar impact craters where the local slope is sufficiently small to maximize the transfer of kinetic energy from LCROSS impact into the lunar regolith target. LRO's LEND neutron spectrometer provided maps of enhanced hydrogen concentrations that could indicate water ice embedded in the upper meter of the lunar regolith. LRO Diviner temperature measurements of the South Polar region were used to reveal the extremely low temperatures of cold traps that can potentially preserve volatiles

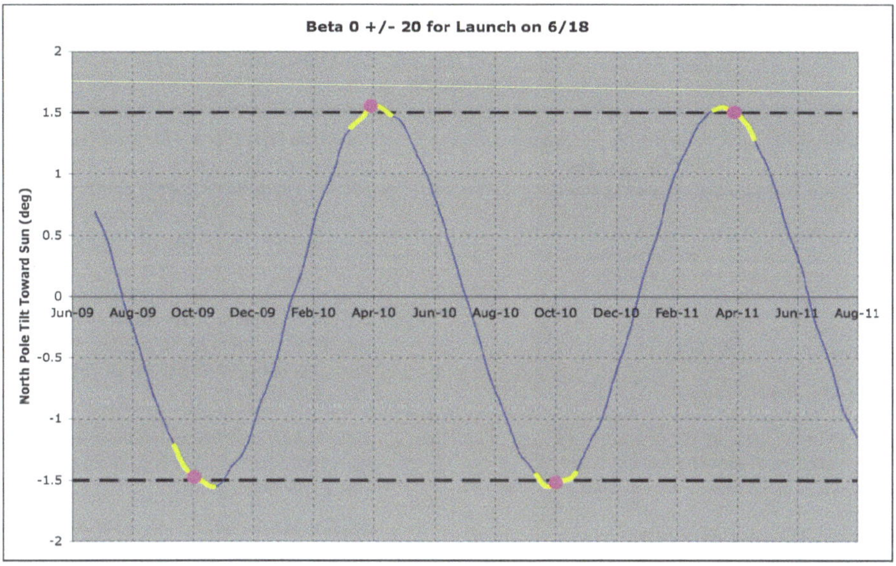

Fig. 5 Variation of the angle between the LRO orbit and the direction of the sun. South Pole summer occurs on October 22

in ice form. Mini-RF dual-frequency radar polarization imaging provided information that indicates the blockiness of the impact site and which can be used to further test for the presence of significant water ice on the basis of anomalous scattering behavior. These combinations of LRO measurements as well as other factors led the LCROSS team to choose the south polar crater Cabeus for its impact target. Before the impact occurs the Cabeus target area will be exhaustively observed by the LRO LROC, LOLA, and Mini-RF instruments in detail to characterize the pre-impact geology.

The LRO spacecraft orbit will be adjusted to pass at closest approach to the Cabeus target site approximately 90 seconds after the LCROSS Centaur impact. At impact and after the impact the LRO LAMP far UV spectrometer will search for evidence of significant volatile species and how they evolve in the Moon's tenuous atmosphere. LRO's Diviner radiometer will peer into the impact site to measure the heating effects of the impact and how it changes over time. The LRO mission plans to undertake a campaign of extensive observations of the post-impact surface geological effects by all instruments over the course of the days and weeks following the LCROSS impact event. Such LRO-based observations will provide important measurements in support of the LCROSS' team analysis of the physics of the impact and how volatile materials may have been mobilized.

An important objective during the primary LRO Exploration Mission is intensive measurements of specific sites, chosen by NASA's Constellation program, as representative of a broad distribution of lunar geological features important for both their value to human exploration as well as for scientific objectives. An LRO Science Targeting Meeting was held at Arizona State University on June 9–11, 2009 to solicit ideas and priorities from the scientific community for LRO targeting, both in terms of focused science themes and specific features on the Moon. Additional areas for intensive measurements will be identified during the LRO mission.

Table 4 Objectives are met by LRO instrument investigations, some with complementary data products

LRO Measurement Objective	LRO Instrument	Data Product
The LRO shall characterize the deep space radiation environment at energies in excess of 10 MeV in lunar orbit, including neutron albedo	LEND	Radiation Data Product for global distribution of neutrons at Moon's orbit with spatial resolution of 50 km at different energy ranges from thermal energy up to >15 MeV separately for periods of quiet Sun and for periods of Solar Particle Events
	CRaTER	Provide Linear Energy Transfer (LET) spectra of cosmic rays (particularly above 10 MeV), most critically important to the engineering and modeling communities to assure safe, long-term, human presence in space
The LRO shall measure the deposition of deep space radiation on human equivalent tissue while in the lunar orbit environment	CRaTER	Provide LET spectra behind different amounts and types of areal density, including tissue-equivalent plastic
The LRO shall measure lunar terrain altitude to a resolution of 10 cm and an accuracy of 1 m for an average grid density of approximately 0.001 degrees latitude by 0.04 degrees longitude	LOLA	Provide a global digital elevation model of the Moon with 10 cm vertical resolution, 1 m vertical accuracy and 50–100 m horizontal resolution with 1 km average cross track sampling at the equator
The LRO shall determine the horizontal position of altitude measurements to an accuracy of 100 m	LOLA	Provide global topography of the Moon with 10 cm vertical resolution, 1 m vertical accuracy and 50–100 m horizonal resolution with 1 km average cross track sampling at the equator
	LROC	For areas of high interest (targets), collect multi-look NAC data reducible to 2 m scale Digital Elevation Models for 25 km^2 areas
	LROC	Acquire 100 m/pixel global stereo imaging reducible to 1 km/pixel global topography in EDR format
The LRO shall obtain temperature mapping from 40–300 K in the Moon's polar regions to better than 500 m spatial resolution and 5 K precision for a full diurnal cycle	Diviner	Temperature maps at better than 500 m spatial resolution from 40–300 K over an entire diurnal cycle to enable the detection and characterization of cold traps in polar shadowed regions
The LRO shall obtain landform-scale imaging of lunar surfaces in permanently shadowed regions at better than 100 m spatial resolution	LOLA	Provide global topography with 10 cm vertical resolution, 1 m vertical accuracy and 50–100 m horizontal resolution with 1 km average cross track sampling at the equator
	LAMP	Albedo maps of all permanently shadowed regions with resolutions down to 300 m
The LRO shall identify putative deposits of water-ice in the Moon's polar cold traps at a spatial resolution of better than 500 m on the surface and 10 km subsurface (up to 2 m deep)	LOLA	Provide Reflectance data from the permanently shadowed regions to identify surface ice signatures at a limit of 4% ice surface coverage by area
	LEND	Develop maps of putative water-ice column density on polar regions of the moon with spatial resolution of 10 km
	LAMP	Develop water-frost concentration maps of the lunar polar regions Mapping resolutions as good as 3 km for frost abundances down to 1.5%

Table 4 (*Continued*)

LRO Measurement Objective	LRO Instrument	Data Product
The LRO shall assess meter-scale features of the lunar surface to enable safety analysis for potential lunar landing sites over targeted areas of 100 km^2	LROC	Provide up to 50 Mosaics of selected potential landing sites with one meter scale resolution
	LROC	Provide crater size density and size distribution maps of up to 10 potential landing sites (100 km^2/site)
	Diviner	Provide rock ($\leq$0.5 m diam) abundance percentages for up to 50 selected potential landing sites by measuring the ratio of high thermal to low thermal inertia material
	LOLA	Provide topography, surface slopes, and surface roughness at 25-m spacing over a 70 m wide FOV swath at up to 50 selected potential landing sites
The LRO shall characterize the Moon's polar region illumination environment to a 100 m spatial resolution and 5 Earth hour average temporal resolution	LROC	Provide uncontrolled illumination movies, 1 each of North and South Lunar Poles over the course of 1 lunar year at an average time resolution of 5 hours or better (Wide Angle Camera)
	LROC	Provide meter scale resolution summer (uncontrolled) mosaics of the lunar poles ($\pm$4 degrees). (Narrow Angle Camera). Some gores in the data beyond the polar regions are acceptable
	LOLA	Polar region maps of latitudes 86°–90° with a vertical resolution of 10 cm and a spatial resolution of 25 to 35 m after one year. This will help to identify potential sites of optimal solar power generation
	Diviner	Provide polar illumination map at better than 500 m spatial scales over a full diurnal cycle
The LRO shall characterize lunar mineralogy by mapping the thermal properties of regolith and characterizing UV, visible, and infrared spectral differences and variations at km scales globally	LROC	Global imaging 400 m/pixel in the ultraviolet bands and 100 m/pixel in the visible bands, ten uncontrolled demonstration multi-spectral mosaics for high priority targets
	Diviner	Global fine-component thermal inertia, silicate mineralogy and Lambert albedo from thermal emission, solar reflectance and topography measurements with greater than 50% spatial coverage at the equator
The LRO shall perform hydrogen mapping of the Moon's surface with a sensitivity of 100 ppm or better, a SNR of 3, and 10 km resolution in the polar regions	LEND	Determine hydrogen content of the subsurface at the polar regions with spatial resolution of 10 km and with sensitivity to concentration variations of 100 parts per million at the poles

4 LRO Science Mission Objectives

Although LRO was conceived and designed to be an Exploration Mission of primary benefit to ESMD (i.e., for providing boundary conditions in support of engineering decisions pertaining to the return of human explorers to the Moon), the science value of the LRO observations was recognized by NASA's Science Mission Directorate (SMD) from the outset (i.e. the Announcement of Opportunity was issued jointly by ESMD and SMD, and SMD was jointly responsible for the selection of LRO instruments). Most of the competitively-selected instruments have a significant heritage in remote sensing devices that have been flown to other planetary objects for scientific purposes. In order to fully realize the scientific potential expected from the wealth of LRO data, NASA's SMD in partnership with ESMD competitively selected in 2007 a team of Participating Scientists (PS) to work closely with the LRO instrument teams and their Principal Investigators (PIs). The Participating Scientists help plan observations during the LRO Exploration Mission and the extended Science Mission and aid in the analysis of LRO data, especially correlating results from multiple instruments to develop improved models of the lunar surface and geological evolution.

At the end of the ESMD Exploration Mission phase, LRO will undertake a minimum two-year Science Mission sponsored by the Science Mission Directorate (SMD). The science objectives of this extended mission will reflect the themes outlined by the National Academy of Science's report on the Scientific Context for Exploration of the Moon (SCEM). Table 5 lists the SCEM themes and an initial formulation of the LRO science objectives that might comprise the core investigations during the Science Mission. This initial formulation will be refined and a proposal for the Science Mission will be submitted to SMD early in 2010.

LRO has several orbital configurations to consider in executing the extended Science Mission, depending on the amount of fuel left at the end of the 1-year baseline mission. Figure 6 shows the fuel required for several types of orbits, expressed in terms of velocity change (m/s) from the 50-km circular orbit. It is expected that about 220 m/s of velocity change capability will remain at the end of the Exploration Mission.

The most compelling options for the LRO Science Mission include:

- continue the 50-km altitude orbit (MO, or Exploration Mission Orbit) for an additional to-be-determined number of months depending on available station-keeping fuel,
- return to the 30 × 216 km commissioning orbit (FO, or Frozen Orbit) with a periselene at the southern hemisphere for several additional years—this orbit requires minimum fuel to maintain,
- move first to a transition mission orbit (TMO) with mean altitude of 110 km (20 × 220 km elliptical), with the periselene initially at 40 degrees south of the equator (argument of periapse ∼220 degrees) moving north towards the equator, and over the north pole (with a 90 × 150 km altitude), and returning to an altitude range of 20 × 220 km at 40 degrees south of the equator after a total of about 350 days, and finally, to the commissioning orbit (30 × 216 km, periapse over the southern hemisphere) for remainder of mission life.

Table 6 summarizes these various orbits and required propulsion fuel capacity. They are organized according to Science Mission duration (Short, Medium, Long). In early 2010 a decision will be made as to the orbital strategy for the Science Mission.

5 Data Accessibility

To make the LRO observations accessible to both human exploration planners as well as the science community, calibrated LRO data will be rapidly deposited into the Planetary Data

Table 5 LRO measurement capabilities can accomplish many NRC SCEM science objectives

SCEM Theme	LRO Approach/SCM Objective	Measurements	Contributing Instruments
Bombardment History of the Inner Solar System	Characterize Recent Impact Flux Crater age studies Role of volatiles on crater morphology Basin formation and modification	Topography, High Resolution & Multispectral Imagery Reimaging of Apollo era sites	LROC, DLRE, LAMP, LOLA
Lunar Interior	Thickness of the Lunar Crust Stratification of the Mantle Heat Flow Tidal Deformation	Global Geodetic Grid Multispectral Imagery Temperature Measurements of PSRs	LROC, DLRE, LOLA
Compositional Diversity of the Lunar Crust	Determine the extent of planetary differentiation, the age, distribution, and origin of lunar rock types, and the composition of the lower crust both globally and regionally	Topography, High Resolution & Multispectral Imagery Neutron albedo Radar images and polar	LROC, DLRE, Mini-RF, LOLA, LEND
Polar Volatiles	Sources, composition, physical properties of locations	Neutron albedo maps Temperature maps of PSRs Reflectance data Surface UV spectra	DLRE, Mini-RF, LOLA, LEND, LAMP, LOLA
Lunar volcanism	Origin, compositional variability, age of mare basalts and pyroclastic deposits	High resolution topography Dual band radar imaging High-resolution and multispectral imaging Day/night thermal emission	DLRE, LOLA, LROC, LEND, Mini-RF

System (PDS) by each instrument Science Operations Center (SOC). Figure 7 illustrates the locations where the various LRO observational data and data products will be located.

Each instrument team is required to submit to the PDS several types of data products. Typically Level 0 data products are raw data in the form of counts accumulated during specific time intervals of measurements, with orbital information and engineering parameters. Higher level data products consist generally of count rates converted into physical units and projected onto the lunar surface as maps of geophysical quantities.

Each instrument SOC is required to deliver validated and calibrated data to the PDS every three months, starting six months after the beginning of the primary Exploration Mission (i.e., March 15, 2010). The initial delivery will include all measurements made from the time of launch through the first three months of the primary orbit (e.g., June 18 to December 15, 2009). Subsequent deliveries will be made every three months after the first delivery and will include all data that is no more than three months old. After the Exploration Mission ends, some investigations plan to reprocess data and develop higher level composite data products. These will be delivered to the PDS no later than six months after end of the Exploration Mission.

Table 5 (*Continued*)

SCEM Theme	LRO Approach/SCM Objective	Measurements	Contributing Instruments
Impact processes	Melt sheet differentiation Multi-ring basin structure Lateral and vertical mixing of compositional units	Regional and high resolution topography of craters and basins Spectral reflectance and thermal emission properties of multi-ring basins	LROC, DLRE, Mini-RF, LOLA
Regolith processes and weathering	Identify ancient regolith Modification processes Rare earth minerals (Gd and Sm)	High resolution slope characterization Physical properties High-resolution and multispectral imaging Day/Night thermal emission S and X band synthetic aperture radar	DLRE, Mini-RF, LOLA. LEND, LAMP, LOLA
Atmosphere and dust environment	Density and composition of exospheric species Size, charge and distribution of electrostatically levitated dust	Fluorescence and light scattering in exosphere from atomic species and dust	LAMP, LROC
Other Lunar and non-Lunar Science	Novel heliophysics and astrophysics studies Solar cycle dependence of the lunar radiation environment	Galactic cosmic rays and solar energetic particles Earth's aurora, plasma-sphere, and astronomical objects in the UV	CRaTER, LEND

Table 6 Options for orbits during the science mission

	Option	Mission Orbit (MO) + Transition Orbit (TO) Duration (Months)	50 × 50 km Nominal Mission Orbit ΔV (m/s)	20 × 140 km Transition Orbit ΔV (m/s)	16 × 200 Maintain Frozen Orbit (FO) ΔV (m/s)	Total ΔV (m/s)
L	6-months MO 2 × 350-days TMO about 4-years FO	30	75	118	19	212
M	10-months MO 350-days TMO 4-years FO	22	125	67	20	212
S	11-months MO MO to FO (costs 50 m/s) About 5-years-FO	11	138		24 + 50 (MO to FO)	212

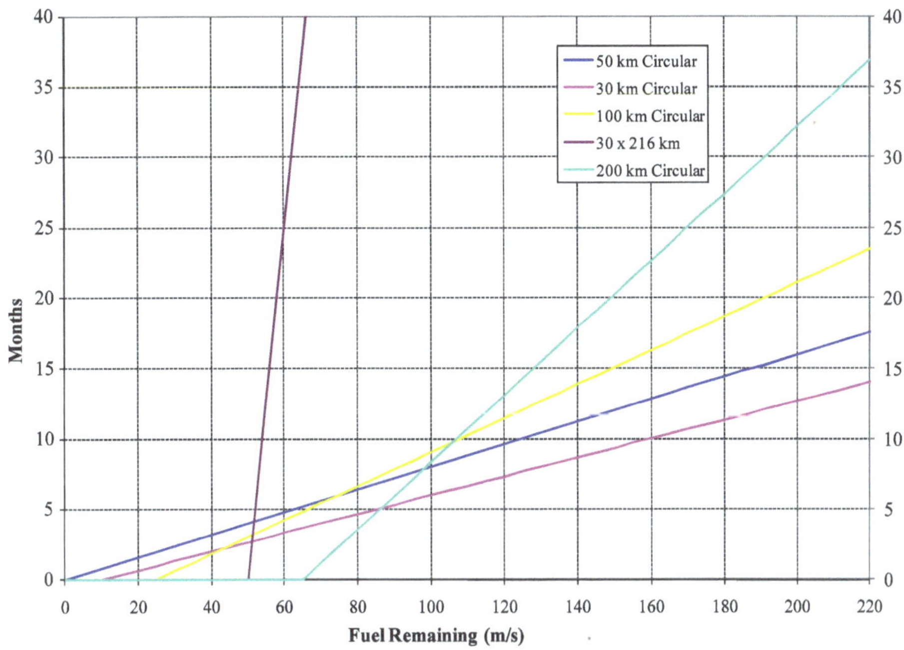

Fig. 6 LRO Extended Science Mission Options: Lifetime vs. Fuel Available at the Start of the Extended Mission for Various Orbits

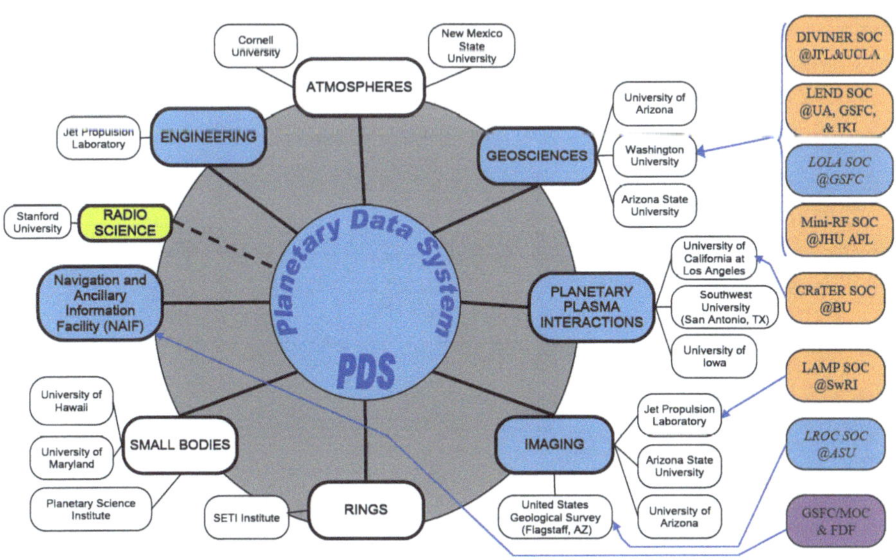

Fig. 7 LRO SOCs and PDS Nodes and Data Nodes

6 Conclusions

LRO is in orbit around the Moon, has successfully completed its commissioning phase and has been inserted into the 50 km mission orbit. The spacecraft and instruments are performing well and there appears to be no barrier toward fulfilling the objectives of the mission. After the one year ESMD mission and especially after an extended mission LRO will have provided a comprehensive data set that will guide future exploration and scientific discoveries. LRO is the first dual-purpose mission conducted by NASA, with a unique capability to accomplish both paradigm-shifting science observations together with critical engineering measurements necessary for cost-effective and safe return of human explorers to the lunar surface.

References

Chin et al., Lunar reconnaissance orbiter overview: the instrument suite and mission. Space Sci. Rev. **129**, 391–419 (2007)
vBOOK0052

C. Tooley et al., Lunar reconnaissance orbiter mission and spacecraft design. Space Sci. Rev., this issue (2010). doi:10.1007/s11214-009-9624-4

M. Zuber et al., The lunar reconnaissance orbiter laser ranging investigation. Space Sci. Rev., this issue (2010). doi:10.1007/s11214-009-9511-z

Space Sci Rev (2010) 150: 23–62
DOI 10.1007/s11214-009-9624-4

Lunar Reconnaissance Orbiter Mission and Spacecraft Design

Craig R. Tooley · Martin B. Houghton · Richard S. Saylor Jr. · Cathy Peddie · David F. Everett · Charles L. Baker · Kristina N. Safdie

Received: 10 July 2009 / Accepted: 10 December 2009 / Published online: 14 January 2010

Abstract Launched June 18, 2009, with its primary mission scheduled to end September 2010, NASA's Lunar Reconnaissance Orbiter will be the first observatory ever to spend an entire year orbiting and observing the Moon at a low altitude of just 50 km. The spacecraft carries a wide variety of scientific instruments and will provide an extraordinary opportunity to study the lunar landscape at resolutions and over time scales never achieved before. This paper is intended as a companion to the series of papers released simultaneously in this journal detailing LRO's instruments and their planned measurements. The paper describes the design and key performance drivers of the LRO spacecraft and overall mission design. It presents a comprehensive description of the operation of the various systems that comprise the spacecraft and illustrates how these systems enable achievement of the mission requirements.

Keywords Lunar · Spacecraft · Mission design

1 Introduction and Overview

In January 2004 the President of the United States unveiled the Vision for Space Exploration which charted the path for the return of humans to the Moon and deep space. The first milestone in that plan was an unmanned lunar orbiter to be launched in late 2008. By late 2004 that first objective had become the Lunar Reconnaissance Orbiter (LRO) mission and as 2005 began development had started at NASA's Goddard Space Flight Center. As the precursor to the manned flights to the Moon the primary LRO mission objective was to

C.R. Tooley (✉) · M.B. Houghton · C. Peddie · D.F. Everett · C.L. Baker
NASA's Goddard Space Flight Center, Greenbelt, MD 20771, USA
e-mail: craig.r.tooley@nasa.gov

R.S. Saylor Jr.
Honeywell Technology Solutions Inc., Greenbelt, MD, USA

K.N. Safdie
Alliant Techsystems Inc., Beltsville, MD, USA

collect the data necessary to support the design and development of the systems and missions which would return human beings to the Moon. The LRO mission requirements fulfill three broad objectives:

(1) Identify Safe Landing Sites This is accomplished via precision altimetry and high resolution imaging. The resulting global maps will be anchored with an improved global geodetic grid. The detailed topography and rock abundance information will enable the selection of safe landing sites.
(2) Locate and Characterize Potential Resources The primary focus is on Hydrogen/water at the lunar poles, regions of continuous solar illumination, and the overall mineralogical composition of the Moon.
(3) Characterize the Lunar Space Radiation Environment Principally the energy level and flux intensity of high energy particles, data which will be used to aid in the design of mitigation techniques to protect human explorers.

To meet the mission objectives NASA solicited proposals for instruments that could contribute to them. While the solicitation (NASA Announcment of Opportunity 2004) was fairly detailed in the types of data desired, the scope and top level requirements for the mission were not fully defined until after the selection of the instruments and their associated measurement proposals. The solicitation did not specify specific instrument types but did favor instrument designs with heritage from other planetary instruments in recognition of the aggressive LRO schedule. The suite of LRO instruments consists of six competitively selected instruments and a seventh directed technology demonstration instrument (Mini-RF). The instruments and key aspects of their design and data products are presented in Table 1. For more information on the LRO instruments and their planned measurement investigations, see Chin et al. (2007).

The Orbiter is illustrated in its operational configuration in Figs. 1 and 2 is a photograph of LRO in the open Atlas launch vehicle fairing atop the LCROSS spacecraft prior to final encapsulation. The LCROSS lunar mission (NASA: LCROSS Pages. www.nasa.gov\lcross) was LRO's sister lunar exploration mission and was launched with LRO on the Atlas rocket.

The fundamental mission and spacecraft performance requirements were defined by the requirement to accommodate the selected instruments and provide for the collection and communication of their measurement data products. Together with the programmatic constraints these resulted in a set of key driving requirements which led to the overall design of the mission and spacecraft. These are summarized and discussed below:

Discovery-Class Mission with an Accelerated Development Schedule LRO was a NASA directed, rather than competed, mission and it was stipulated when the project was initiated that a Discovery-class mission was envisioned and it was to be launched before the end of 2008 to support the NASA Exploration Program's requirements. This translated into an approximately $450M budget and four years to develop the mission from concept to launch. These realities influenced all aspects of the LRO design as it was clear that the success or failure of the mission would be judged not only on technical performance but also on meeting the schedule and budget. As NASA's first act in the Exploration Initiative it was imperative the mission was a complete success. Thus the use of available technologies and in-production hardware and an overall simplicity of design became central to the design effort to enable the rapid pace of development.

Orbit and Mission Duration To meet data resolution requirements and ensure data was collected across all seasonal lunar lighting conditions a 50 km polar orbit and a primary

Table 1 LRO instruments

LRO Instrument Summary

Instrument		Primary Measurements	Key Characteristics
CRaTER—Cosmic Ray Telescope for the Effects of Radiation *Boston University/MIT*		• Lunar & Deep Space Radiation Environment	– Measures LET Spectra behind Tissue Equivalent Plastic (TEP) – Nadir FOV 70 deg. – Zenith FOV 35 deg. – Operates autonomously over entire orbit
DLRE—Diviner Lunar Radiometer Experiment *UCLA/JPL*		• 300–500 m resolution scale maps of surface temperature • Characterizes rock abundance and albedo • Silicate mineralogy	– 40–400 K 9 channel radiometer – 21 pixel push-broom line detector – 3.15 km swath – Near copy of MCS on MRO – Operates autonomously over full orbit
LAMP—Lyman-Alpha Mapping Project *Southwest Research Institute*		• Maps of frosts and landforms in permanently shadowed regions using Lyα albedo	– 465–1880 Å pass-band – 0.3 × 6.1 deg. Slit – 1.8 Å full slit spectral resolution – Near copy of ALICE on New Horizons – Operates primarily on night side of orbit
LEND—Lunar Exploration Neutron Detector *Russian Space Agency*		• Maps Hydrogen in upper 1 m of regolith at 10 km scale using neutron albedo	– Collimated neutron telescope – Measures and differentiates thermal, epithermal, and energetic neutrons – Derived from HEND on Mars Odyssey – Operates autonomously overfull orbit
LOLA—Lunar Orbiter Laser Altimeter *Goddard Space Flight Center*		• Global geodetic Topography with 10 cm vertical and 1 km (equator) to 25 m (poles) horizontal resolution • Characterize slopes, roughness, and brightness	– 1064 nm laser pulsed at 28 Hz – Beam split into 50 m/5 spot pattern – 4.4B measurements/year – Operates autonomously over full orbit

mission duration of 1 year was required. This in turn was a design driver for the spacecraft propulsion system design and reliability requirements.

Large On-board ∇V *Capability* In order to perform the necessary Lunar Orbit Insertion (LOI) and the subsequent regular orbit adjustments a large propulsion system was required.

Table 1 (*Continued*)

LRO Instrument Summary			
Instrument		Primary Measurements	Key Characteristics
LROC—Lunar Reconnaissance Orbiter Camera **NAC**—Narrow Angle Camera (2) **WAC**—Wide Angle Camera *ASU/MSSS*	 	• Target imagery at 0.5 m resolution at poles and selected sites. • Global imagery in visible and UV at 100 m resolution • Characterize lighting conditions • Multispectral imagery maps ilmenite and other minerals	NACs: - f3.59 Cassegrain (Ritchey-Chretien) - FL = 700 mm - FOV 2.86 deg per NAC - Push-broom line imager WAC: - f/5.1(vis)/f/5.5(UV), - FL = 6 mm (vis)/4.7 mm (UV) - FOV = 90 deg (vis)/60 deg (UV) - Push-broom frame imager - LROC is timeline driven and operates on day side of every orbit
Mini-RF—Technology Demo of Small Synthetic Aperture radar (SAR) *DOD/NAWC*	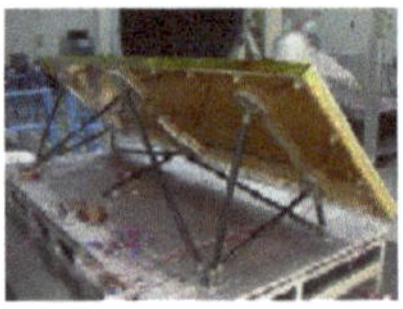	• Radar imagery and interferometry • Flown as demo • Can contribute to topography and resource identification	- X & S band radar - 15–150 m resolution (mode dependent) - Timeline driven when opportunities are available. Can operate over entire orbit

Almost half of LRO's mass at launch was fuel. Accommodation of the large fuel tanks was defining characteristic of the overall spacecraft physical configuration.

Near Continuous Nadir Pointed Operation To fulfill the science data product requirements LRO must be operated in a nadir pointed science operation mode at least 95% of the time during the year. This necessitated designing relatively complex continuously articulated antenna and solar array systems.

High Rate Data Downlink and Large On-board Data Storage The high resolution imagery data products required a high data downlink rate (100 Mb/s) and high capacity on-board data storage to store data when the link is not available. Meeting this requirement led to the selection of a high bandwidth Ka-band science data communication system paired with a new dedicated ground station and a large capacity on-board solid state recorder integrated into the avionics.

Capability for Extended Mission The LRO primary mission requirement was 1 year, as noted above, but in addition LRO was required to have the capability of performing an extended mission of up to 3 additional years. This requirement had to be designed into the consumables budget. More significantly, this required LRO to be designed to survive the most severe lunar eclipses although they did not occur during the primary mission. These

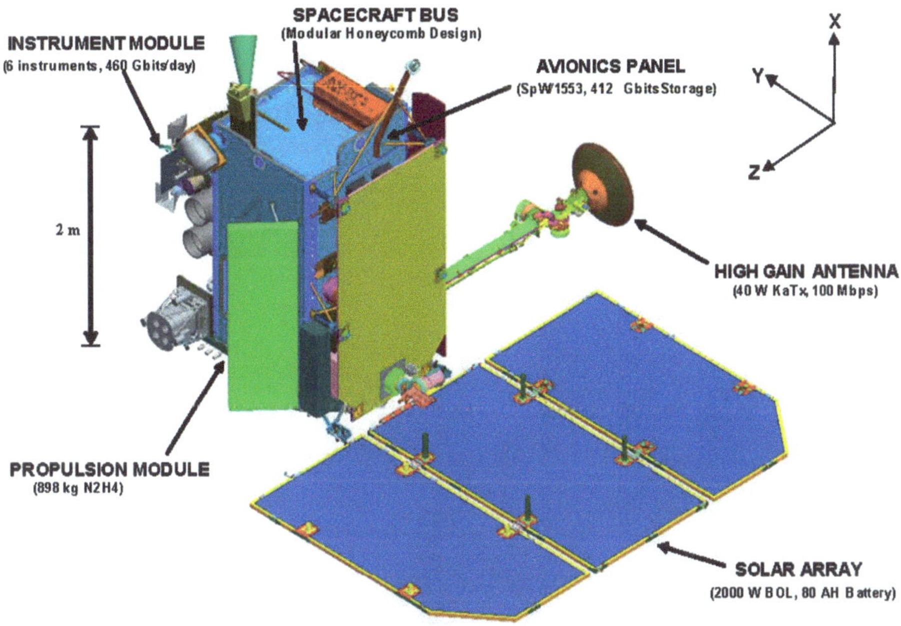

LRO Orbiter Characteristics		
Mass (CBE)	1916 kg	Dry: 1018 kg, Fuel: 898 kg (1313 m/sec)
Orbit Average Bus Power	647 W @ Beta 0	
Data Volume, Max Downlink rate	461 Gb/day, 100 Mb/sec	
Pointing Accuracy, Knowledge	60,30 arc-sec	

Fig. 1 Lunar Reconnaissance Orbiter—deployed configuration with key attributes given

eclipse periods became the worst-case design points for the spacecraft power and thermal systems.

Harsh Lunar Thermal Environment Designing a mission for a long-duration, low-altitude lunar mission with the added constraint that spacecraft must essentially point the instruments nadir continuously proved to be more challenging from a thermal control standpoint that initially anticipated. This reality manifested itself during the design and development in two areas: the overall configuration of the spacecraft which was dominated by the need to radiate almost all heat from a zenith face of the spacecraft (discussed in more detail in the thermal section that follows) and the extensive efforts required to adapt the heritage instrument designs for the environment.

This paper is intended as a companion to the series of papers published simultaneously in this journal which describe each of the LRO instruments and planned investigations in detail. It provides a detailed presentation of the LRO design with the goal, together with the instrument papers, of enabling a comprehensive understanding of how LRO functions and its capabilities. The remainder of this paper is divided into two main sections, the first presenting an overview of the mission design and operation, and the second presenting a description of the design and functional operation of the LRO spacecraft.

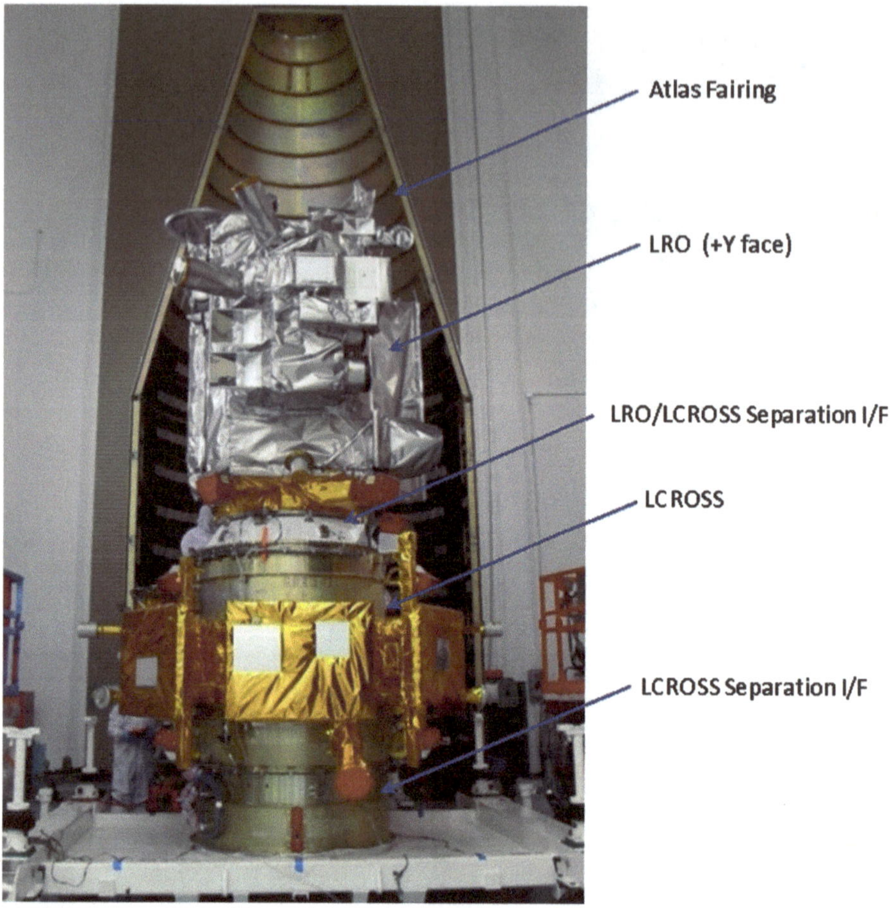

Fig. 2 LRO in Atlas Fairing with LCROSS Spacecraft

2 Mission and Operations Design

2.1 Baseline Mission

LRO was launched on an Atlas V 401 Evolved Expendable Launch Vehicle (EELV) from the Kennedy Space Center (KSC), located on the east coast of Florida, USA at 5:32 pm EDT June 18, 2009. It followed a minimum energy transfer trajectory to the Moon (see Fig. 3), taking four and a half days to complete the journey.

Once in the vicinity of the Moon, LRO began a sequence of Lunar Orbit Insertion (LOI) maneuvers, with the first maneuver (capture) nominally lasting roughly 40 minutes. Subsequent LOI maneuvers were performed over the next several days (see Fig. 4), culminating in a low maintenance, 30 × 216 km quasi-frozen orbit (Folta and Quinn 2006) that LRO used during its 60 day commissioning period (see Fig. 5). After completing its commissioning activities on September 15, 2009, LRO moved to its nominal 50 km polar mapping orbit (Fig. 6), where it will remain for a minimum of 1 year, collecting data over the entire lunar surface under all possible lighting conditions.

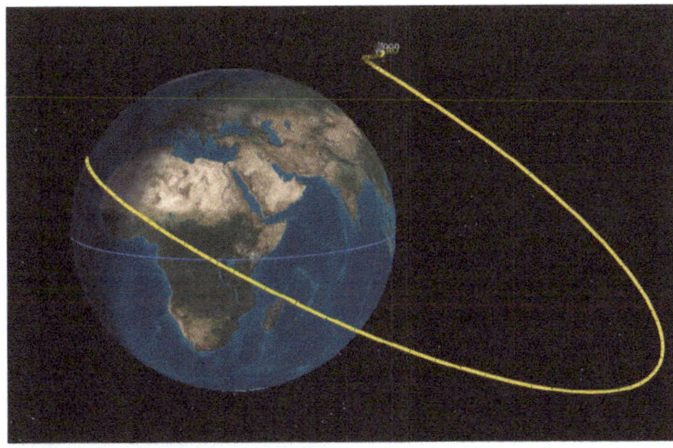

Fig. 3 Minimum energy transfer to the Moon (4–5 days)

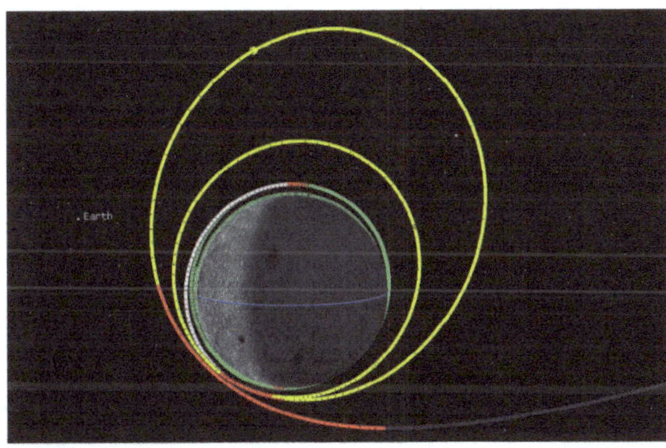

Fig. 4 Lunar orbit insertion sequence (4–6 days)

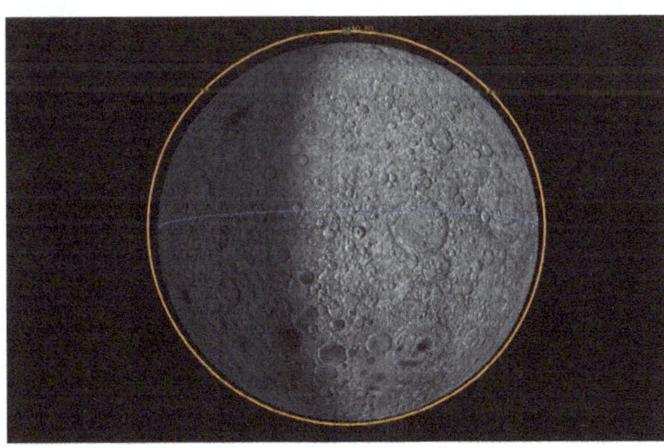

Fig. 5 30 × 216 km quasi-frozen orbit (up to 60 days)

2.2 Launch Windows

LRO launch window selection was primarily driven by the seasonal lighting conditions on the lunar surface and the desire to definitively identify any permanently lit or permanently shadowed areas near the lunar poles. To achieve these goals, LRO needed to be oriented in a particular orbit plane relative to the solar cycle, to maximize observation of the most extreme polar lighting conditions. Specifically, the LRO orbit plane must be oriented such

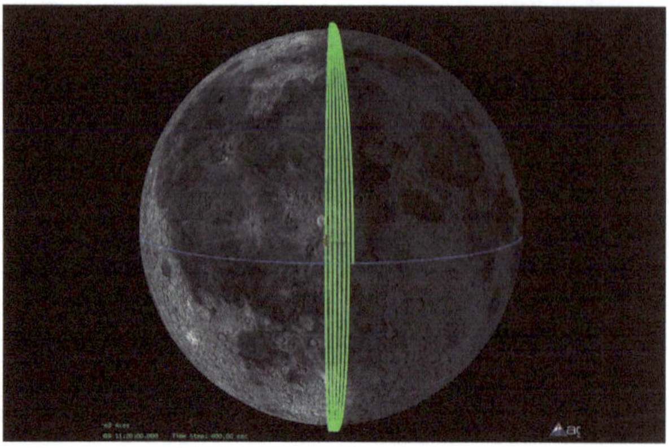

Fig. 6 50 km polar mapping orbit (at least 1 year)

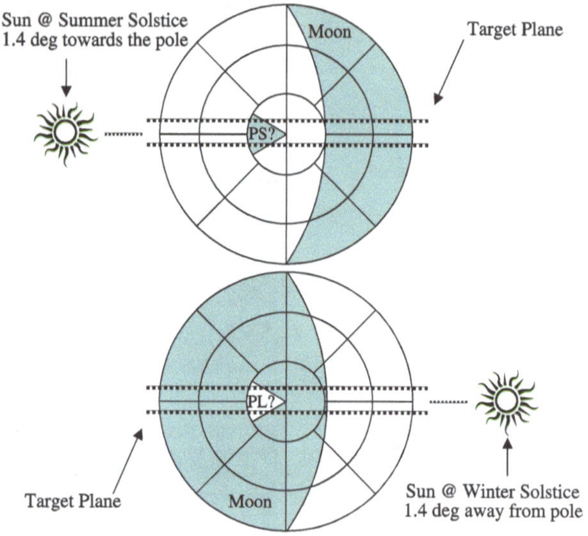

Fig. 7 The search for permanent light (PL) and shadow (PS) regions

that it is near edge-on to the Sun (0 deg beta-Sun angle) during the lunar solstice periods (see Fig. 7).

At insertion (LOI), LRO's orbit plane is fixed with respect to the Earth (~85 deg beta-Earth angle), regardless of the relative geometry between the Earth, Moon, and Sun (see Fig. 8). Therefore, since the lighting constraint drives the target plane to a *particular inertial* orientation (low beta-Sun angle at the solstices), it constrains the Earth departure (launch) to be within a few days of the point at which the natural insertion plane coincides with the inertial target plane. Forcing the difference between the two to be below 20 deg constrains the launch window to 2–3 day periods every two weeks (see Fig. 9 for examples of launch windows).

2.3 Lunar Orbit Insertion

LRO's launch enabled it to reach the Moon in roughly four and a half days, at which time a successful Lunar Orbit Insertion maneuver (LOI-1) was required to capture into a stable orbit around the Moon. The LOI-1 burn and the subsequent LOI maneuvers were mission

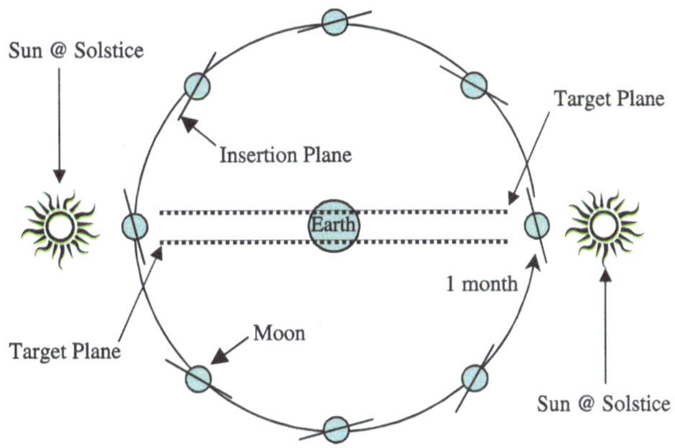

Fig. 8 LRO insertion plane relative to Earth and target plane

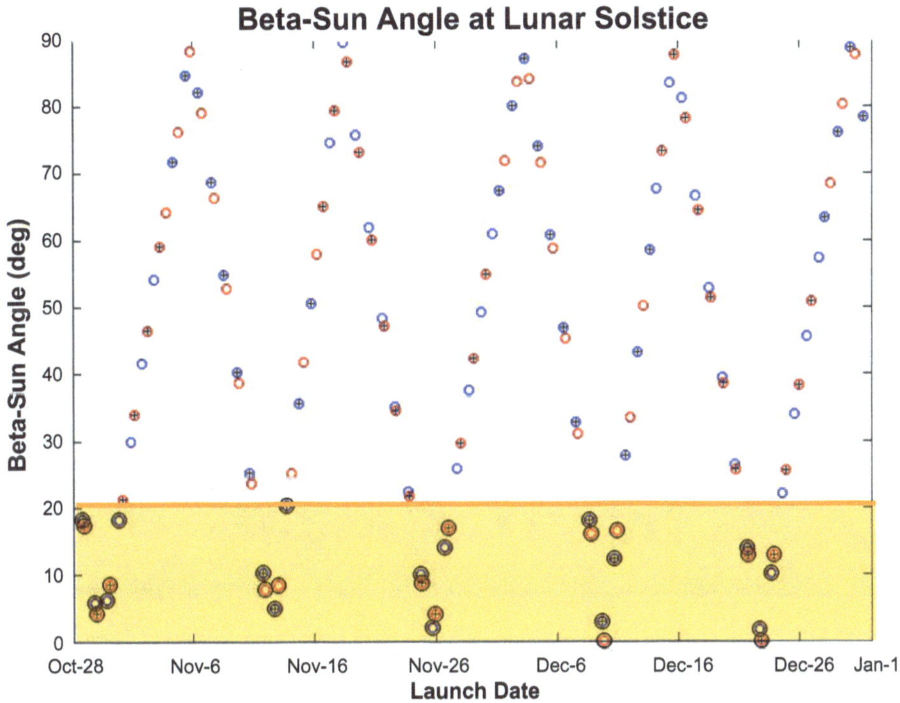

Fig. 9 LRO launch windows; 2–3 days every 2 weeks meet the beta-sun angle requirement

critical, and many steps were taken to ensure their flexibility and probability of success. In addition, these maneuvers determined the fuel budget for the entire LRO mission.

The first of these maneuvers (LOI-1) provided the necessary change in velocity (∇V) to allow LRO to be captured by the Moon's gravitational field. The magnitude of this maneuver is a function of the exact geometry of the transfer trajectory and the allocation of an adequate amount of fuel to the LOI-1 maneuver in turn establishes the fuel budget for the mission (see Table 2) (Beckman 2007). Although LRO's launch vehicle imposed a 2000 kg mass limit on LRO, the orbiter carried a ∇V requirement of 1270 m/s and had a maximum fuel load of

Table 2 LRO ∇V and fuel budget pre-launch

Mission plan	ΔV (m/s)	Fuel (kg)	Source
Mid-Course Correction	30	28.6	3σ LV errors
LOI-1 (with Checkout)	591	454.5	Deterministic
LOI-2 thru LOI-4	362	226.3	Deterministic
Mission Orbit Insertion	56	31.9	Deterministic
Orbit Station-Keeping	162	91.5	Deterministic
Ext. Mission/Margin	69	33.1	3 yrs Frozen
Momentum Unloading	–	17.0	4 years Total
Other (Residuals)	–	15.4	Conservative
Total	1270	898	

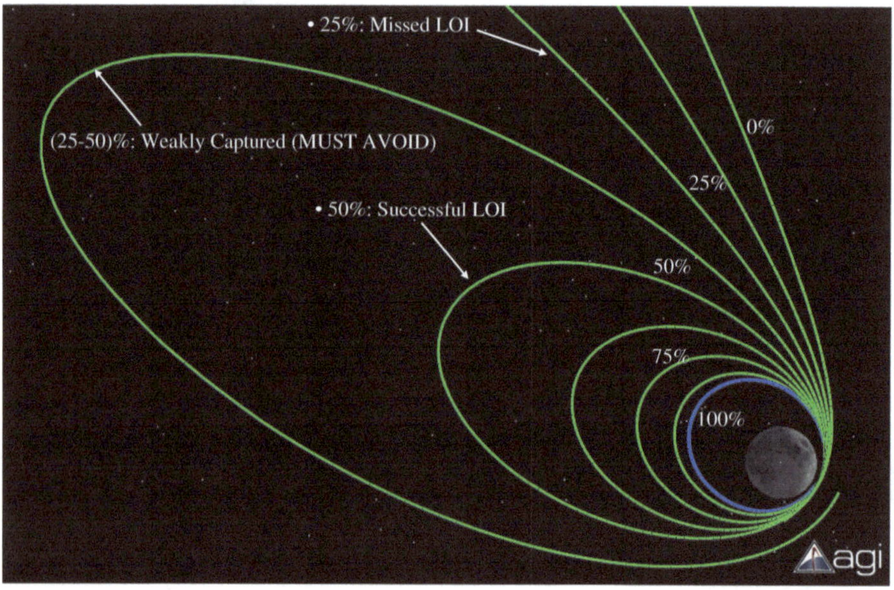

Fig. 10 Lunar Orbit Insertion (LOI) orbits shown as a function of percent completion of LOI maneuver

895 kg (hydrazine). The maximum fuel load was set by the capability of the available tanks LRO was designing with. The required ∇V, in combination with the expected effective specific impulse (Isp) of LRO's propulsion system (212.2 s), set the maximum allowable total launch mass for LRO at 1965 kg. LRO's final measured mass was 1915 kg, with the difference contributing to a significant increase in reserve ∇V.

Many measures were implemented to ensure the success of this critical maneuver. Firstly, the burn (lasting approximately 40 minutes), was designed such that only 50% of the ∇V would be required to capture into orbit (see Fig. 10). In addition, it was designed such that capture could be achieved either through thrust or duration, defending against possible failures in the propulsion system, as well as possible interruptions caused by software and/or processor faults.

The propulsion system, described in a following section, is fully redundant with two banks of two 88 N insertion thrusters and two banks of four 20 N attitude control thrusters. The insertion thrusters deliver a total force of ~350 N, twice the required thrust needed for lunar capture. In the event of any thruster failure, either of the insertion thruster banks could have been used with either of the attitude thruster banks in order to successfully execute an insertion maneuver. With half the thrust, the burn duration would slightly more than double (due to increased finite burn losses), but the resulting orbit would be stable, and the mission could continue.

Similarly, at full thrust, a 40 minute burn duration is twice what is needed for a successful lunar capture. If the burn was interrupted for less than 20 minutes, for any reason, the resulting orbit would still be stable, and LRO would be able to continue with its mission. Restart procedures were created for all conceivable scenarios, and the LRO operations team was thoroughly trained in executing these procedures.

In the event of a catastrophic failure or interruption (one that results in less than 25% of the required ∇V), LRO would not be captured by the Moon's gravity (see Fig. 10), but would have one final chance of getting into lunar orbit. It would have required a deep space maneuver within 10 days of the first lunar encounter, used to target a second lunar encounter about 90 days later (see Fig. 11). This would have depleted significant amounts of fuel, but LRO would still have been able to get into lunar orbit, albeit with limited options in terms of orbits and durations. It would have been able to fly in a higher (~215 km), circular orbit for up to a year, or go into and stay in the low maintenance, 30 × 216 km, quasi-frozen orbit originally meant for commissioning for up to 3 years.

Finally, if, as a result of failures and/or interruptions, the delivered ∇V was in the 25–50% range, LRO would have been weakly captured by the Moon's gravity. This would have resulted in chaotic behavior and have had an irrecoverable impact on LRO's orbital inclination. If this happened, LRO would not have been able to achieve any sort of polar lunar orbit, and its primary objectives could not be met. This catastrophic threat factored into all contingency planning, and was a prime focus of early mission rehearsals and training.

LRO successfully completed LOI maneuvers, without the necessity of exercising any contingency procedures, in late June 2009 and achieved the planned commissioning orbit. Commissioning took roughly 60 days, and LRO was then transitioned into the primary mission 50 km polar mapping orbit on September 15, 2009.

2.4 On-Orbit Operations

Once in orbit around the Moon, LRO's universe became essentially Moon-centered (see Fig. 12). From this perspective, the Earth circumnavigates the Moon once a month, and the Sun circumnavigates the Moon once a year. During normal operations, the $+Z$ axis of the Orbiter is pointed toward the Moon, with the Sun remaining on the $-Y$ side of the X–Z plane. The solar arrays track the Sun in two axes, and the high-gain antenna points toward Earth whenever it is in view. LRO's orbit has a mean period of 113 minutes and a maximum eclipse time of 48 minutes, occurring at 0° Sun-beta angle. As previously discussed, LRO's orbit is targeted so that the lunar solstices occur near these 0° Sun-beta angle periods. These also mark the points at which LRO executes a 180° yaw flip to keep the Sun on the solar-array-side ($-Y$) of the spacecraft.

LRO's orbit will be in full view of the Earth for ~2 days twice per month, and in full view of the Sun for ~1 month twice per year. LRO's momentum management and station-keeping maneuvers are executed while the orbit is in full view of the Earth.

LRO will make use of a global network of S-band ground stations for nominal spacecraft tracking (at least 30 minutes per orbit) and one Ka-band station for downlink of all the stored

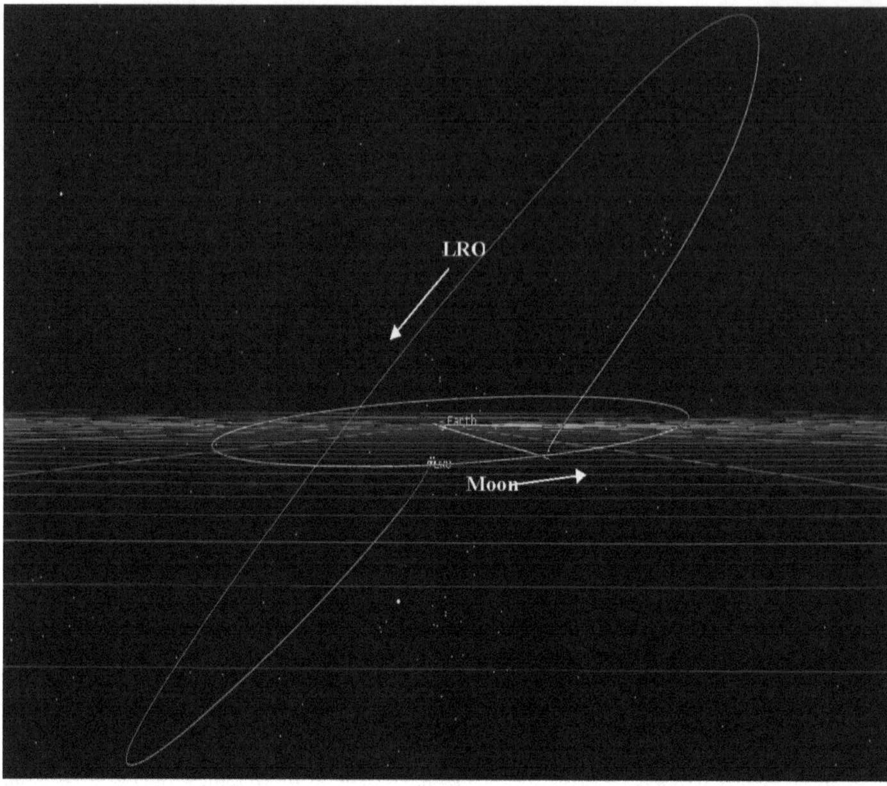

Fig. 11 Deep space recovery trajectory in the event of a missed LOI or less than 25% complete LOI propulsion burn

instrument and spacecraft data. Nominally, LRO will never be out of contact with the ground for more than one hour at a time. A depiction of the LRO ground system is given in Fig. 13.

A snapshot of the nominal on-orbit operations over 3 different time scales is shown in 14. Station-keeping maneuvers and instrument calibrations occur once a month. Momentum management maneuvers occur every 2 weeks. There is an S-band pass every orbit (12 per day), and 4 (on average) Ka-band passes every day. Most of the instruments operate autonomously over the course of a single orbit, while two (LROC and Mini-RF) require daily tailored command timelines. Nominally, LRO will receive a new command timeline from the ground once per day.

2.5 Data Downlink

On a given day, about 460 Gbits of data are generated on-board the LRO spacecraft (see Table 3). Data is downlinked at 100 Mbps through a single Ka-band ground station at White Sands, New Mexico, USA (designated WS1). On average, there are four passes between LRO and WS1 every day, each lasting 45 minutes, but the actual number fluctuates between 2 and 6, as the Moon moves through its entire declination range each month (as seen from the Earth). Table 4 shows the effect that this has on the Ka-band link utilization. Even in the worst case (2 passes), there is sufficient time to downlink the entire day's data volume.

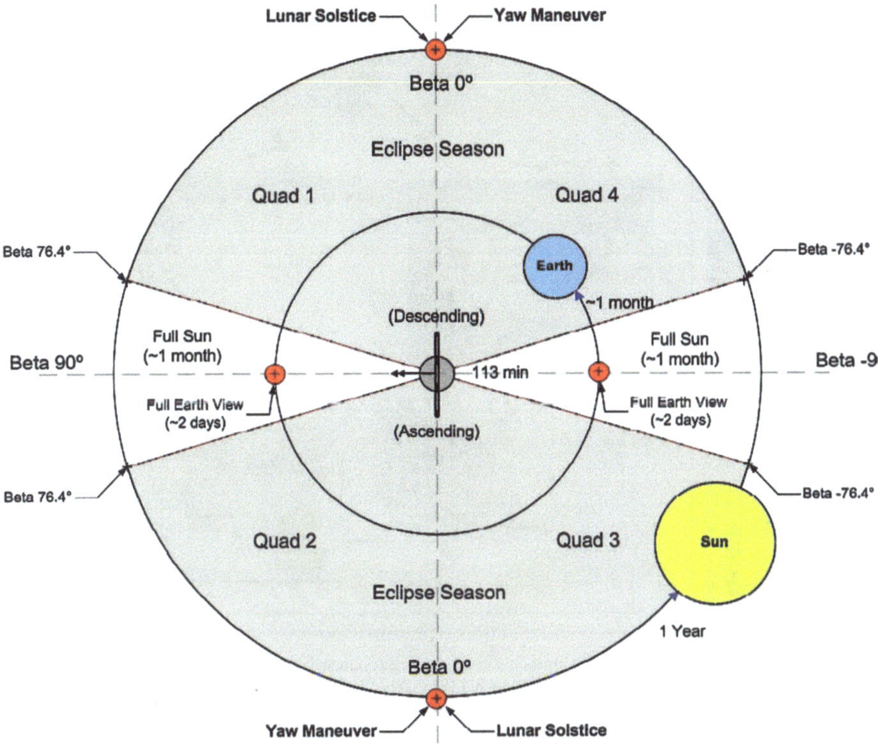

Fig. 12 "Moon Centered Universe" provides useful diagram for depicting the relationship of the LRO orbit to the Earth and Sun

Table 3 Breakdown of LRO's daily data volume

LRO Data Volume Budget				
Type	Data per Orbit (Mbits)	Data per Day (Gbits)	Files per Orbit	Files per Day
CRaTER	610.20	7.78	2.00	25.49
Diviner	180.52	2.30	22.60	288.00
LAMP	168.22	2.14	2.00	25.49
LEND	20.52	0.26	2.00	25.49
LOLA	226.29	2.88	2.00	25.49
LROC	34,659.16	441.67	28.00	356.81
Spacecraft Housekeeping	379.65	4.84	2.00	25.49
Total (Gbits):	36.24	461.87	61	772

2.6 Lunar Orbit Station Keeping

Lunar orbits can be characterized by the evolution of their eccentricity and argument of periapsis over time. The Moon's non-uniform gravitational field causes significant perturbations to these two parameters (Beckman 2007). Figure 15 illustrates the evolution of these parameters, over time, for LRO's two main lunar orbits (the 30 × 216 km quasi-frozen com-

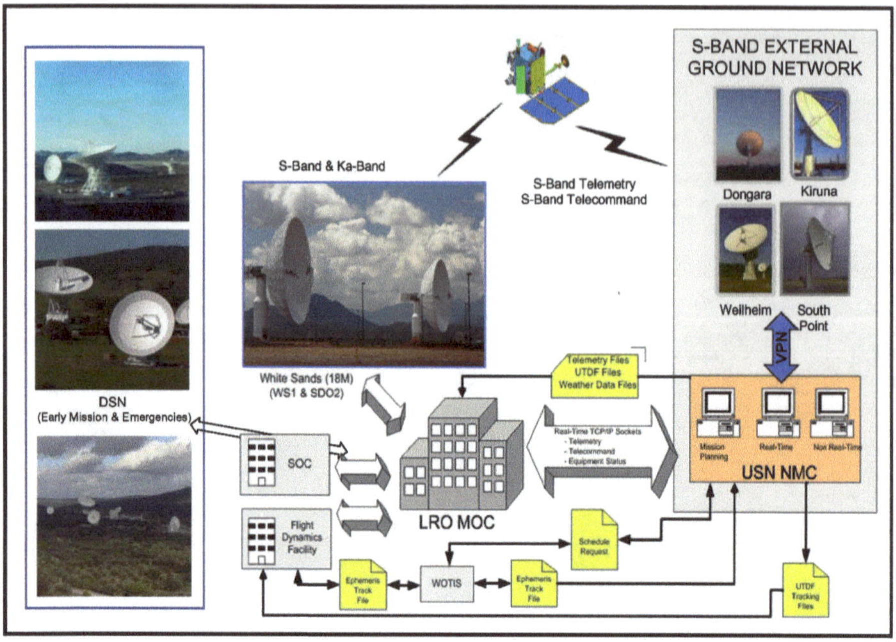

Fig. 13 Representation of the LRO Ground System showing relationship between Mission Operations Center (MOC) and Flight Dynamics Facility at NASA GSFC, the ground stations, and the Instrument Science Operations Centers (SOC). Note only one of the seven SOCs is shown

Table 4 Ka-band downlink utilization

Ka-band Dowlink Utilization*					
Passes	2	3	4	5	6
No.	Pass Utilization minutes)				
1	45.0	45.0	45.0	45.0	45.0
2	40.1	33.4	26.7	20.0	13.3
3	–	6.7	6.7	6.7	6.7
4	–	–	6.7	6.7	6.7
5	–	–	–	6.7	6.7
6	–	–	–	–	6.7
Used	94.6%	63.0%	47.3%	37.8%	31.5%
Margin	5.4%	37.0%	42.7%	62.2%	68.5%

*Based on D/L rate of 100 Mbps—10% overhead

missioning orbit and the 50 km polar mapping mission orbit). The quasi-frozen orbit shows virtually no secular growth in eccentricity or argument of periapsis over time. The altitude is bounded, and the periselene remains over the South Pole. This orbit requires very little station-keeping fuel (about 5 m/s per year).

In contrast, the 50 km mission orbit shows significant evolution in eccentricity and argument of periapsis from month to month. In Fig. 15 this is most clearly seen in the increasing eccentricity as the orbit evolves. If left uncorrected, these perturbations would cause LRO

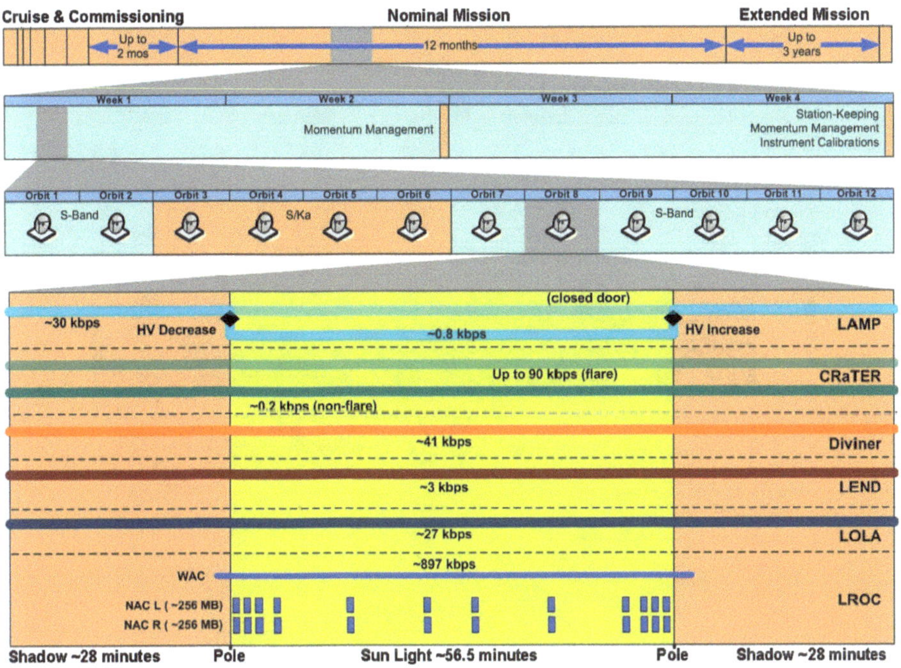

Fig. 14 Typical LRO on-orbit operations

Fig. 15 Evolution of eccentricity and argument of periapsis

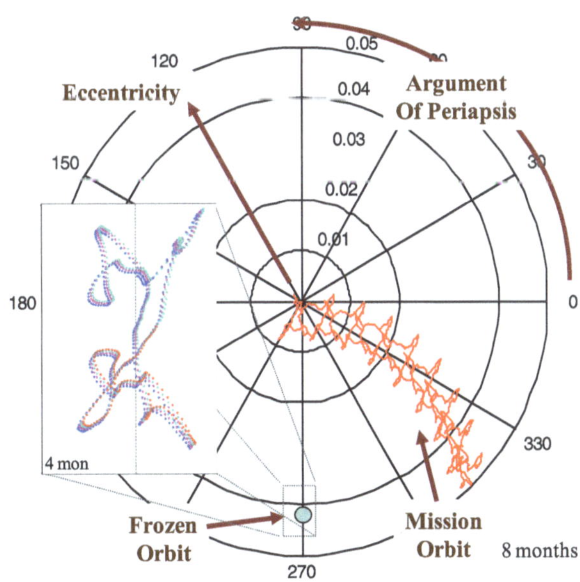

to hit the lunar surface within roughly 68 days. The variation in altitude with and without station-keeping is shown in Fig. 16. The LRO station-keeping strategy makes use of the repeating pattern that can be seen in the eccentricity and argument of periapsis parameters (see Fig. 17). The goal is to precisely reset the pattern at the end of each month, so that the

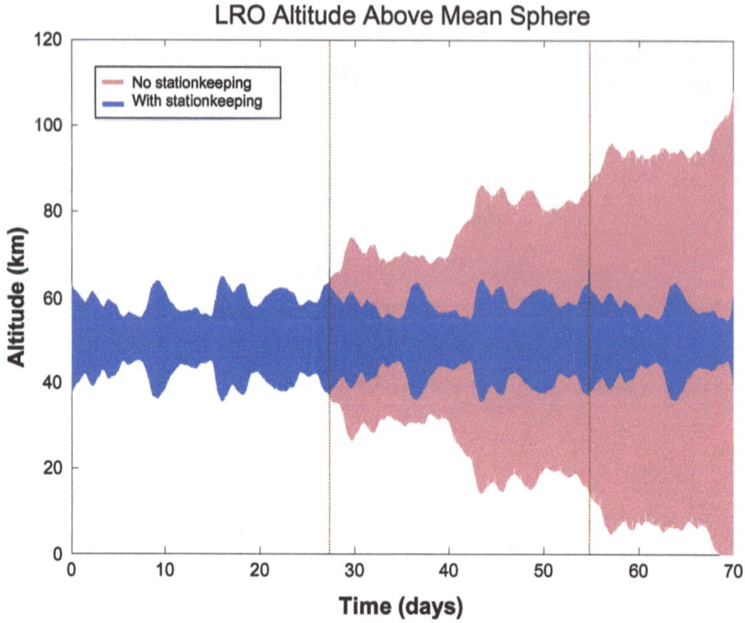

Fig. 16 Simulated LRO mission orbit altitude variation with and without station-keeping, thus showing the necessity of the monthly station-keeping maneuvers

evolution is bounded. This is accomplished with a 2-burn sequence that first circularizes the orbit and then de-circularizes it in the proper direction so as to center the repeating pattern around zero eccentricity. These maneuvers keep LRO's altitude within 15 km of the target 50 km orbital altitude and a safe distance from the lunar surface. This is an unavoidably fuel intensive procedure, consuming approximately 150 m/s of fuel per year.

2.7 Lunar Eclipses

Looking back at Fig. 12, it is clear that the Earth will pass between the Sun and the Moon every month. Although Earth will usually pass far enough above or below the Sun-Moon line that its shadow will not fall on the Moon, twice a year (on average) it will pass close enough to the line that it will cast a significant shadow on the Moon. These lunar eclipses vary in severity over a cycle of roughly four years; with a peak occurring in the middle of 2011 (see Table 5). For descriptive purposes, these eclipses have been labeled as Type 1 (Moon partly in the penumbra only), Type 2 (Moon fully in the penumbra), Type 3 (Moon partly in the umbra), and Type 4 (Moon fully in the umbra), based on existing data like that shown in Fig. 18 (NASA: Eclipse Pages. sunearth.gsfc.nasa.gov/eclipse).

In order to assess the effects of these lunar eclipses on the LRO spacecraft, geometric models were developed to estimate the amount of solar input that would be received during each event (see Fig. 19). These models show LRO coiling through space as it orbits the Moon, while it (with the Moon) passes through the Earth's shadow. These results were used to size the spacecraft battery and design the power management schemes to ensure LRO

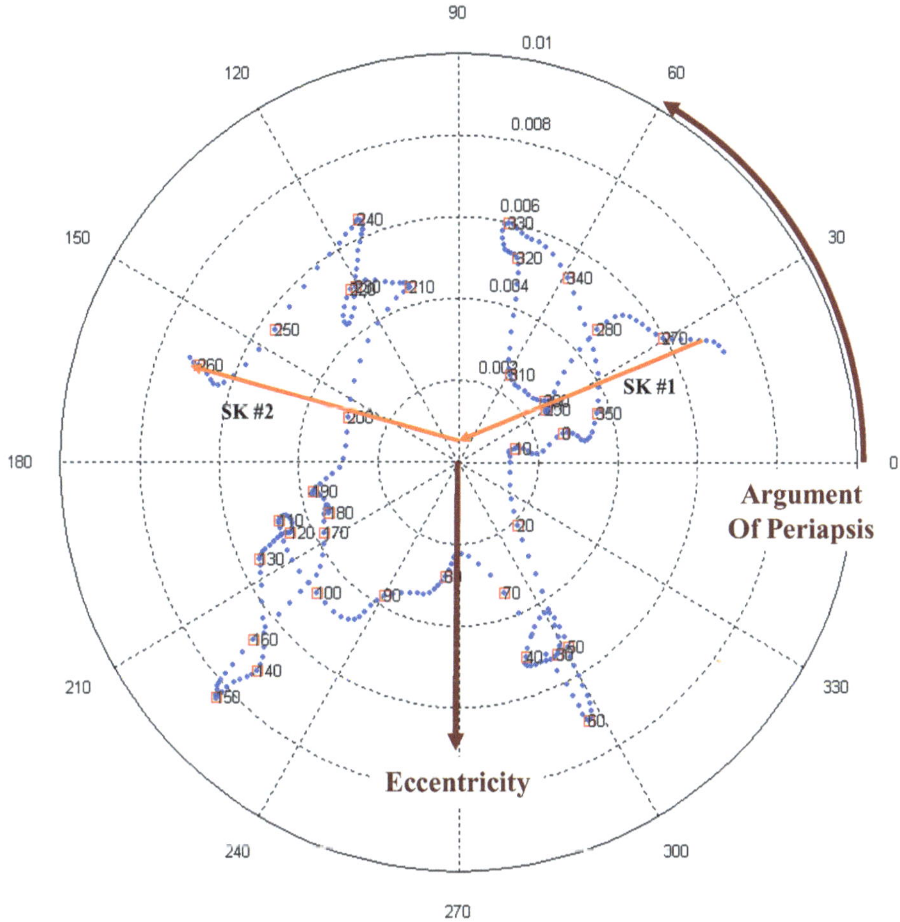

Fig. 17 Mission orbit station-keeping strategy—each month two station-keeping (SK1 & SK2) maneuvers reset the perturbation driven evolution of the orbit eccentricity and argument of periapsis to maintain the orbit at 50 km ±15 km

can survive the eclipses. During its nominal mission, LRO will encounter Type 1, 2, and 3 lunar eclipses. The worst of these will occur in June 2010 resulting in a worst case battery depth of discharge of ~30%, which is within the normal operating range for the spacecraft. During the potential extended mission timeframe, lunar eclipses become increasingly severe culminating with the worst of the cycle in June 2011. This event will result in a worst case depth of discharge of almost 60% and in addition requires the spacecraft to undergo a preheat operation prior to entering the eclipse in order to survive. The spacecraft was designed and qualified to survive this worst case event.

2.8 Extended Mission

Once its one year nominal mission is complete, LRO will, by design, have a significant amount of fuel remaining. Fuel to provide 65 m/s was allocated in the fuel budget for an extended mission and an additional 165 m/s worth of fuel was gained largely due to the

Table 5 Tabulation of lunar eclipses during LRO mission timeframe

Lunar Eclipses: 2009–2013				
Date	Type	Penum.	Partial	Total
2009 Feb 09	(2)	4:03	–	–
2009 Jul 07	(1)	2:12	–	–
2009 Aug 06	(1)	3:16	–	–
2009 Dec 31	(2)	4:15	1:02	–
2010 Jun 26	(3)	5:26	2:44	–
2010 Dec 21	(4)	5:38	3:29	1:13
2011 Jun 15	(4)	5:39	3:40	1:41
2011 Dec 10	(4)	6:00	3:33	0:52
2012 Jun 04	(3)	4:33	2:08	–
2012 Nov 28	(2)	4:41	–	–
2013 Apr 25	(2)	4:12	0:32	–
2013 May 25	(1)	0:54	–	–
2013 Oct 18	(2)	4:04	–	–

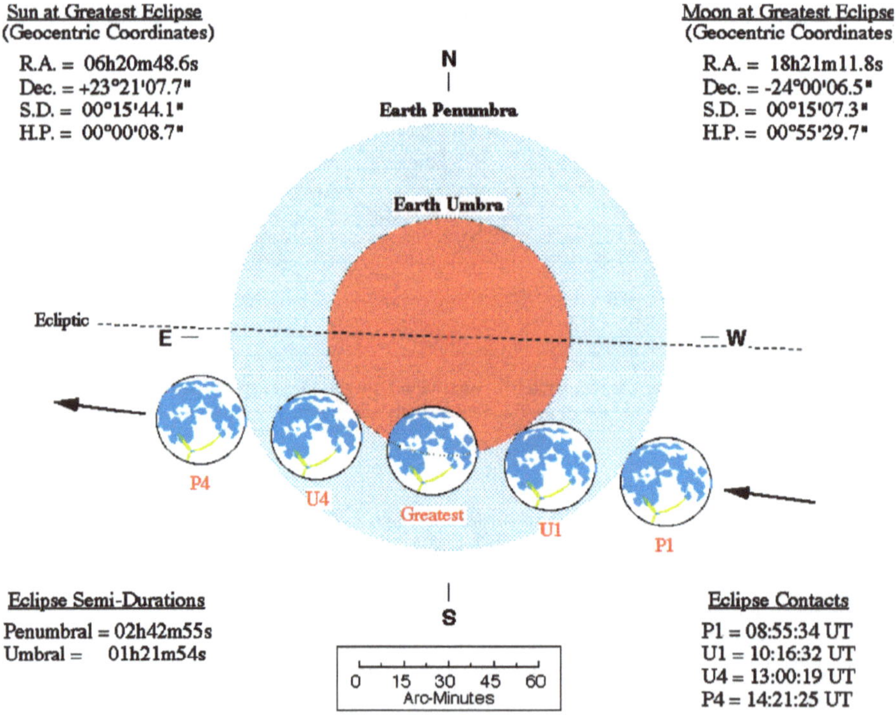

Fig. 18 Example eclipse data—June 26, 2010 Lunar Eclipse

positive mass margin at launch, better-than-budgeted thruster performance, and the very accurate trans-lunar injection by the launch vehicle (thus requiring very little correction). The extended mission is likely take on one of two forms. With the currently projected fuel reserve LRO could remain in the nominal 50 km mission orbit for an additional 18 months.

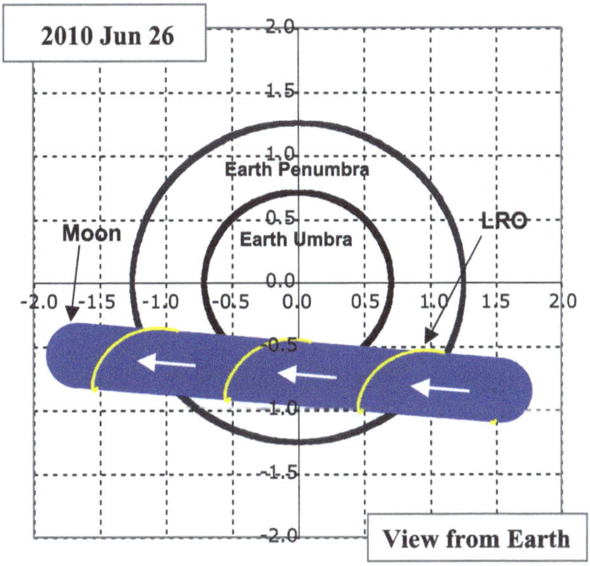

Fig. 19 Geometric model of 26-Jun-2010 Lunar Eclipse showing LRO's path (*spiral path*) about the Moon as it passes through the eclipse. This model was used to design the LRO power and thermal systems to survive the eclipses

Alternatively, that fuel could be used to transition back to the 30 × 216 km quasi-frozen commissioning orbit (∼50 m/s) and remain there for well over 10 years (5 m/s per year), although this is well beyond the design life of the spacecraft. At present NASA has committed to an extended LRO science mission of at least 2 additional years. The desired orbit has not yet been determined.

Regardless of the extended mission selected, LRO will eventually impact the lunar surface. Once its fuel is depleted, it will no longer be able to maintain its orbit, and the perturbations caused by the non uniform gravitational field (even in the frozen orbit) will eventually result in an impact.

3 Spacecraft Design

LRO's spacecraft design was driven by the very aggressive schedule, diverse suite of seven instruments selected for the mission, and unique challenges of a long duration, low altitude lunar mission. These key drivers manifested themselves in LRO's highly modular physical design. Key aspects of the modularity are listed below:

- Propulsion module includes all plumbing and thermal control, enabling independent assembly and proof testing.
- Propulsion module includes two available from the NASA X-38 program, enabling early assembly of propulsion module.
- Spacecraft avionics attached to a single panel, containing embedded heat pipes, enabling a table-top integration and thermal coupling of most heat sources.
- Optical instruments mounted on a separate instrument module, composed of aluminum honeycomb panels with graphite-composite face skin, decoupling high-precision pointing instruments from the rest of the structure.
- Single, zenith-facing radiator, decoupled from large swings in lunar heat load and coupled to most of the structure to provide a large thermal mass to damp variations induced by the environment.

The Orbiter is shown in Figs. 1 and 2. Figure 20 shows an expanded representation of the Orbiter which shows the major components and illustrates the modularity of the spacecraft. Figure 21 shows multiple views of the actual Orbiter to enable key components to be seen. In response to both schedule and cost constraints, GSFC drew upon highly successful in-house experience with rapidly executed single-string design missions and designed LRO as a high-reliability largely single-string spacecraft (Lee 2007). The LRO system block diagram is shown in Fig. 22 which serves as a roadmap for the descriptions of the major subsystems that follow.

3.1 Command and Data Handling Subsystem

The LRO Command and Data Handling (C&DH) subsystem is the centerpiece of the LRO control and data flow architecture. The C&DH subsystem is housed in a single enclosure, see Fig. 23, and includes a Single Board Computer (SBC), a data storage system, a SpaceWire interface (SpW) (ESA Standard 2008) which serves as a high speed connection for some of the instruments and C&DH dataflows, a MIL-STD-1553B low-speed bus controller (Department of Defense Interface Standard for Digital Time Division Command/Response Multiplex Data Bus 1978), interface card to the communication subsystem, analog telemetry conditioning cards and two ultra-stable oscillators (USO). Each of these has heritage from previous NASA and/or DOD missions (Nguyen et al. 2008).

The primary functions of the C&DH system are:

- Hosting the Guidance, Navigation and Control (GN&C) and C&DH Flight Software (FSW)
- Command and data handling functions (command acceptance and distribution, telemetry collection, science data storage) for the instruments
- Telemetry, command and control interface for spacecraft subsystems
- Interface to the spacecraft communications transmitter and transponder for high speed telemetry
- Science data formatting

A block diagram of the C&DH system and its relationships to adjacent systems and instruments is shown in Fig. 23.

The LRO C&DH consists of seven major the elements which are briefly described below.

3.1.1 Single Board Computer (SBC)

LRO's Single Board Computer (SBC) is the processing platform for the Flight Software (FSW). It utilizes the BAE RAD750 133 MegaHertz (MHz) processor and carries 36 Megabytes (MB) of Static Random Access Memory (SRAM) storage to store the executable code, and provide backup data storage. In addition, it provides 64 KB of Start-Up Read Only Memory (SUROM) storage for Boot Code and 4 MB of Electrically Erasable Programmable Read Only Memory (EEPROM) of storage for Application Code. The spacecraft's MIL-STD-1553B Low Speed Bus (LSB) used to communicate with the majority of subsystems and instruments is integrated into the SBC. The Single Board Computer supports SpaceWire interfaces to communicate with LROC, Mini-RF, the HK/IO card, and the Communication cards. It communicates with the Data Storage Board (DSB) cards via a Compact Peripheral Component Interface (cPCI) interface on the backplane

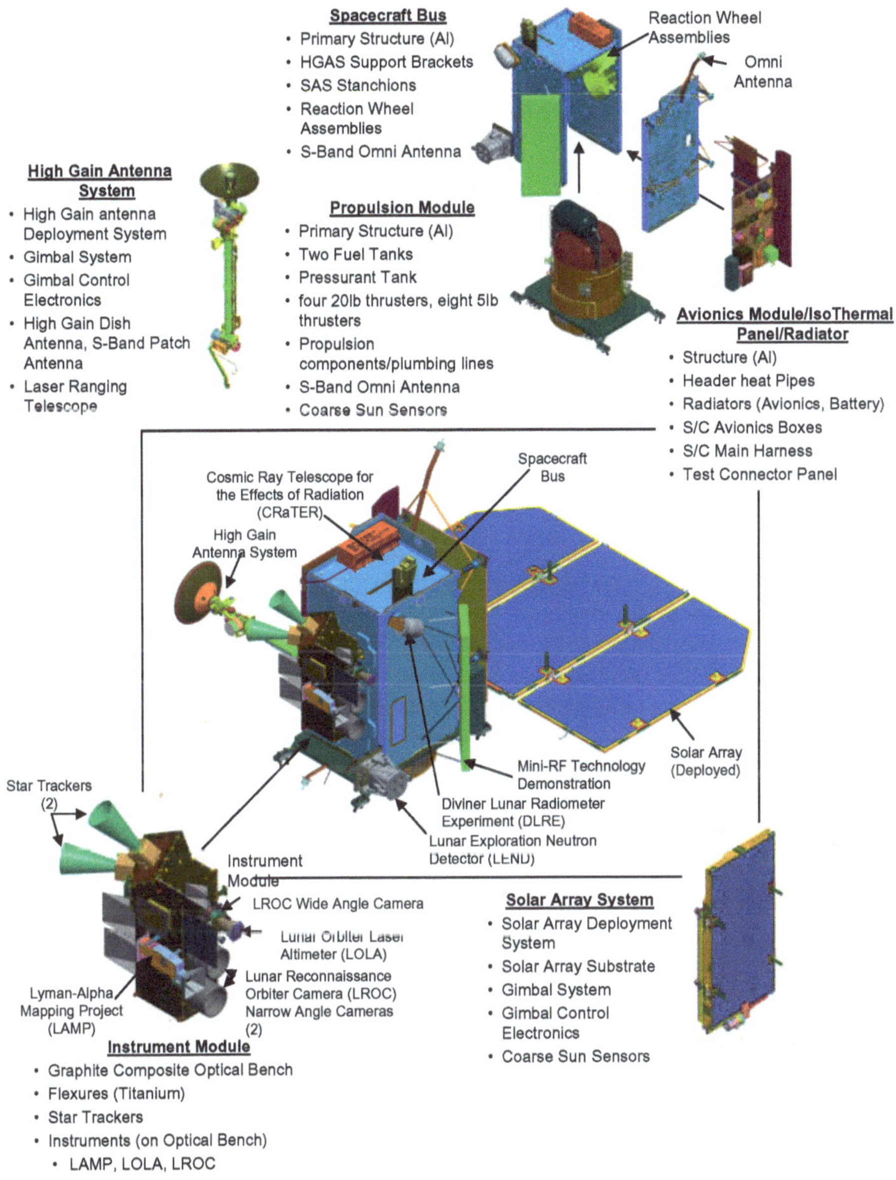

Fig. 20 LRO design overview showing modular construction

3.1.2 Communication Cards (Ka-Comm & S-Comm)

LRO's C&DH subsystem includes two communication cards, the Ka-Comm and S-Comm, which enable spacecraft communications with the ground.

The Ka-Comm card is designed to accommodate a high-speed telemetry interface to Earth using the Ka-Band frequencies. It is connected to the SBC via a SpaceWire link.

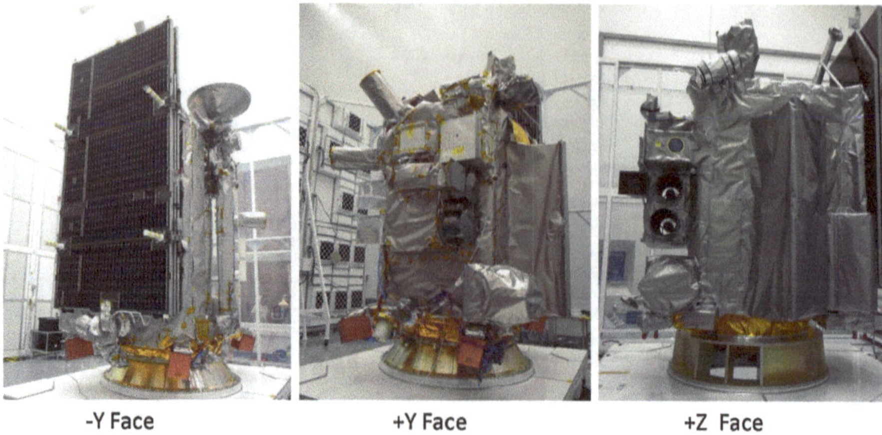

Fig. 21 LRO in launch configuration

During a Ka-Band pass, the data from the SBC flows directly into the Ka-Comm at the SpaceWire link rate of 132 Mega bits per second (Mbps), where the link rate equals the data rate plus the SpaceWire overhead. The data rate is a maximum 100 Mbps. The data is encoded for transmission per Consultative Committee for Space Data Systems (CCSDS) (Consultative Committee for Space Data Systems (CCSDS). http://public.ccsds.org/) recommendations for telemetry using concatenated encoding. Prior to sending on to the Ka-band system for modulation onto the RF carrier the data stream is also split into two streams for Offset Quadrature Phase Shift Keying (OQPSK) modulation.

The S-Comm card is designed to accommodate a telemetry interface to Earth using S-Band frequencies. The S-Comm card is connected directly to the SBC via a 10 Mbps SpaceWire link. During a ground station pass the data from the SBC flows directly into the S-Comm card and is encoded for transmission per CCSDS recommendations for telemetry encoding using concatenated encoding. The S-Comm provides one telemetry stream at up to 1.093 Mbps to the transmitter for Bi-Phase Shift Keying (BPSK) modulation onto the RF carrier.

The S-Comm accepts uplinked CCSDS telecommands from the S-Band transponder at 4 Kbps. Normally commands are forwarded to the SBC for further processing via the SpaceWire interface unless they are tagged as hardware-decoded commands. Hardware-decoded commands are performed directly by the S-Comm card. The S-Comm card can support up to 8 hardware-decoded commands, 4 of which are RS-422 outputs. Hardware-decoded commands are only used in contingency situations.

Both the S-comm and Ka-comm card provide control of the respective Communication Subsystem transmitters via asynchronous low rate serial interfaces.

3.1.3 HK/IO Board

The Housekeeping Input/Output (HK/IO) board maintains the spacecraft mission elapsed timer (MET), maintains the time of received uplink commands, provides timing synchronization to the instruments through a once per second pulse (1 PPS), and provides a unique high speed science data and a low speed command/telemetry interface to the LAMP instrument (all other instruments communicate via SpaceWire or MIL-STD-1553). The HK/IO board interfaces with the C&DH system via SpaceWire thought the Ka-comm card.

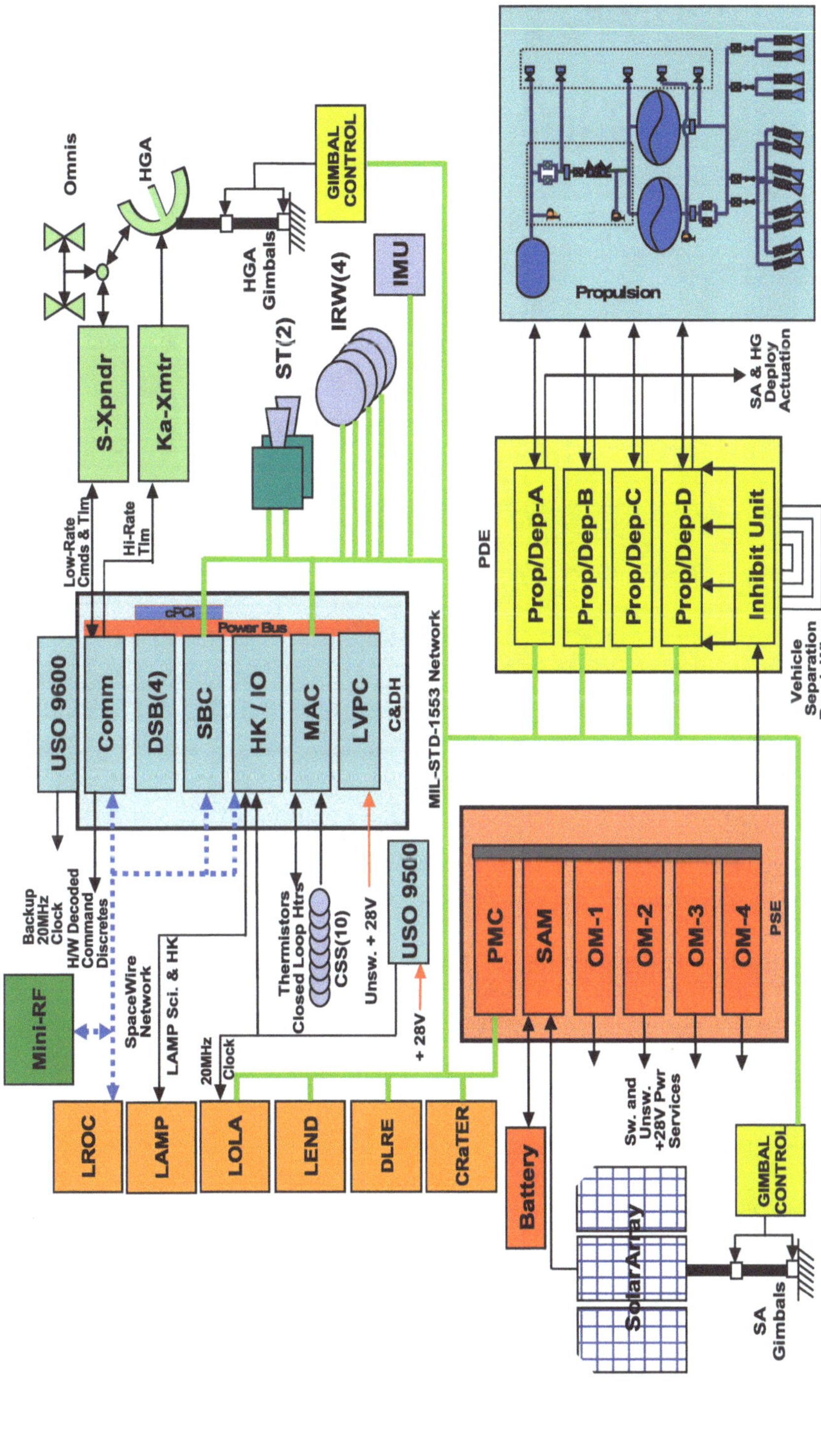

Fig. 22 LRO simplified system block diagram

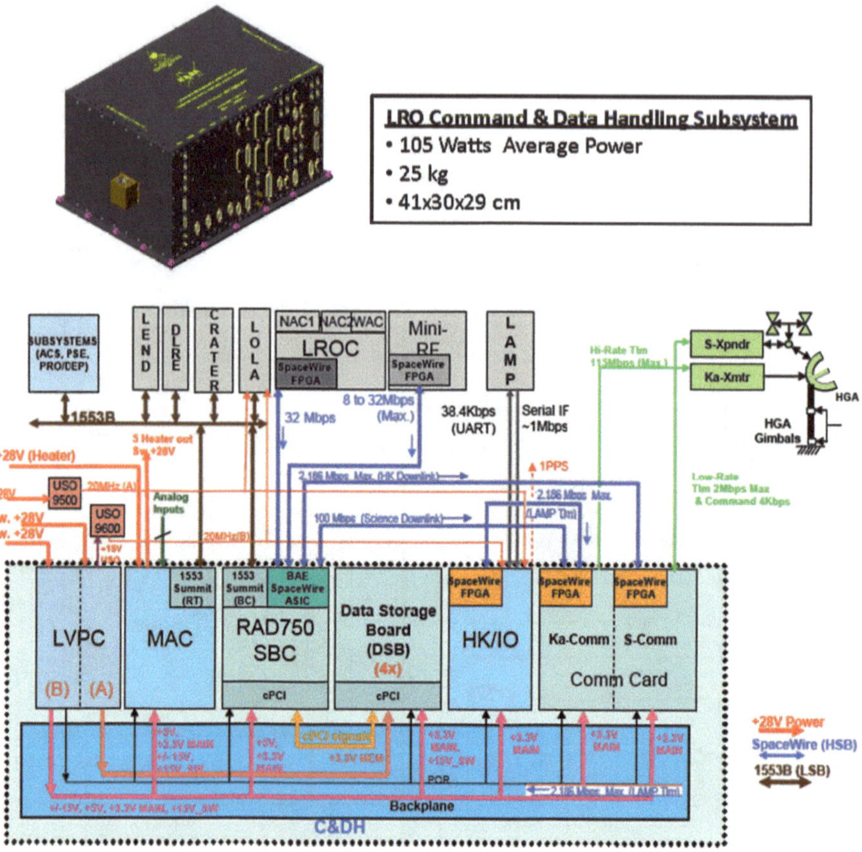

Fig. 23 LRO C&DH system

3.1.4 Multifunction Analog Card

The Multifunction Analog Card (MAC) is a set of interface electronics that samples and conditions a variety of spacecraft housekeeping inputs, including thermistors, hinge potentiometers, coarse sun sensors (CSS), analog voltage inputs from the Ka-band Travelling Wave Tube Amplifier (TWTA), Analog to Digital (A/D) converter scale reference voltages, and C&DH subsystem LVPC outputs. The MAC also controls the transmission of +31 VDC power to switched heater services. All MAC data transfer to/from the SBC is over the MIL-STD-1553 bus.

3.1.5 Data Storage Board (DSB)

The data recorder on LRO is a solid state design consisting of four DSB cards designed as a file system to handle the storage and retrieval of files. The DSBs provide 384 Gigabits (Gbits) of memory capacity for incoming data files for a minimum of 17.5 hours of science data and HK data collection. Data transfer to and from the SBC is via a cPCI interface on the backplane. A robust error detection and correction scheme will correct up to 2 nibbles of 4 bits each in the 32-bit words.

3.1.6 Low Voltage Power Converter (LVPC)

The primary task of the LVPC is to provide power to the subsystems within the C&DH enclosure, including the four DSB boards, SBC, MAC, HK/IO Board, and Communication Cards. The LVPC converts an unswitched primary input voltage with an operational range of +21 Volts Direct Current (VDC) to +35VDC into secondary power outputs. The LVPC also houses the circuitry that drives the magnetic relays to provide power for the RF systems.

3.1.7 Ultra Stable Oscillators (USO)

The C&DH subsystem includes both a primary and redundant USOs to provide precision timekeeping onboard the spacecraft. Each oscillator provides two 20 MHz signals. One signal goes to the C&DH HK/IO card and the second goes directly to the LOLA instrument.

Only one of the oscillators will be powered at any given time. The primary USO has a frequency stability factor of 10 parts per billion (ppb) over one millisecond (ms). It was chosen to provide sufficient stability to meet the needs of the laser ranging system. It receives a +31VDC switched service from the Power Subsystem Electronics (PSE).

The redundant USO has a frequency stability factor of 0.3 parts per million (ppm) over one millisecond (ms). It was chosen to provide sufficient stability to meet the needs for LOLA reconstruction of the orbital ephemeris. It receives a +15VDC switched service from the LVPC.

3.2 Attitude Control Subsystem

The Attitude Control System (ACS) controls the pointing of the LRO spacecraft. ACS control algorithms are implemented in flight software hosted by the LRO C&DH and control the distributed sensors and actuators via the MIL-STD 1553 bus. Through the use of Star Trackers and Coarse Sun Sensors, the ACS determines where the spacecraft is pointing in fine and course accuracies, respectively. A three axis, ring laser gyro (called an Inertial Measurement Unit or IMU) measures the rate at which the spacecraft is rotating. The use of Reaction Wheels allows the spacecraft to smoothly point in any desired direction, as well as compensate for disturbance torques (such as High Gain Antenna movement and Solar Array movement). Eight small thrusters (20 N) are used to provide steering (attitude control) during large thruster (80 N) firings for Lunar Orbit Insertion (LOI). Additionally, the Propulsion System 20 N thrusters are used every two weeks to zero out the momentum buildup in the Reaction Wheels (primarily due to gravity gradient effects) and perform station keeping maneuvers while in the mission orbit. The ACS subsystem components are described in Table 6 and the Reaction Wheels and Star Trackers are shown in Fig. 24.

The Attitude ACS also determines spacecraft attitude, provides guidance to reach the desired inertial pointing vector, and uses actuators to achieve the desired pointing vector. Additionally, the ACS provides pointing support for High Gain Antenna and Solar Array. The nominal ACS control flow is illustrated in Fig. 25.

There are four primary LRO ACS control modes: Sun Safe, Observing, Delta-V, and Delta-H which are illustrated in Fig. 26. Each is described in detail below.

Sun Safe is the mode in which the Orbiter is starts in and was used to acquire the Sun after separation from the launch vehicle. It is the mode that Orbiter will eventually return to in the event of anomalies detected on-board. This mode moves the solar array to the "index" position (with the panel against the Orbiter facing $-Y$), damps any body rates, and points the

Table 6 LRO ACS components

ACS Component	Power (W)	Mass (kg)	Qty.	Performance
Star Tracker (ST)	11.2	4.2	2	20 arc-sec, 90 arc-sec about boresight, 16 deg field of view
Inertial Measurement Unit (IMU)	25	4.5	1	0.2 arc-sec resolution up to 375 deg/s, 100 ppm linearity, 5 ppm stability
Coarse Sun Sensors (CSS)	0	0.011	10	170 deg field of view
Reaction Wheels	16.6	11.9	4	~90 N m s at 24 V, ~120 N m s at 31 V, >0.16 N m torque
Propulsion-Deploy Electronics (PDE)	13.2	9.4	1	FPGA based electronics that control the thrusters, valves, and deployment initiators. Provides safety inhibits

LRO Reaction Wheels mounted to spacecraft structure

LRO Star Trackers mounted on Instrument Module

Fig. 24 LRO reaction wheels & star trackers

$-Y$ axis toward the Sun. Sun Safe uses the coarse Sun sensors and the IMU to sense attitude and rates, and it uses the Reaction Wheels to lower body rates and point the Orbiter. This mode is capable of operation without the IMU in the event it is off or producing bad data by using the Sun sensors to derive rates. Sun Safe mode can also be used to accommodate a stuck solar array.

Observing mode is the primary mission control mode. It uses the star trackers and the IMU, combined with spacecraft ephemeris to determine the Sun, Earth, and Moon vectors in inertial space. The ground operators select solar inertial or lunar nadir pointing, and the Reaction Wheels provide the control. In Observing mode the Orbiter can point to and track any commanded inertial target. This also mode controls the 2-axis gimbals on both the solar array and the high-gain antenna, for tracking of the Sun and Earth respectively.

Delta-V mode provides changes in the Orbiter velocity for orbit control. This mode holds the commanded attitude by propagating rates measured by the IMU. It uses the thrusters for both the velocity change and for attitude control. The 80 N insertion thrusters are only used for the lunar orbit insertion maneuvers, and, in that configuration, the 20 N attitude control thrusters are on-pulsed for control. All other maneuvers only use the 20 N attitude control thrusters, operating in an off-pulse configuration.

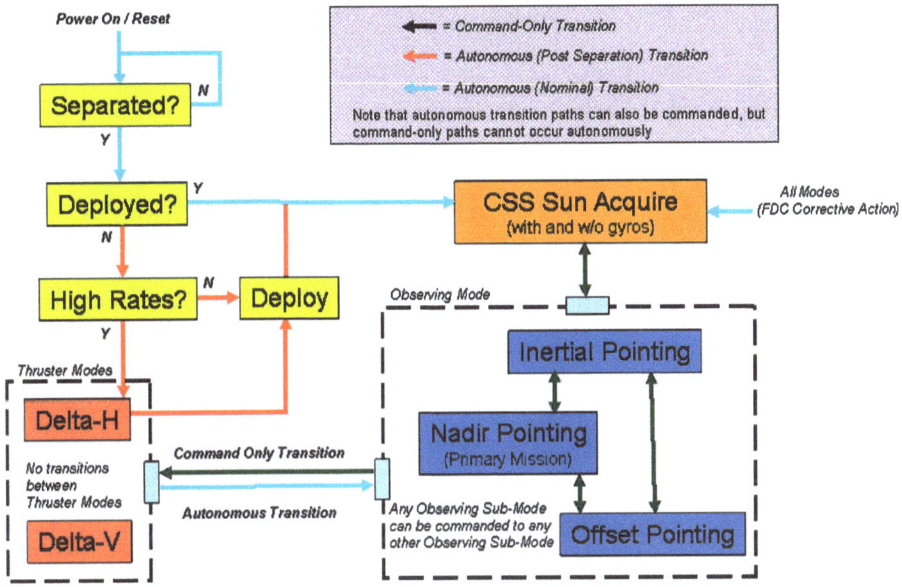

Fig. 25 ACS nominal control mode flow

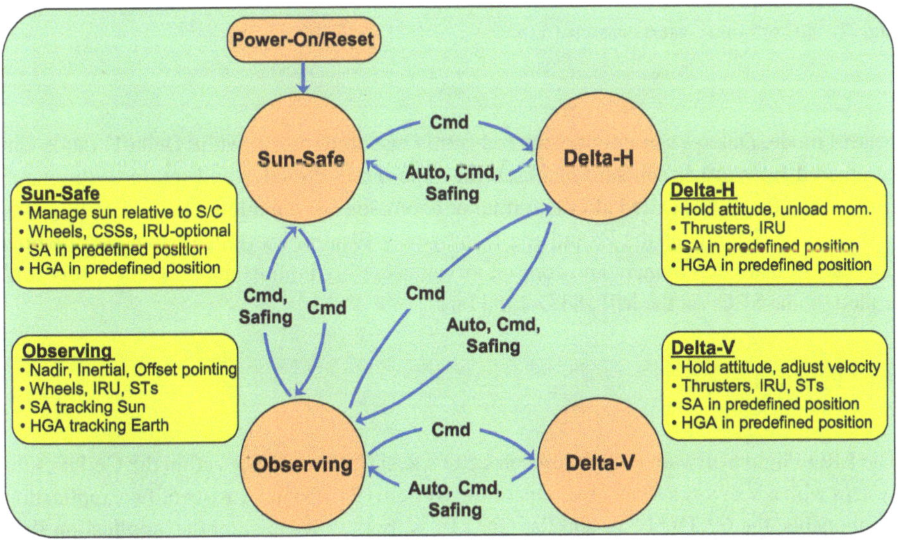

Fig. 26 LRO control modes

Delta-H mode is used to reduce the Orbiter system momentum. This mode holds the commanded attitude by propagating rates measured by the IMU. The 20 N attitude control thrusters fire to maintain attitude as the reaction wheels are commanded to the speed indicated by the operators.

Although the Orbiter autonomously enters a lower-state control mode when an anomaly is detected, ground commands must be provided to move the Orbiter into a higher-state

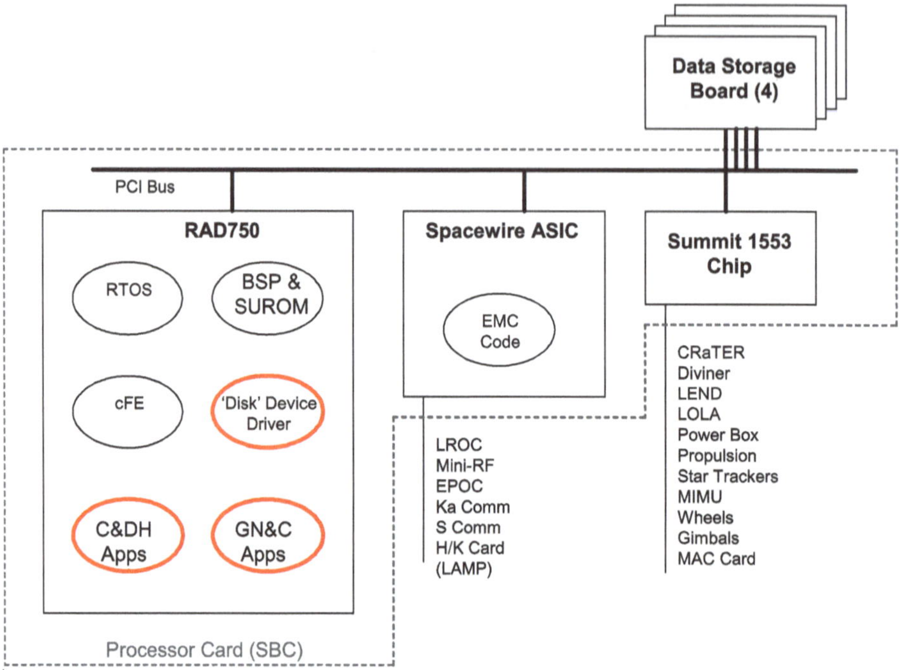

Fig. 27 SBC software context diagram

control mode. Delta-Vcan only be entered from Observing mode, while Delta-H mode can be entered from either sun safe or observing. The thruster modes will return to the mode from which they were called after a normal or anomalous termination.

LRO's ACS includes the Propulsion Deployment Electronics (PDE) component, which provides thruster deployment and control, as well as safety inhibits during launch. It is controlled by the SBC via the MIL-STD-1553 bus.

3.3 Flight Software

The LRO Flight Software (FSW) runs on LRO's RAD750 processor within the C&DH subsystem with a VxWorks Real-Time Operating System (RTOS). It consists of two application components, the C&DH FSW and the GNC FSW, both developed on the Application Program Interface (API) provided by the GSFC Core Flight Executive (cFE) (Wilmot 2006). The cFE provides a mission independent flight software operational environment with a set of services with functional building blocks used to create and host applications. In addition to the RTOS, the cFE, and the applications, the RAD750 also includes a software Board Support Package (BSP) which provides the software interface between the RTOS and the hardware located on the SBC. The software components running within the SBC are illustrated in Fig. 27 and the overall FSW architecture in shown in Fig. 28.

The primary functionality of the LRO FSW can be broken down into 5 areas or Tasks to better understand its overall operation, these which are described below:

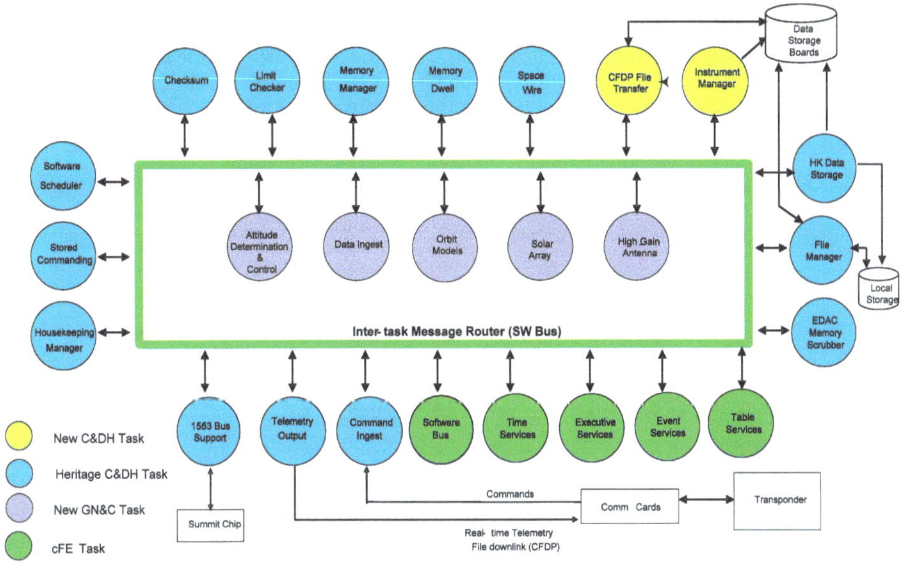

Fig. 28 Flight software architecture with heritage prior to LRO shown

3.3.1 Software Scheduling and Housekeeping

Predictable system activities, such as housekeeping telemetry collection, operate on a schedule which is controlled by the software schedule table. The table allows up to five activities in any 10 ms slot across the one-second scheduling interval and is designed to ensure that tasks can be accomplished in the time available. The schedule begins at the interrupt from the 1 pulse per second (PPS) signal and continues to the next one second tick.

The housekeeping task collects housekeeping data from all of the other software tasks, and it services the processor watchdog. If one of the critical tasks fails to report back in two consecutive cycles, the software will initiate a processor reset.

3.3.2 Commanding and Command Sequences

The majority of LRO's commands are executed by one of two variants of stored commands, although LRO can also execute individual commands from the ground. Absolute Time-tagged Sequence (ATS) commands are executed with reference to Universal Coordinated Time (UTC), and Relative Time-tagged Sequences (RTS) execute stored commands with timing relative to the start of sequence execution. Both types of stored command use a one second resolution. Ground commands (an individual command or a command to start an RTS or ATS) are ingested by the S-Comm card in the C&DH then forwarded to the flight software in the SBC for processing via the SpaceWire interface. Up to eight commands can processed each second.

3.3.3 Limit Checking

The Limit Checker (LC) implemented in the FSW provides the safing functions for the Orbiter. A table defines watch points and action points. The watch points monitor telemetry and are considered "true" when the referenced telemetry data has exceeded a limit. Action points initiate an RTS when one or more watch points meet the criteria indicated by the action point table. The action point table defines which RTS to execute for which watch points, how the watch points should be logically combined, and how long the entire condition should be true prior to execution. This provides great flexibility to define and adjust safing as the Orbiter performance changes with time. The detailed architecture and implementation of LRO's safing design is detailed in Andrews et al. (2008).

3.3.4 Instrument Control and Data Handling

The FSW provides Orbiter level control of each of the instruments, including the handling of commands, housekeeping telemetry, science data and the distribution of time information. In order to take advantage of as much instrument heritage (from previous missions) as possible, each interface has unique characteristics, such as the time message format, command structure, and housekeeping telemetry definition. The FSW handles these unique interfaces and provides the translation to the standard CCSDS interface used by the Orbiter and the ground system. Low-data-rate instruments (CRaTER, DLRE, LEND, and LOLA) communicate with the FSW over the MIL-STD-1553 interface, while SpaceWire is used for the high-rate instruments (LROC and Mini-RF) and for LAMP via the custom serial interface on the HK/IO card. The LROC's high data rate, combined with minimal buffering, requires the FSW to prioritize LROC science data collection and precludes the ability to operate both Mini-RF and LROC simultaneously. The FSW sets up a Direct Memory Access (DMA) transfer to move LROC data to a special location in RAM and then to the DSBs. A similar process is used to transfer data from the DSBs to the Ka-comm card.

All on-board data is stored in files, including configuration tables. The FSW handles the creation of instrument data and housekeeping files. Files are stored in a directory structure and are downlinked by command.

3.3.5 File Transfers

Files are transferred between the Orbiter and ground (in both directions) using CCSDS File Delivery Protocol (CFDP). CFDP code must run on both the Orbiter and the ground system to reliably transfer files. The file sender divides the file into Protocol Data Units (PDU), which are sent to the receiver. The PDUs include information so that the receiver can determine if any are lost. The receiver requests retransmission of any lost PDUs and then provides positive acknowledgment when the entire file is received. LRO will delete a file when positive acknowledgment is received from the ground; otherwise files remain on board, guaranteeing that science data is reliably transferred to the ground. This strategy allows for a large amount of compression of LROC images, without fear of the bit error rate corrupting large portions of an image. LRO also has the capability to downlink data without a corresponding uplink, using a different class of CFDP but the capability will only be used in a contingency.

3.3.6 Time Keeping

LRO's flight software works with the hardware to maintain Orbiter time to within 100 ms of UTC. Post-processed knowledge of time will be accurate to 3 ms. A hardware counter on the HK/IO card counts continuously from the time the card is powered on. This Mission Elapsed Timer (MET) contains 32 bits of seconds and 16 bits of sub-seconds (>136 years before rolling over and a precision of better than 16 μs). The software adds an offset to the MET so that the total corresponds to the time since the LRO epoch (January 1,

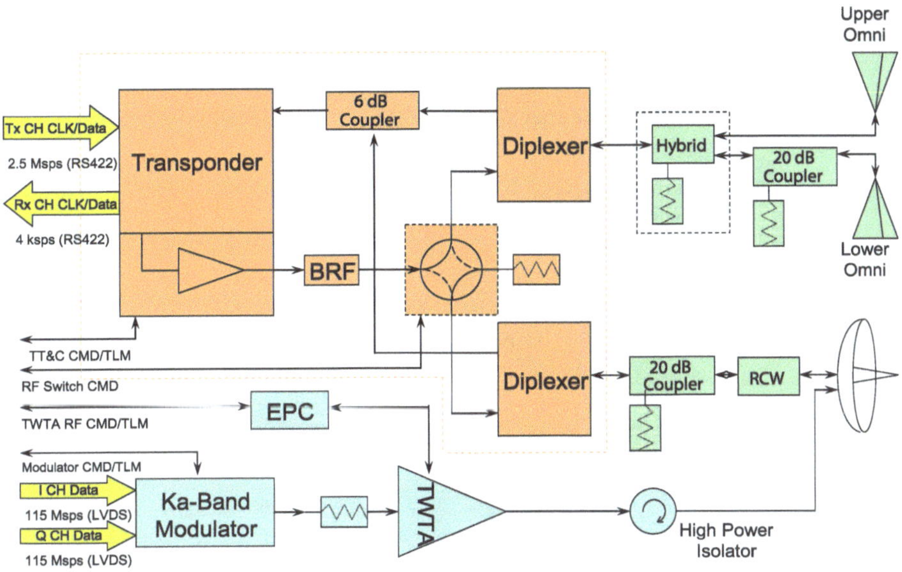

Fig. 29 LRO communications system block diagram

2001 00:00:00 UTC). This time is inserted in the time field of housekeeping and science data packets. The software distributes time with a "time at the tone" message via the instrument data interfaces. The tone is either a 1 pulse per second hardware signal for the 1553 instruments or a SpaceWire time tick for the SpaceWire instruments. The ground adjusts the time on the Orbiter using a feature on the S-Comm card which pulses the HK/IO card and latches the MET when a command is received. The ground compares this MET to the time the command was sent, compensates for the propagation delay, and then sets the software offset accordingly to ensure the correct time on board.

3.4 Communication Subsystem

LRO's communications system consists of an S-band system to provide tracking, telemetry and commanding (TT&C) and a high data rate Ka-band downlink only system for telemetry and science data transfers. The system design is shown in Fig. 29.

The S-Band system has a fixed forward link data rate of 4 kbps, and a selectable on orbit return link data rate between 125 bps and 1093 kbps. It consists of one Spacecraft Tracking and Data Network (STDN) compatible transponder, an S-band Radio Frequency (RF) Switch, and the RF paths to and from the two Omni-Directional antennas and the S-band feed on the High Gain Antenna (HGA). The transponder downlinks at a frequency which is phase-locked to the uplink, providing two-way Doppler tracking information to the ground with an accuracy of better than 1 mm. It also repeats uplinked range tones with a fixed delay, allowing the ground to determine the distance to the Orbiter within 15 m.

The Ka-band system includes a Ka-Band modulator, a 40 W Traveling Wave Tube Amplifier (TWTA) consisting of a traveling wave tube (TWT), an Electronic Power Conditioner (EPC), and a High Power Isolator. The Ka-band return link is also selectable on orbit and varies from 25 Mbps to 100 Mbps

Although the Ka-band system uses only the High Gain Antenna, the S-band system can utilize either the Omni-Directional or the High Gain Antenna for transmit and always uses

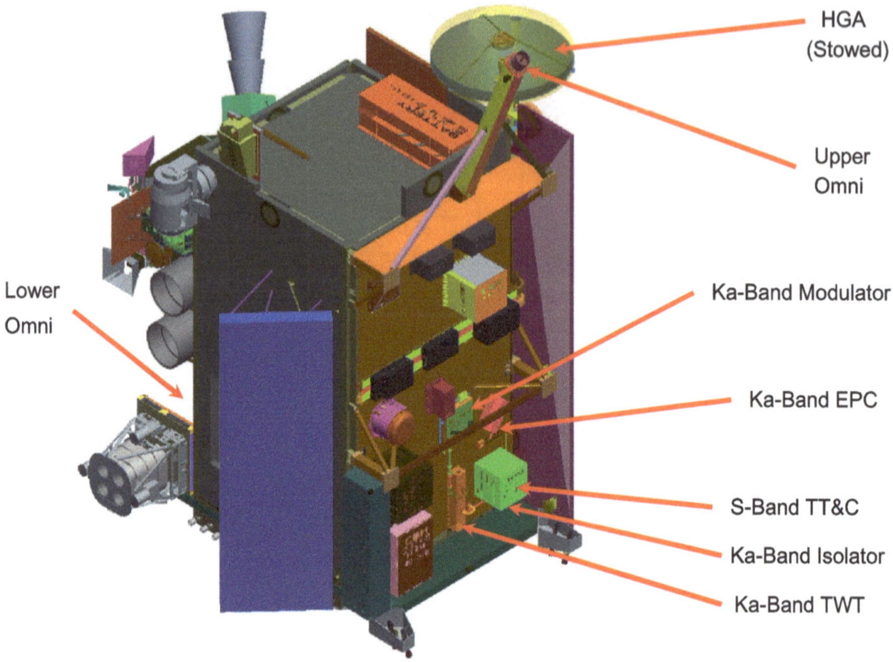

Fig. 30 Communications system components (Avionics radiator not shown)

Table 7 LRO communication system summary

LRO Communication Subsystem Characteristics		
Parameter	S-Band System	Ka-Band System
Frequency	2271.2 ± 2.5 MHz (Transmit) 2091.3967 ± 2.5 MHz (Receive)	25.65 GHz ±150 MHz (Transmit)
Mass	8.8 kg	8.4 kg
Polarization*	LHCP (Omni), RHCP (HGA)	LHCP
RF transmit power	5.8 Wat Diplexers	41.9 Wat TWTA Output

*L/RHCP—Left/Right Hand Circular Polarization

both paths for receiving. The ground selects the uplink path via polarization, as the HGA has opposite polarization from the omni antennas for this reason.

Major components of the communications system are shown in their installed location on the avionics panel of the spacecraft bus in Fig. 30. Specifications for each of the S- and Ka-Band subsystems are shown in Table 7.

3.5 Propulsion Subsystem

LRO's propulsion system was designed to provide mid-course transit corrections after separation from the launch vehicle, lunar orbit capture and then momentum dumping and station keeping for the remainder of the mission. Figure 31 illustrates the overall design and shows

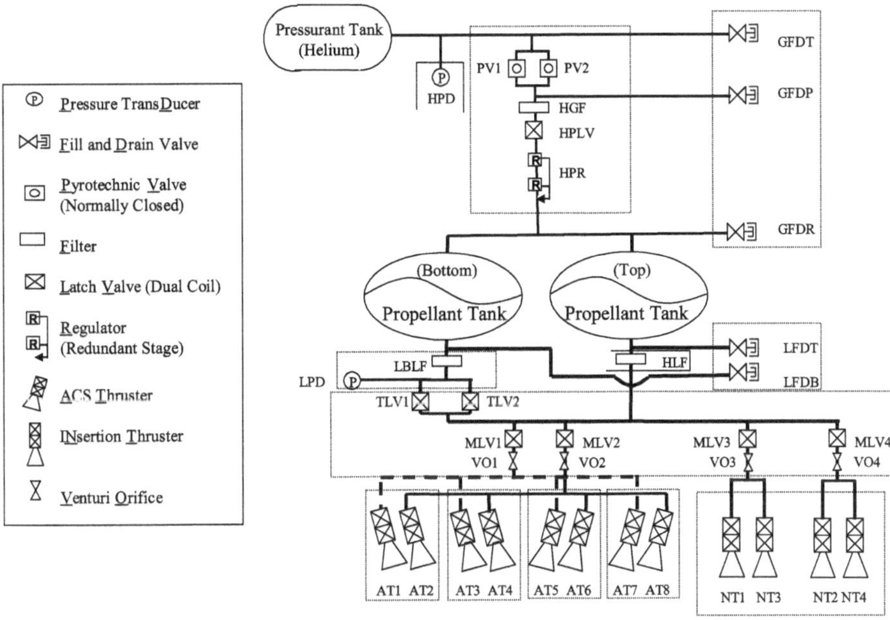

Fig. 31 Diagram of propulsion system

the redundancy of the propulsion system, and Fig. 32 shows the flight Propulsion Subsystem prior to the buildup of the Orbiter around it.

The propulsion system is a monopropellant hydrazine (N_2H_4) system, chosen for its simplicity. It carries a hydrazine fuel load of 895 kg (>1270 m/s ∇V capability) in two identical titanium diaphragm type 28,144 in^3 oblate spheroid propellant tanks. These tanks were specifically chosen because of their availability from a previous program, the cancelled NASA X-38 mission. The system is Helium pressure regulated, with the pressurization provided by a single composite overwrapped tank filled to 4200 psi at launch.

The system includes twelve dual coil catalytic hydrazine thrusters, four of which are on-axis 80 Newton-class insertion thrusters located around the spacecraft center of gravity (in the x-axis). Eight canted 20 Newton-class attitude control thrusters provide attitude control, lunar orbit maintenance maneuvers, and momentum dumping. Isolation valves with redundant coils are used to isolate thruster banks in the event of a thruster failure. The LRO Propulsion subsystem forms the physical core of the LRO spacecraft and the integration of the spacecraft was essentially the process of building up the Orbiter around it.

3.6 Power Subsystem

The LRO Electrical Power System (EPS) employs Direct-Energy-Transfer (DET) architecture and consists of three sub-elements: the Solar Array, Power System Electronics (PSE), and a Lithium-Ion Battery. The LRO Power System block diagram is shown in Fig. 33.

3.6.1 Solar Array

LRO's solar array unfolds to a single wing of three panels as shown in Fig. 1, and is continuously gimbaled in two axes to track the Sun. The array configuration consists of three

Fig. 32 LRO propulsion subsystem (on support stand)

aluminum honeycomb panels populated with 3,696 Advanced Triple Junction solar cells, for a total area on the order of ten square meters, and a power generating capability of 1,849 W at end of life.

3.6.2 Power Subsystem Electronics (PSE)

LRO's PSE performs the functions of power distribution and battery charging. The basic design and operation of the LRO PSE architecture is modeled after the architecture used on the Microwave Anisotropy Probe (MAP) and uses the solar arrays to convert sunlight energy into electrical energy while in the sunlight. The electrical power is then transferred to the PSE, where it is conditioned and directed to all of the electrical loads connected to the spacecraft bus. During the eclipse seasons, the PSE will also direct a portion of the sunlight-generated electrical power to the battery for energy storage recharging. During the eclipse portion of the orbit the battery will provide all of the energy to the spacecraft. Any excess power from the solar arrays not needed for battery charging or spacecraft loads is shunted back to the solar array. The PSE architecture can be seen within Fig. 33. The PSE is designed for single fault tolerance while still being capable of meeting all mission requirements.

3.6.3 Lithium-Ion Battery

A single Lithium-Ion battery provides power to the spacecraft when illumination of the solar array is insufficient. In order to minimize electrical losses due to power converters

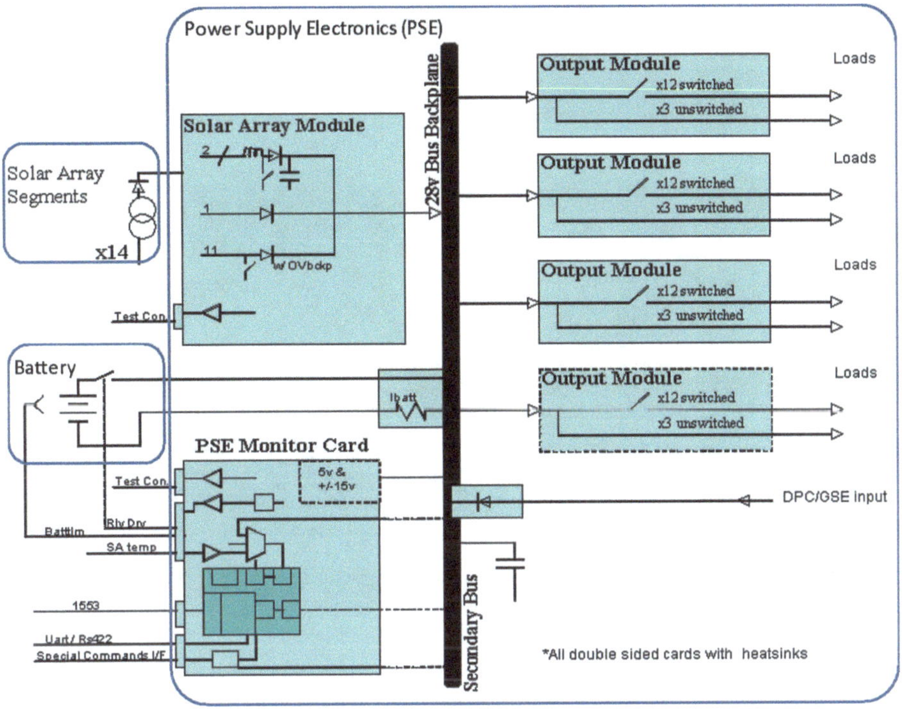

Fig. 33 Power system block diagram

and maintain a stable voltage spacecraft bus range, the battery is connected directly to the electrical bus. The battery energy storage capacity is 100 Ampere hours at Beginning of Life (BOL) within a 24–34 V voltage range. The battery was selected to meet a requirement of storage capacity of more than 80 A-hr at End of Life (EOL) and keeping the Depth of Discharge (DOD) to less than 30% during normal operations. This limited DOD enables the battery to survive over the 5000 cycles expected during the nominal mission with a significant margin. Thermal control of the battery is provided at the spacecraft level though a heat pipe assembly which connects the battery directly to a zenith facing radiator.

The battery is constructed from small cells arranged in strings and blocks. The basic block size is 12 strings of 8 cells each and seven blocks are used, to give a total of 672 cells. Within each block, the cells are connected in strings with eight cells in each string in series. This design ensures that cell failures (both open-circuit and short circuit) result in benign events. An open circuit failure results in the loss of a cell string, and a battery capacity loss of ~1% of capacity. A short circuit failure activates a cell internal protection device leading to an open-circuit failure and a similar loss in capacity. The battery weighs 36.4 kg and is shown in Fig. 34.

3.7 Thermal Control Subsystem

The LRO Thermal Control subsystem basic design accommodates four major elements: the Spacecraft Bus (Avionics, Reaction Wheels, and Battery), Instrument Module, Propulsion Module, and Deployables (high gain antenna and solar array). Because the lunar surface temperature varies from less than −140°C to greater than 125°C, the heat load on Orbiter $+Z$ face (nadir) varies from only a few W/m^2, when flying over lunar midnight, to

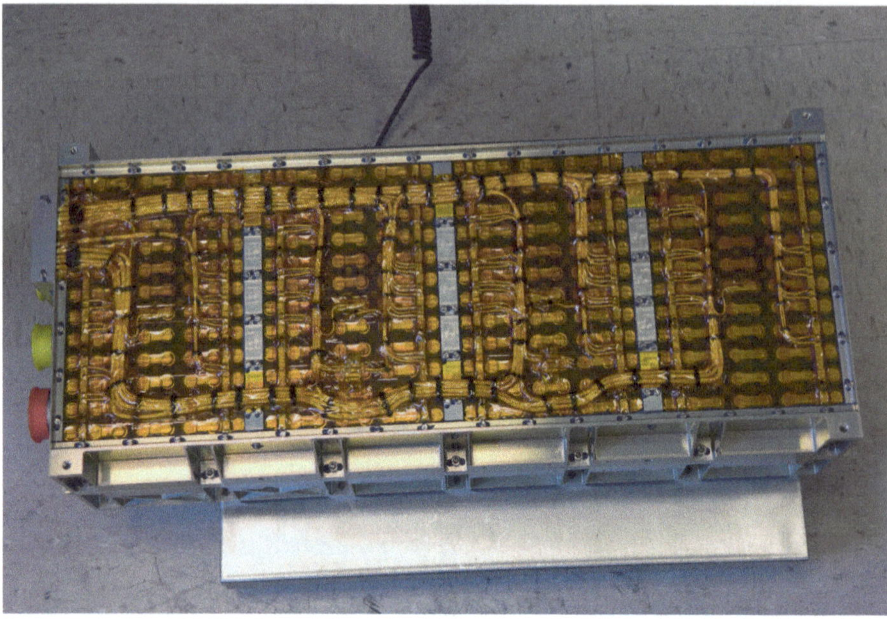

Fig. 34 LRO battery, top view showing cell interconnections

1330 W/m^2 from the Moon at lunar noon. A similarly high load is experienced from the Sun at lunar noon near the equator on the Orbiter $-Z$ face (zenith), further details are given in Fig. 35 and Table 8. This variation occurs in the span of less than an hour when the Sun-line is near the plane of the orbit. When the Sun is nearly perpendicular to the orbit plane, LRO experiences its nominal (routine) cold operational case, even though it does not pass through the lunar shadow, since the Moon surface at the terminator is near $-140°$C or colder. The most extreme cold case occurs during Type 4 lunar eclipses, discussed in the mission design section earlier in the paper in that in addition to a cold moon, the solar input is blocked by the earth for extended periods causing the thermal system to drive the battery sizing.

The LRO Thermal Subsystem was designed and qualified to ensure the survival of the Orbiter in the most extreme lunar eclipse cold case.

The overall orbiter is designed primarily with zenith directed radiators providing minimal exposure to the lunar environment (Baker et al. 2009). These radiators are covered with Optical Surface Reflectors (OSR), which reflect most of the optical energy of the Sun, but still emit in the infrared. These mirror-like OSR's allow LRO's radiators to run cold, even in the direct Sun.

Most of LRO's avionics are thermally coupled into an embedded Constant Conductance Heat Pipe (CCHP) aluminum honeycomb panel. Dual bore header heat pipes couple the isothermal panel to a CCHP avionics radiator that is separately mounted on the zenith surface of the spacecraft. The reaction wheel assemblies have a dedicated heat pipe system that couples the assemblies to the zenith avionics radiator. Coupling all of the avionics and the wheels to a common radiator and wrapping that around the propulsion system provides a large thermal mass that minimizes temperature variation over the span of an orbit. To survive long passes through the Earth's shadow during lunar eclipses, this thermal mass is pre-heated, using software-controlled heaters, so that minimal heater power is required during eclipse scenarios. The Lithium-Ion battery is maintained on a separate heat pipe network

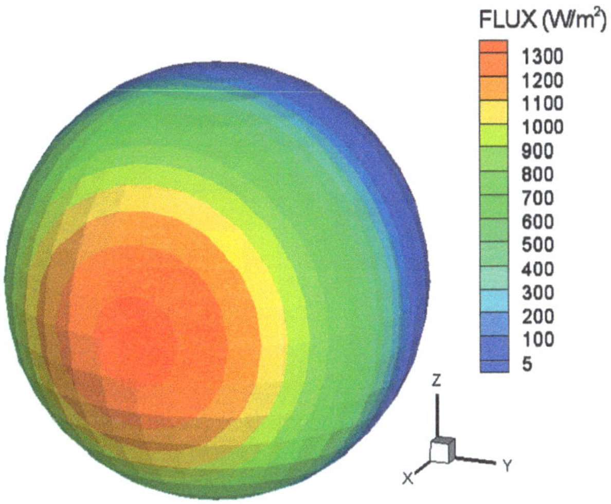

Fig. 35 Modeled lunar thermal environment

Table 8 LRO design lunar thermal environment

LRO Design Lunar Thermal Environmental Parameters		
Parameter	Hot Case	Cold Case
Solar Flux (Q_{sun})	1420 W/m^2	1280 W/m^2
Lunar Albedo Factor (F_{albedo})	0.13	0.06
Peak Lunar IR Flux ($Q_{\text{lunar-max}}$)	$Q_{\text{lunar-max}} = Q_{\text{sun}}(1 - F_{\text{albedo}})$	
(lunar sub-solar point)	1335 W/m^2	1114 W/m^2
Minimum Lunar IR Flux ($Q_{\text{lunar-min}}$) (dark lunar surface)	5 W/m^2	5 W/m^2
Lunar IR Flux to Spacecraft ($Q_{s/c\ \text{lunar}}$)	$Q_{s/c\ \text{lunar}} = [(Q_{\text{lunar-max}} - Q_{\text{lunar-min}})\cos(\beta)\cos(\theta)] + Q_{\text{lunar-min}}$ where β = solar beta angle & θ = angle from sub-solar point	

and a dedicated radiator on the zenith face of the Orbiter. An oversized radiator combined with heaters ensures tight control of its temperature.

A decoupled instrument optical bench (the Instrument Module) accommodates the instruments requiring high accuracy alignments (LOLA, LROC, LAMP, and the Star Trackers). It is constructed of low coefficient of thermal expansion (CTE) composite faced honeycomb material, is fully blanketed, and is heated with low density heaters to maintain cold limit temperatures. The CRaTER, Diviner, Mini-RF and LEND instruments are mounted directly to the main Orbiter structure and are individually controlled with heaters and radiators.

Within the Propulsion Module, most components are attached to a temperature-controlled cylinder that is part of the spacecraft bus structure including the upper propellant tank, pyrotechnic valves, and high and low pressure panels. Thruster valves are coupled to the spacecraft and tailored to survive soak-back after burn sequences. The lower propellant tank is thermally coupled to the aft deck for structural reasons. Remaining lines not on the structural cylinder are independently heated.

The High Gain Antenna uses multi-layer insulation (MLI), thermal isolation, and numerous heaters and radiators to maintain temperatures within allowable limits. The solar

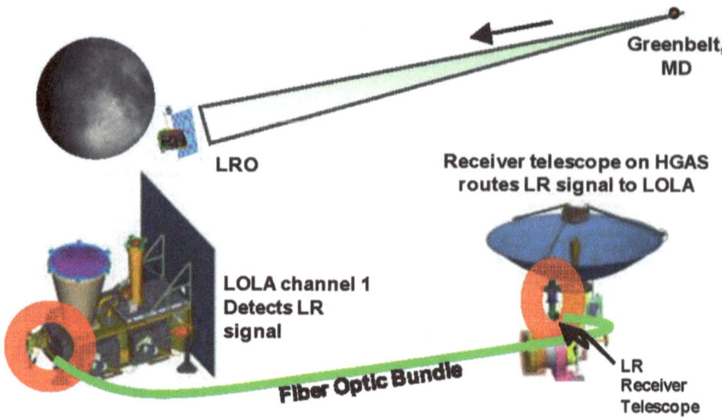

Fig. 36 LRO active one-way laser ranging system elements

array assembly uses MLI, dedicated radiators and heater control to maintain actuators and dampers.

Prior to launch the Thermal Subsystem performance was verified and optimized during Orbiter level Thermal-Vacuum testing which simulated the expected lunar environment and has meet all requirements since launch.

3.8 Laser Ranging Subsystems

3.8.1 Active One-Way Laser Ranging System

During the preliminary design phase an active on-way (uplink) laser ranging system was added to LRO to improve the tracking, and thus geodetic accuracy of the data products, over that possible with the S-Band tracking system alone. Figure 36 shows the elements of the active laser ranging system, beginning with the ground station at GSFC, which transmits a 532 nm pulse at 28 Hz, this pulse is captured by the laser ranging telescope mounted on the high gain antenna and then transmitted via fiber optic cable to the detector assembly located within the LOLA Instrument. Time-stamped arrival and departure times allow for precise range measurements from Earth to the LRO spacecraft, thus helping to determine precise position relative to the Earth's surface. That data, combined with known positioning of the spacecraft and LOLA data gathered from the Moon, will provide enough information to improve the lunar gravity modeling which in turn will be used to improve the orbit determination.

3.8.2 Passive Two-Way Laser Ranging System

In addition to the active laser ranging system LRO carries a retro-reflector assembly consisting of twelve 1″-diameter solid retro-reflective prisms in a square array mounted on the $-Z$ face (zenith radiator). This retro-reflector will allow LRO to perform two-way laser ranging with high-power terrestrial laser ranging sites. This ranging will be conducted as an experiment of opportunity late in the mission. The retro-reflector assembly is visible on the Orbiter in Fig. 37.

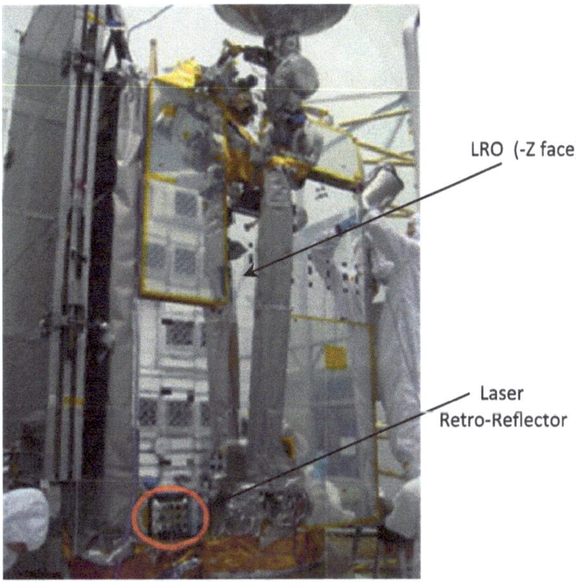

Fig. 37 Laser retro-reflector corner cube array (indicated in oval) on LRO

3.9 Mechanical Subsystem

The spacecraft structure and mechanical subsystem is of modular design, with four (4) distinct structural units (modules) for propulsion (PM), spacecraft bus (SB), avionics (AM), and instruments (IM), and two deployable systems with mechanisms for deployment and articulation, these being the Solar Array System (SAS) and the High Gain Antenna System (HGAS).

LRO structural construction is primary a combination of machined metal elements and metal honeycomb panels with bonded-in bolted fittings. The exception is the IM which utilizes composite face-sheet honeycomb construction for thermal stability. The major structural elements can be seen in Fig. 20.

Both the SAS and HGAS are deployed via spring-hinge-damper mechanisms with release controlled by the activation of Non-Explosive Actuators (electrical melt wire devices). Both systems have two-axis gimbal systems that employ stepper motors paired with Harmonic-Drive gear reducers. Position and rate telemetry from the gimbal assemblies is used by the GNC FSW to drive the motors and control the pointing of the Solar Array and High Gain Antenna.

4 Summary

LRO is an ambitious mission engaged in exploring the lunar surface and environment like never before in preparation for the human return. Its data products will enable future lunar exploration for decades to come. It is a challenging mission with several key driving factors, all of which have been met with robust design margins and innovative operations concepts. The mission development was completed within budget and on schedule. LRO is currently in lunar orbit executing its primary exploration mission with all of the spacecraft systems and science instruments performing flawlessly. Additional information about the LRO mission and its current status can be found at the NASA LRO website (www.nasa.gov\lro).

References

S. Andrews, M. Houghton, R. Saylor, Building fault tolerance into NASA's Lunar Reconnaissance Orbiter. AAS 08-036, 2008

C. Baker et al., Lunar Reconnaissance Orbiter (LRO) rapid thermal design development, in *International "Heat Pipe for Space Application" Conference*, 2009

M. Beckman, Mission design for the Lunar Reconnaissance Orbiter, in *AAS Guidance & Control Conference*, 2007

G. Chin et al., Lunar Reconnaissance Orbiter: The instrument suite and mission. Space Sci. Rev. **129** (2007)

Department of Defense Interface Standard for Digital Time Division Command/Response Multiplex Data Bus, MIL-STD-1553B, 21 September 1978

ESA Standard, SpaceWire—links, nodes, routers and networks. ECSS-E-50-12C, 31 July 2008

D. Folta, D. Quinn, Lunar frozen orbits, in *AAS/AIAA Astrodynamics Specialist Conference*, 2006

L. Lee, Utilizing reliability predictions in space flight electronics designs, in *International Military & Aerospace/Avionics COTS Conference*, 2007

NASA Announcement of Opportunity, Lunar Reconnaissance Orbiter Measurement Investigations, NNH042SS00O, 18 June 2004

Q. Nguyen et al., A high performance command and data handling system for NASA's Lunar Reconnaissance Orbiter. AIAA-2008-7926, 9 Sept. 2008

J. Wilmot, Implications of responsive space on the flight software architecture, in *AIAA-RS4 2006-6003; AIAA Responsive Space Conference*, 2006

Space Sci Rev (2010) 150: 63–80
DOI 10.1007/s11214-009-9511-z

The Lunar Reconnaissance Orbiter Laser Ranging Investigation

Maria T. Zuber · David E. Smith · Ronald S. Zellar · Gregory A. Neumann · Xiaoli Sun · Richard B. Katz · Igor Kleyner · Adam Matuszeski · Jan F. McGarry · Melanie N. Ott · Luis A. Ramos-Izquierdo · David D. Rowlands · Mark H. Torrence · Thomas W. Zagwodzki

Received: 20 December 2008 / Accepted: 15 April 2009 / Published online: 16 May 2009

Abstract The objective of the Lunar Reconnaissance Orbiter (LRO) Laser Ranging (LR) system is to collect precise measurements of range that allow the spacecraft to achieve its requirement for precision orbit determination. The LR will make one-way range measurements via laser pulse time-of-flight from Earth to LRO, and will determine the position of the spacecraft at a sub-meter level with respect to ground stations on Earth and the center of mass of the Moon. Ranging will occur whenever LRO is visible in the line of sight from participating Earth ground tracking stations. The LR consists of two primary components, a flight system and ground system. The flight system consists of a small receiver telescope mounted on the LRO high-gain antenna that captures the uplinked laser signal, and a fiber optic cable that routes the signal to the Lunar Orbiter Laser Altimeter (LOLA) instrument on LRO. The LOLA instrument receiver records the time of the laser signal based on an ultrastable crystal oscillator, and provides the information to the onboard LRO data system for storage and/or transmittal to the ground through the spacecraft radio frequency link. The LR ground system consists of a network of satellite laser ranging stations, a data reception and distribution facility, and the LOLA Science Operations Center. LR measurements will enable the determination of a three-dimensional geodetic grid for the Moon based on the precise seleno-location of ground spots from LOLA.

Keywords Moon · Laser ranging · Lunar Reconnaissance Orbiter · Orbit determination

M.T. Zuber (✉)
Department of Earth, Atmospheric and Planetary Sciences, Massachusetts Institute of Technology, Cambridge, MA 02139-4307, USA
e-mail: zuber@mit.edu

D.E. Smith · G.A. Neumann · X. Sun · J.F. McGarry · D.D. Rowlands · T.W. Zagwodzki
Solar System Exploration Division, NASA Goddard Space Flight Center, Greenbelt, MD 20771, USA

R.S. Zellar · R.B. Katz · I. Kleyner · A. Matuszeski · M.N. Ott · L.A. Ramos-Izquierdo
Advanced Engineering Technology Directorate, NASA Goddard Space Flight Center, Greenbelt, MD 20771, USA

M.H. Torrence
Stinger Ghaffarian Technologies, Greenbelt, MD 20770, USA

1 Introduction

1.1 Motivation

The Lunar Reconnaissance Orbiter[1] (LRO) mission (Chin et al. 2007) will perform precise geophysical, geological and geochemical mapping of the Moon in order to provide an observational framework for future orbital, landed and human exploration. The mission represents the first lunar exploration component of the Vision for Space Exploration (Bush 2004). A high-priority component of the mission's exploration objectives is an accurate lunar topography model that will be derived from the Lunar Orbiter Laser Altimeter (LOLA) (Smith et al. 2009). LOLA's lunar topography model will be used to construct a high-accuracy, global geodetic grid that will provide the foundation for positioning data sets from LRO as well as other lunar missions.

1.2 Tracking of LRO

The LOLA instrument has a ranging precision of about 10 cm (Smith et al. 2009), which is considerably better than the orbit determination capability originally baselined for the LRO mission. Uncertainties in spacecraft position represent the limiting error source for both LOLA's topographic model and geodetic grid. The Moon's synchronous rotation presents a special challenge in determination of the spacecraft orbit because the spacecraft cannot be tracked directly while orbiting above the lunar far side as viewed from Earth (cf. Lemoine et al. 1997; Konopliv et al. 1998, 2001). The spacecraft must be tracked as precisely as possible on the near side so that far side errors do not accumulate above the level that would violate the orbit positioning requirement.

The baseline tracking system for LRO is an S-band (2.2–2.3 GHZ) radio frequency link for ~20 hours per day (Chin et al. 2007). A commercial network, the Universal Space Network (USN), will provide tracking with stations at Dongara, Australia; Kiruna, Sweden; Weilham, Germany; and South Point, Hawaii. The Doppler accuracy of the USN is $\sim 1\ \mathrm{mm\,s^{-1}}$ (1σ) averaged over 10 s, which for the tracking time allocated will permit LRO orbits to be determined to ~10 m radially and 300 m along-track and across-track. The expected orbit accuracy is based upon simulations (Smith et al. 2008) using the NASA/GSFC GEODYN/Solve orbit analysis software (Pavlis et al. 2001).

Altimetric crossovers from the LOLA instrument will also be used in the LRO orbit determination process (Smith et al. 2009). Crossovers are points on the surface of the target body where spacecraft orbital trajectories cross. For places where surface topography does not change with time (in contrast, for example, to Earth's ocean surface), surface positions reconstructed from a combination of altimetric ranges and orbital tracking on different passes are indicative of orbit error. A global database of crossover mismatches can be minimized in a least square sense to improve orbital knowledge. The use of orbit crossovers has been successfully demonstrated to improve orbits of the Mars Global Surveyor spacecraft (Rowlands et al. 1999), the gravity field of Mars (Lemoine et al. 2001), and the radial accuracy of topography (Neumann et al. 2001) using observations from the Mars Orbiter Laser Altimeter (Zuber et al. 1992; Smith et al. 1999, 2001). Simulations of orbit determination have also been performed for the terrestrial Vegetation Canopy Lidar (VCL) and ICESat missions (Luthcke et al. 2000). The Lunar Reconnaissance Orbiter will be the first to use

[1] All acronyms are defined in the Appendix.

crossovers from a multi-beam altimeter to improve spacecraft orbits for a planetary body with synchronous rotation, where direct tracking of the target body's far side is not possible.

To take maximum advantage of the 10-cm range resolution of LOLA and the 50-cm pixel resolution of the Lunar Reconnaissance Orbiter Camera (LROC) (Robinson et al. 2008), the radial component of the orbit of LRO must be reconstructed to the sub-meter-level. This requirement dictates the need for higher-precision tracking on the lunar near side than is provided by the spacecraft's baselined S-band tracking system, or even by a combination of S-Band tracking and altimetric crossover analysis (Smith et al. 2008). Attainment of the precision tracking requirement on LRO will be achieved by introducing a laser ranging (LR) system in which an Earth-based laser will be used to range to LRO whenever in view.

1.3 LR System Overview

As illustrated schematically in Fig. 1, the LR is a one-way, time-of-flight measurement system that uses laser pulses to determine the range between an Earth-based satellite laser ranging station and the LRO spacecraft in orbit around the Moon. The system is an innovative combination of proven flight hardware components and satellite laser ranging technology. Stable clock oscillators on the spacecraft and at Satellite Laser Ranging (SLR) stations on Earth enable a precise time of flight measurement to be obtained whenever the LRO spacecraft is in view from a station. The system is designed to measure centimeter-level orbit perturbations over a few seconds of flying time and meter-level perturbations from pole to pole. The LR system has three main elements: the flight segment, the ground station and the data processing system. Tables 1 and 2 list key parameter values for the flight system and ground system elements.

As illustrated schematically in Fig. 2, the LR flight element includes a laser ranging telescope (LRT) attached to the LRO high-gain antenna (HGA), a fiber optic bundle to transmit the pulses from the LRT to the LOLA detector, and timing electronics on the LOLA instrument to time stamp the pulse arrival times. LOLA's Channel 1 detector assembly is designed to receive the LR signal through wavelength multiplexing, with the LR pulses at 532 nm and the lunar surface returns at 1064 nm. The LOLA timing electronics is designed to time stamp both the LR signal from Earth and the LOLA signal from the lunar surface through time multiplexing.

The LR ground element consists of SLR stations that track the LRO spacecraft as it orbits the Moon and synchronously transmit laser pulses to the spacecraft. The primary ground station for the LR system is NASA's Next Generation Satellite Laser Ranging (NGSLR) station, located at Goddard Space Flight Center in Greenbelt, Maryland, shown in Fig. 3. In addition, stations from the International Laser Ranging Service (ILRS) have been invited to partner with the NGSLR facility and participate in ranging activities. While the single

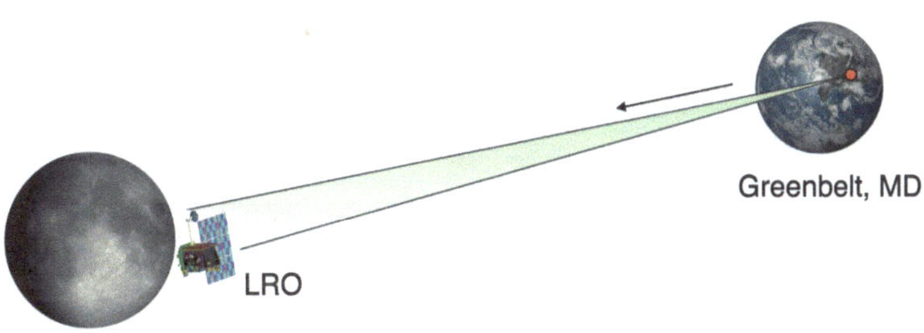

Fig. 1 Schematic illustrating the geometry of the LR uplink

Table 1 LRO LR flight system parameters

Parameter	Unit
Receiver aperture diameter	1.9-cm clear aperture, 3.0-cm outer diameter
Field of view	30 mrad
Fiber optic bundle	400-μrad core diameter, 7 each, 0.22 NA, step index 40% transmission, including fill factor and connector losses
Optical system transmission	27% transmission, including LOLA aft optics
Optical bandwidth	0.3 nm FWHM
Wavelength	532 nm
Detector (LOLA Ch. 1)	
Quantum efficiency	45%
Avalanche gain	120
Dark current	83 pA
Preamp noise	1.5 pA/Hz$^{1/2}$
Impulse response pulse width	5 ns FWHM
Timing electronics	
Timing resolution	<0.2 ns
Clock stability	2×10^{-12} over 1 hour
Pulse energy monitor	10% single shot
Earth window	8 ms

Table 2 NGSLR ground station characteristics

Parameter	Unit
Laser pulse energy	50 mJ
Wavelength	532 nm
Pulse width and rate	<10 ns FWHM, 28 Hz, synchronized to MET
Beam divergence (full angle at $1/e^2$ points)	55 μrad (11.3 arcsec)
Pointing uncertainty (3σ)	10 μrad (2 arcsec)
Spacecraft position prediction uncertainty	10 μrad (2 arcsec)
Laser optics transmission	50%
Transmission through atmosphere	
Minimum elevation	20°
Atmospheric transmission	70% zenith 35% 20° elevation

NGSLR station is sufficient to meet the LR objectives, the participation from the ILRS will broaden ranging coverage and improve the overall LRO tracking coverage and thus data products.

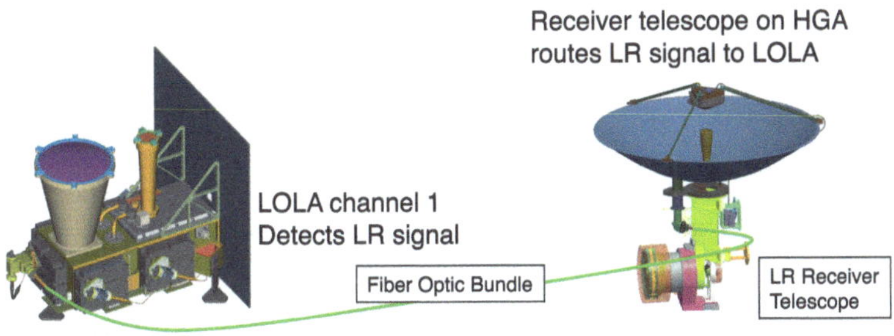

Fig. 2 Schematic showing the LR flight system. The Laser Ranging Telescope (LRT) receives the signal from Earth and transmits it to Channel 1 of the LOLA receiver via the fiber optic bundle (FOB)

Fig. 3 NASA's Next Generation Satellite Laser Ranging System (NGSLR) at NASA/Goddard Space Flight Center, Greenbelt, MD

LR operation is coordinated by the LRO Mission Operations Center (MOC), and the LOLA Science Operations Center (SOC), with the NGSLR facility serving as the single point of contact for ILRS partner stations. The LRO MOC transfers the full set of LOLA science data to the LOLA SOC where the LR pulse transmission and detection times are used to produce ranges to the LRO spacecraft. The ranges are to be archived in the Radio Science Node of the Planetary Data System (PDS).

The LR is required to achieve LRO's precision orbit determination requirement and will allow the production of a precise geodetic grid from LOLA altimetry (Smith et al. 2009) that will enable all LRO and many other archived lunar data to be precisely-located on the lunar surface. In addition, the LR range data along with LRO-supplied S-band ranging data and LOLA altimetry data will also be used to generate a refined gravity model of the Moon on a best-effort basis.

2 Flight Segment

2.1 Laser Ranging Telescope

Laser signals from the ground station are intercepted by the LRT (Ramos-Izquierdo et al. 2008), shown in Fig. 4, which is mounted on and co-bore-sighted with the HGA. Figure 5 shows a schematic model of the spacecraft and points out the location of the LRT, which views the Earth through a 3.81-cm diameter, off-centered hole in the primary reflector of the HGA. This location was chosen for the telescope because the HGA will be constantly pointed toward Earth as LRO orbits the Moon and no additional pointing mechanism(s) or commanding (e.g. spacecraft slewing) are necessary to accommodate the acquisition of LR measurements. The HGA is mounted on two, orthogonal gimbals each capable of rotating 180°, giving the HGA and LRT a hemispherical range of motion. When deployed, the axes

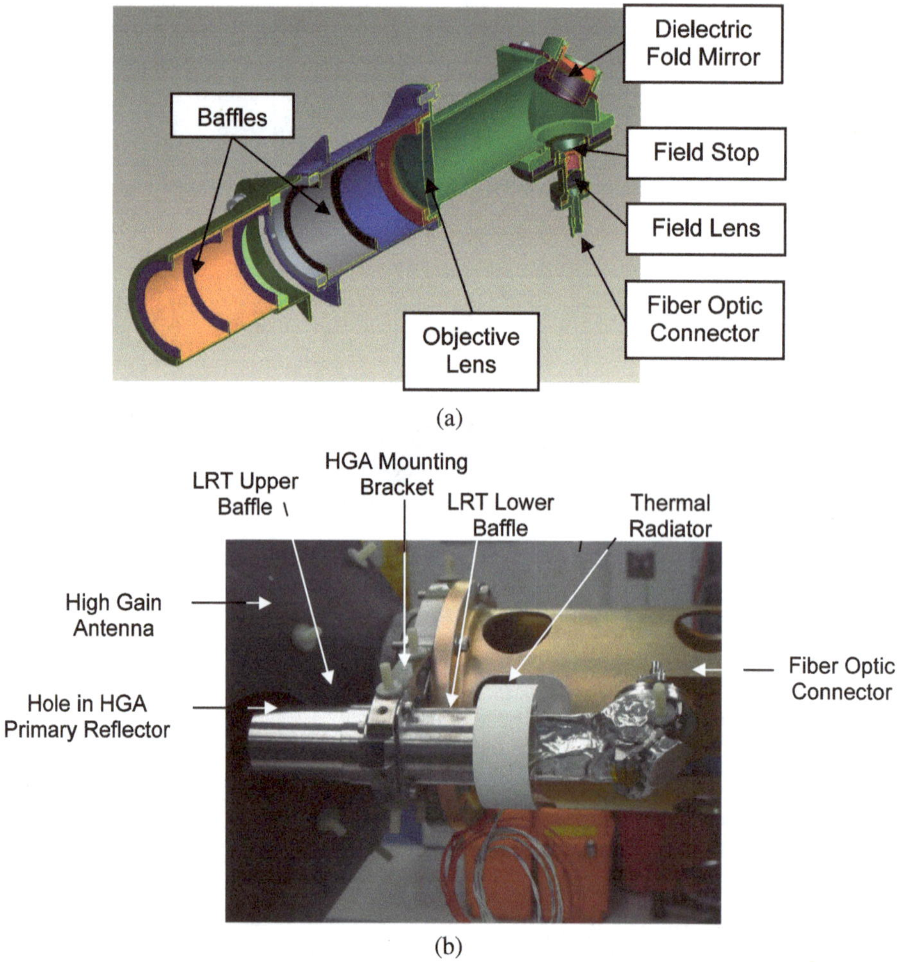

Fig. 4 Laser Ranging Telescope. (**a**) Schematic cutaway. (**b**) Photograph of flight unit. The LRT is attached behind the LRO High Gain Antenna, viewing through a hole in the primary reflector

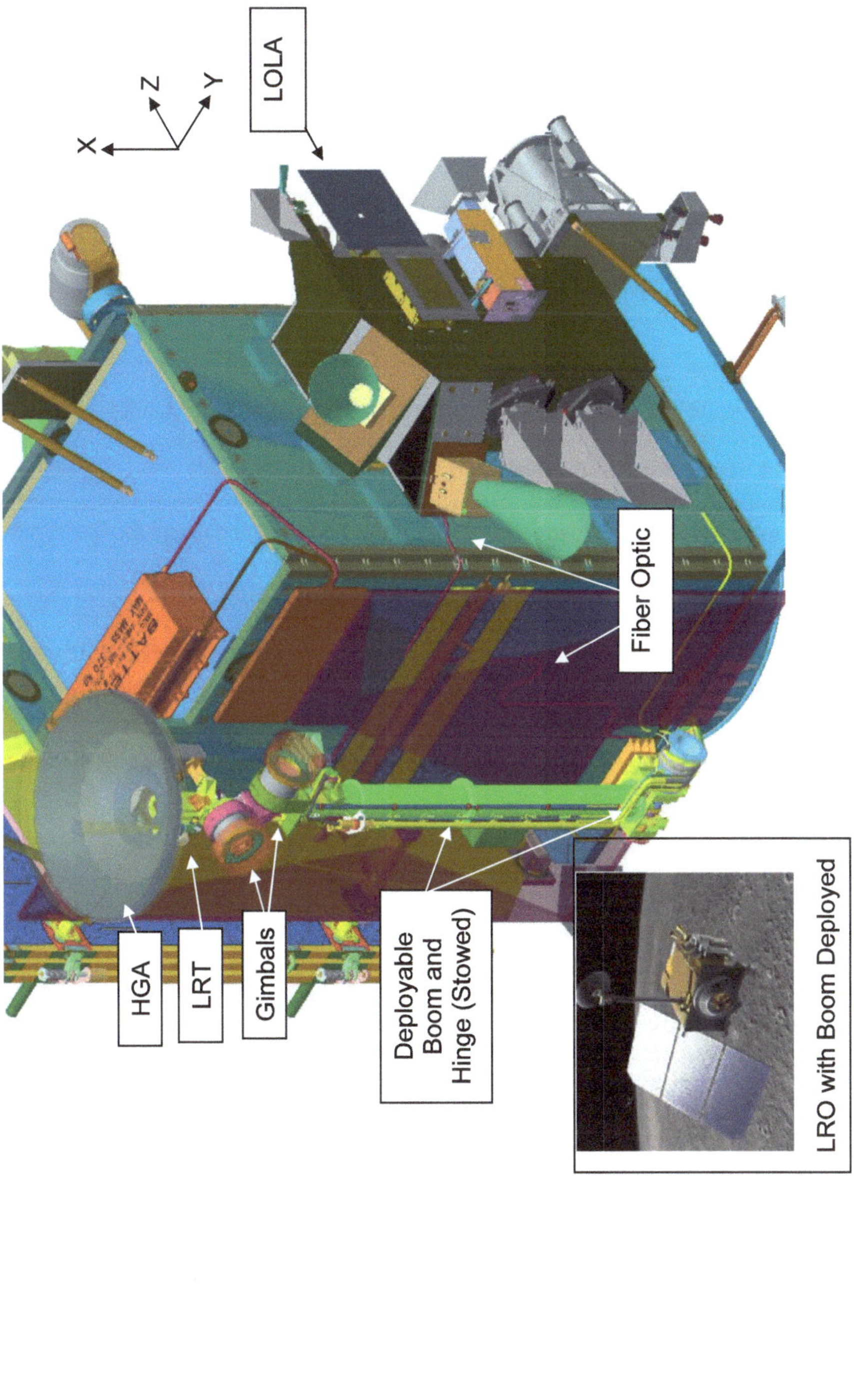

Fig. 5 Schematic showing positions of various components involved in laser ranging on the LRO spacecraft

of the gimbals are oriented parallel to the Y and X spacecraft axes, and are attached to the end of a deployable boom mounted to the $-Z$ side of the LRO spacecraft.

LRO will utilize the HGA Ka-band transceiver to downlink telemetry by pointing the HGA at the White Sands Missile Range, White Sands, NM, whenever it is in view. As observed from LRO, the position of the NGSLR Ground Station at Greenbelt with respect to White Sands will change depending on the orientation of the spacecraft, the position of the HGA gimbals, and the rotation of the Earth. In order to maximize LR ranging opportunities without impacting LRO HGA operations, the LRT field of view (FOV) was designed to cover nearly the entire Earth while centered on White Sands.

The LR telescope is boresighted to the HGA to within 0.01° (cf. Fig. 5). The HGA beam has a divergence of 0.9° and the pointing stability is 0.1° RMS. The telescope has a 30-mrad (~1.7°) field of view with a 19-mm clear aperture, and consists of a sapphire objective lens, a 45° angle of incidence dielectric fold mirror with maximum reflectivity at 532 nm, a 2.74-mm (30-mrad) diameter field stop, and a molded aspheric field lens. A baffle tube assembly limits the telescope's optical acceptance angle.

2.2 Fiber Optic Bundle

A multiple fiber-optic cable (Ramos-Izquierdo et al. 2008) is used to transmit signals received at the LRT to the LOLA channel 1 detector housing. This cable, shown schematically in Figs. 2 and 5, is referred to as the Fiber Optic Bundle (FOB). The FOB is a parallel set of seven individual 400-μm-diameter, 0.22-NA step index fibers. The FOB has three segments, one attached to the antenna, one to the gimbal assembly and boom, and one to the spacecraft body.

As shown schematically in Fig. 5, the fiber optic route passes through each gimbal, down the spacecraft boom, around the deployment hinge and across the spacecraft deck, requiring the fiber to be approximately 10 m in length. The gimbal assemblies contain special channels for FOB and other electrical cables to control the movement and bend curvature when the HGA is in motion. The FOB also has to survive the one time deployment of the boom.

2.3 LOLA Receiver

LOLA is mounted on the LRO Instrument Module attached to the $+Y$ side of the spacecraft. The LOLA instrument Channel 1 aft optic assembly contains a dichoric beam splitter to combine the LR signal and the LOLA signal onto the detector. The LOLA Channel 1 aft optic assembly re-images the light from the FOB into a 600-μm diameter image at the surface of the detector. The aft optic includes a 532-nm bandpass filter with 0.3-nm FWHM to limit the effects of the background light from the sunlit Earth.

To improve the receiver sensitivity and reduce the false detections, the LOLA receiver utilizes range windows. Only signals within the range windows are detected and processed. There are two separate range windows within the LOLA timing electronics. The first, referred to as the Earth window, is 8 ms wide and opened before each LOLA laser pulse emission. The second, referred to as the lunar window, is less than 5 ms wide, and opened after a small delay from the laser pulse emission. The LOLA laser operates at 28 Hz, and the laser emission times and consequently the range windows are synchronized with the LRO Mission Elapse Time (MET). The ground stations will synchronize the laser emission time with the predicted MET for the optimal LR measurement rate.

The LRO MET is based on a ultrastable, oven-controlled crystal oscillator (OCXO). The OCXO is stable to 1 part in 10^{-12} over an hour to enable the measurement of small spacecraft orbit perturbations from pole to pole. The long-term stability of the OCXO is expected to be better than 1 part in 2×10^{-11} per day, which allows estimation of the spacecraft MET to $\ll$1 ms with respect to coordinated universal time (UTC). Together with the orbit prediction, ground stations will synchronize pulse arrival to well within the LR's Earth window.

The LR time stamp precision from the LOLA instrument is about 0.5 ns in standard deviation and depends on the signal pulse energy and pulse width. The equivalent ranging precision for the one-way laser ranging is about 15 cm.

3 The NGSLR Ground Station

3.1 Introduction/Overview

The primary ground station for LRO laser ranging is NASA's NGSLR station. NGSLR is the prototype for NASA's next generation of eye-safe, automated SLR stations. NASA's current SLR Network provides two-way ranging to Earth-orbiting satellites that are equipped with retro-reflector arrays. The NGSLR has recently been modified for dual use of LRO LR and satellite ranging.

The NGSLR was designed as a prototype to replace the current NASA stations with a more automated, less hazardous, easier to maintain system (McGarry and Zagwodzki 2006). Operational capabilities of NGSLR include the traditional two-way laser ranging to retro-reflector equipped satellites, and also the ability to perform one-way and two-way transponder laser ranging. During the LRO mission the NGSLR will be used to track the spacecraft when possible and it will track Earth satellites at other times. Routine satellite tracking observations not only provide data to the SLR data system but also verify the system's pointing accuracy for LR.

NGSLR points its telescope at LRO using ILRS predictions in Consolidated Prediction Format, referred to as CPFs. These are Earth-centered, Earth-rotating vectors spaced at intervals depending upon the orbit. The software interpolates and translates these vectors to the site location and transforms them to the local azimuth and elevation required to drive the mount. The local mount angles are corrected for refraction and mount pointing errors. Measurements taken at the ground station of barometric pressure, temperature and humidity are used in the calculation of the elevation and range refraction. The mount pointing errors are determined by pointing to stars.

3.2 NGSLR Modifications for LRO LR

In order to support the LR experiment, several modifications to the basic NGSLR design were made. Figure 6 shows a block diagram of NGSLR as configured for LRO operations. Modifications included: (1) the addition of a higher power, 28-Hz laser, (2) installation of an aircraft avoidance radar, with a range of 20 km, in support of the higher power laser, and (3) control of the laser fire time by the software.

The laser is a diode-pumped, frequency-doubled (532 nm) Nd:YAG built by Northrop Grumman Space Technology Cutting Edge Optronics of St. Charles, Missouri. The laser pulse width is 5.5 ns (FWHM) and the energy is nominally 50 mJ per pulse. The laser is mounted on an upper level breadboard above the NGSLR host system and is coupled into the telescope with a removable aperture share mirror. Final laser beam divergence is controlled with an external beam expander mounted on the upper breadboard The removable mirror allows for easy switching back and forth between NGSLR operations and LRO

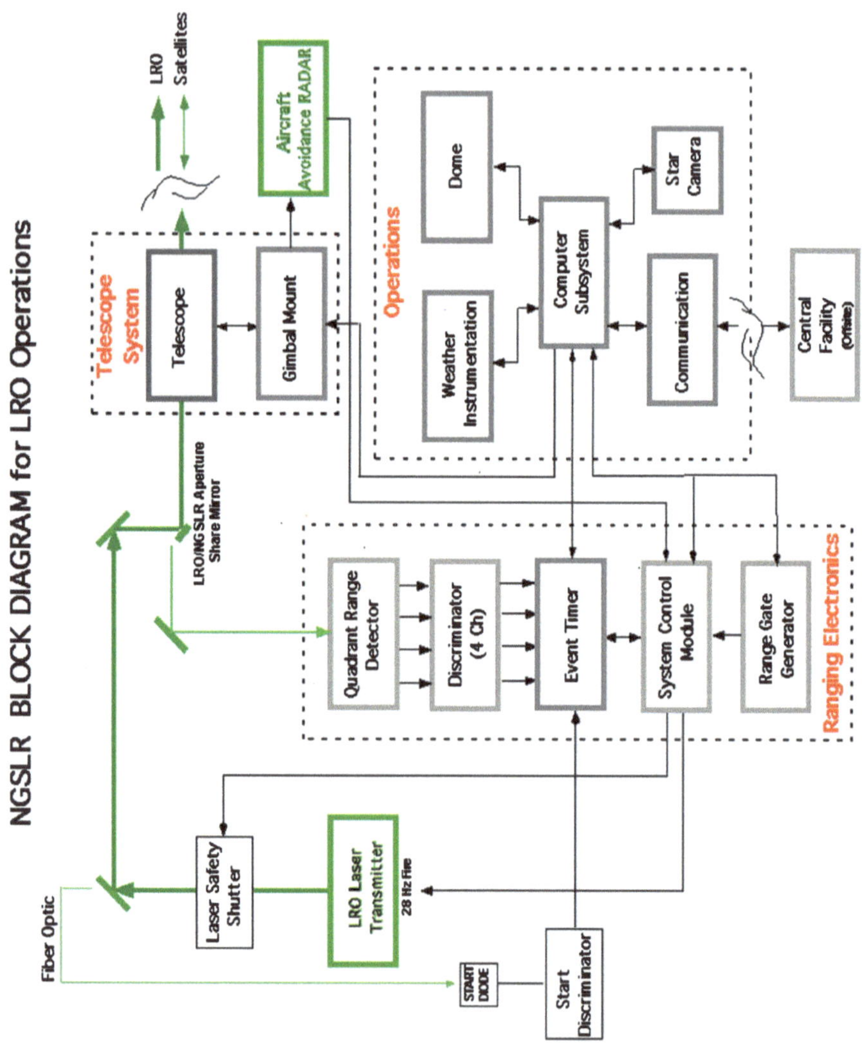

Fig. 6 LRO block diagram utilizing the NGSLR as the host tracking system

tracking and permits SLR tracking utilizing the LRO laser for system check out and verification. In testing, the LOLA LR filter was tilt-tuned to the NGSLR LR laser to optimize its transmission.

The Laser Hazard Reduction System (LHRS) provides a means of detecting aircraft before they intersect the transmitted laser beam out to a range greater than the nominal ocular hazard distance. This function is accomplished by the use of a pedestal-mounted radar that is bore-sighted to the laser-transmitting telescope. Upon detection of an aircraft by the radar, the LHRS provides a signal to activate a laser beam block. The X-band radar is based on a commercial marine radar that is interfaced with several redundant safety systems that constantly monitor system parameters to ensure proper radar operation and pointing along the laser beam axis.

3.3 NGSLR Timing System

The NGSLR station controls the laser pulse fire times to synchronize the pulse arrival with the LOLA Earth window, compensating for spacecraft movement and Earth rotation. Outgoing laser pulses from Earth are time-stamped with <0.1-ns precision for individual laser shots. The ground-based timing system has a drift rate less than 1 part in 10^{-12} per hour and has an absolute time bias of <1 μsec with respect to UTC. The station time is tied to UTC via the TrueTime XL-DC GPS-steered Rubidium oscillator that provides a 1 pulse per second input to the system. A Cesium frequency standard (Symmetricom model 4310), provides the 10-MHz external input to the Event Timer.

3.4 Telescope Pointing

LRO position predictions will be generated by the Goddard Flight Dynamics Facility. These predictions are expected have an accuracy requirement of 4 km. Laser ground stations must be able to open loop point their systems to provide between 1 and 10 fJ cm^{-2} of laser energy density on LRO, while ensuring less than 0.07 mW of peak power on the LOLA detector to avoid damage. NGSLR's peak power will be less than 0.001 mW.

The absolute accuracy of the laser fire measurement must be within 100 ns of UTC with a laser fire inter-arrival time average measurement error of 200 ps or less over a 10-s period. NGSLR's station timing satisfies both of these requirements with the use of the TrueTime Rb Station Timing, the unstirred cesium external source, and the Honeywell Event Timer.

3.5 ILRS Participation

Several ILRS stations (Pearlman et al. 2008) will also be participating in the LR investigation as ground stations. The addition of ILRS stations provides global coverage for LR and increases the laser ranging data set. A few of the participating ILRS stations are also synchronizing their laser fires to the LOLA Earth window in a similar manner to NGSLR, while several others are firing asynchronously at 5 or 10 Hz. At 10 Hz only 2 to 4 pulses per second will fall in the LOLA Earth window. The remaining 6–8 pulses will be treated as noise by LOLA, but these are not enough to cause significant disruptions to the LOLA measurements.

3.6 Feedback to Ground Stations

Since this is an uplink-only ranging measurement, there is no real-time feedback of spacecraft tracking as in normal SLR operations. Instead, LOLA's real-time housekeeping telemetry will be used to provide needed feedback to the stations. LOLA's onboard signal processing indicates whether the Earth pulses are arriving at LRO, and in addition tracks the time at which these pulses occur within the Earth window. The LOLA Earth energy monitor also provides an integrated energy over each Earth window. This information is posted in graphical form to a password protected website that will provide feedback to all participating stations. The delay between spacecraft event and webpage plot should be less than 30 s. NGSLR will be able to bias its fire times from this information to: (1) search for the Earth window if no signal is seen in the LOLA housekeeping telemetry, and/or (2) ensure that its laser pulses are arriving as early as possible in the single-stop Earth window.

3.7 Operations, Data Flow and Scheduling

LRO is visible to Earth for approximately one hour during its two-hour orbit period when the Moon is above the horizon. Laser ranging can only occur at NGSLR above 20° elevation due to regulations by the Federal Aviation Administration and radar viewing limitations. After approximately each pass, the laser emission times and associated information are written to a file that is ultimately transmitted to a computer at the Crustal Dynamics Data Information System (CDDIS) (Noll et al. 2006) at NASA/GSFC, where it is transferred to the LOLA SOC for analysis.

The global network of participating stations will be scheduled from a central facility. Laser stations will be scheduled to range during periods when nearby S-band stations have been scheduled for downlink. This will ensure that the real-time feedback to the stations can occur. Initially only one station will be scheduled to range to LRO at a time, however, as the mission progresses, we anticipate two or possibly three ground stations ranging at overlapping times. In addition to providing redundancy, multiple observations can slightly reduce the error in the observations.

4 Link Margin

The LR receiver uses the same Silicon avalanche photodiode detector as LOLA, but at 532-nm wavelength. Under sunlit Earth background, the detector sensitivity is about

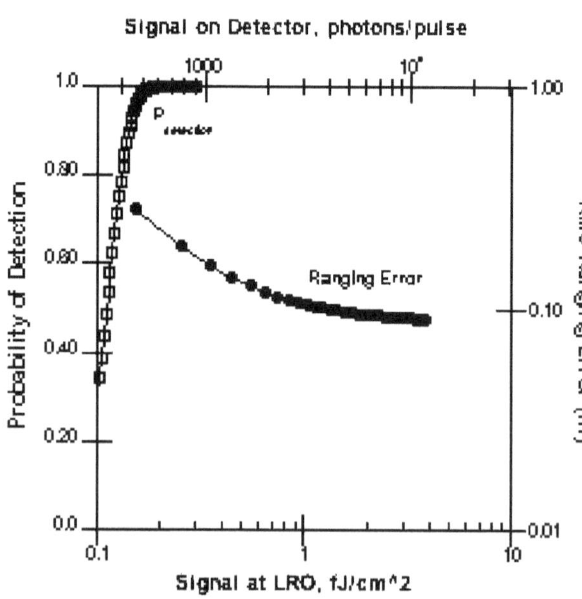

Fig. 7 Link analysis for the LR system. The plot displays the probability of detection and associated range error for signal levels expected during the LRO mapping mission

400 photons pulse^{-1} at 95% detection probability. The optical transmission from the LR telescope to the detector is about 40%, which gives a minimum detectable signal level of about 0.2 fJ cm^{-2} at the entrance of the LR telescope. The ranging error improves with the signal level. Figure 7 shows the receiver probability of detection and the root mean square (RMS) ranging error as a function of input signal level in both cm^{-2} at the entrance of the LRT and photons/pulse on the detector. The desired signal level is 1–5 fJ cm^{-2}, where the ranging error approaches the noise floor of the LOLA receiver timing electronics. The laser pulse width is assumed to be 8 ns at FWHM points. The ranging error generally improves as the laser pulse width gets shorter but is dominated by the receiver electronics to about 20 cm (0.67 ns) from each laser shot. The receiver also measures the pulse energy which can be used to correct any range-walk associated with the pulse amplitude. The mid- to long-term timing accuracy is governed by the OCXO, which is stable to better than 3 ns peak-to-peak over a period of several hours after removing a linear frequency drift rate.

5 Data Processing

LOLA science data and Earth ranges are collected at the SOC following each downlink. Transmit times from each participating Earth station are provided in UTC seconds-of-day, in a common ILRS Consolidated Ranging Data (CRD) format (Ricklefs 2006). The LR range signals are calibrated to pulse centroid time with corrections for fixed system biases in the predicted MET. The LRO timing system (LRO 2006) provides a Spacecraft Time Correction Factor (STCF) that converts the MET counter content to time in UTC. This number of seconds is then used to calculate an approximate UTC arrival time for each range measurement. The difference between the Earth-based laser transmit times and the received time at the spacecraft, times the speed of light, is the nominal one-way range. The STCF, which provides the common time system for LRO instruments, has a 3-ms accuracy requirement, which is sufficient for the initial LR acquisition. Once acquired, the LRO clock frequency and drift rate can be much better predicted from LR data by trending the time bias between the predicted and actual MET.

The production process consists of matching ground fires with Earth ranges using a predicted one-way time-of-flight. Production requires predictions of the spacecraft ephemeris, the exported lunar ephemerides with respect to Earth, and the current station positions and Earth orientation parameters. The light-time-corrected distance between the ground station and LRO, divided by the speed of light, is added to the fire time. LOLA- Earth ranges within an 8-ms window are matched to these predicted times. If the number of matched differences fills a given 200-ns histogram bin to a significant level, the Earth range times in seconds from midnight are merged with the full-rate CRD data.

Normal points consisting of 5-s fits to the valid transmit and receive times in their respective time systems, are provided to the Flight Dynamics Facility and archived with the full-rate data in the ILRS data system.

An example of the value of pulse averaging to produce normal points is given in Fig. 8, which shows observations from the Mercury Laser Altimeter (Cavanaugh et al. 2007). In spite of considerable noise background from sunlit Earth (Fig. 8a, black dots), the signal (red dots) in a 200-ns bin, shown at smaller scale in Fig. 8b, can, after removing three outliers, be fit to a linear function to produce a normal point.

The raw laser ranges consist of pairs of times that will be recorded by separate clocks, and represent only approximate one-way times. The ground station biases and spacecraft clock drift will be solved for in the course of tracking solutions within the GEODYN/Solve orbit

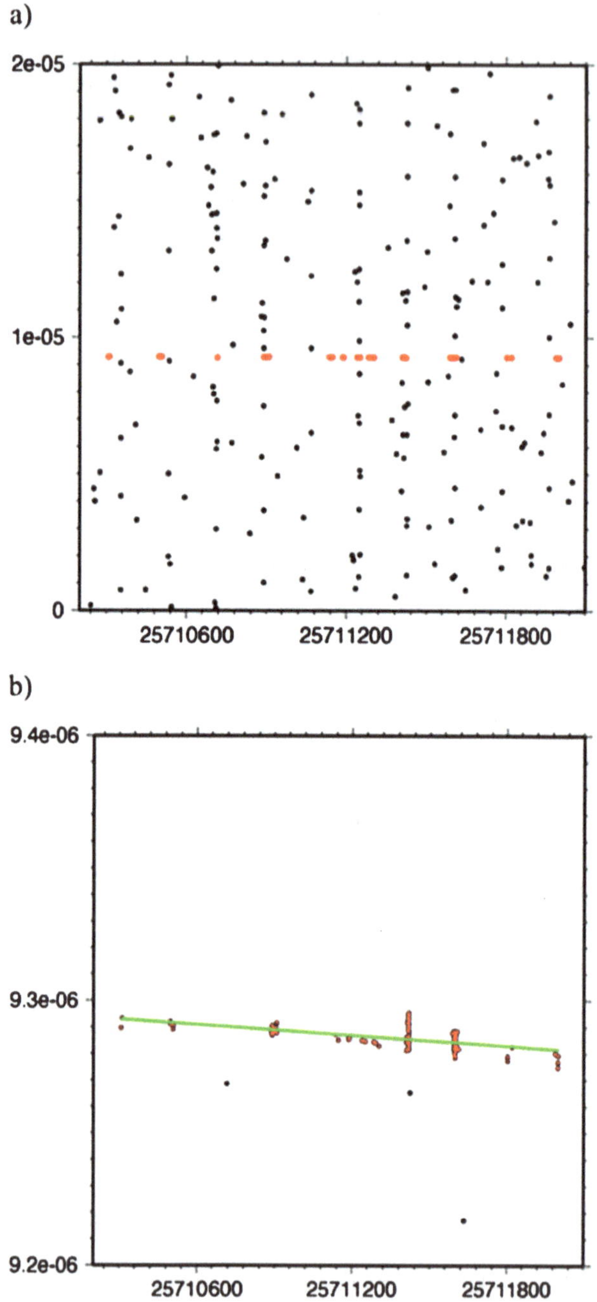

Fig. 8 (**a**) Triggers recorded by the Mercury Laser Altimeter on May 27, 2005 versus time, with laser ranges from Earth in *red*. Ordinates indicate residual times (s) with respect to predicted range. (**b**) Linear fit (in *green*) to residuals (in *red*) in a 200-ns-wide bin, excluding three outliers (in *blue*)

determination software system (Pavlis et al. 2001), applying corrections for time delays due to general relativity. To minimize the range error due to drift inherent in the OCXO, clock drift will be monitored and correlated with variations in temperature and supply voltage from spacecraft housekeeping. These parameters will be monitored from the ground, and their effects will be modeled and calibrated. After the initial calibration period, the long-term frequency drift will also be monitored and modeled. A byproduct of this analysis will be a much more precise clock correction factor for station fire time prediction and altimetry analysis.

The baseline for calibration of the LR measurement is via the process of precision orbit determination. As the LRO spacecraft orbits the Moon the LR measurement will be characterized as a biased range and a range rate coming from the drift of both the spacecraft and ground oscillators. The bias, and perhaps in addition the drift rate of the spacecraft oscillator, should be estimable from the data during orbit analysis.

In a demonstration experiment, a ground-based laser at the Apache Point Observatory in Sunspot, NM plans to range every two weeks to a laser retroreflector array on LRO. The array is comprised of twelve 37.7-mm-diameter solid Surprisil (quartz) cubes, mounted on the $+Y$ panel of the spacecraft. These ranges will be obtained in cooperation with the APOLLO lunar laser ranging experiment (Murphy et al. 2000). Ranging will occur over one or two days during each two-week period, for about 15 minutes per orbit. Although relatively infrequent, and not required for the LR investigation to achieve its objectives, these two-way measurements will provide centimeter-level ranges to the LRO spacecraft that are expected be useful in further calibrating LR ranges. In addition, retroreflector measurements can be used to calibrate the spacecraft oscillator and can conceivably improve the timing for LRO instruments.

6 LRO Orbit Improvement

LR normal points represent a data type in the determination of precision spacecraft orbits using the GEODYN program (Pavlis et al. 2001). The combination of radio tracking combined with orbit crossovers from the LOLA instrument (Smith et al. 2009) and optical tracking from the LR, will be used in a best effort basis to improve the gravity field of the Moon to contribute towards selenocentric location of LOLA altimetric shots and the production of a global geodetic grid. The LR observations are particularly useful in improving orbital knowledge on the Moon's far side that lacks direct tracking. The LR observations limit time drift to <3 ns hour^{-1}, and then provide an additional "anchor point" for timing when the spacecraft re-emerges from the far side.

Orbit determination for the LRO mission (Chin et al. 2007) for the current lunar gravity field (Konopliv et al. 2001) has been studied via simulations (Smith et al. 2008; Rowlands et al. 2008). Table 3 summarizes the various contributions from S-band tracking, LOLA and the LR, to precision orbit determination of LRO on the basis of the analysis by Smith et al. (2008). Unlike the analysis by Smith et al., the analysis by Rowlands et al. (2008) treats the case of multi-beam rather than single beam altimetry, the former of which is better for orbit improvement from crossover analysis than the latter. However, the analysis of Smith et al. incorporates experience from orbit and gravity field improvement in tracking of Mars Global Surveyor and in particular exploits the power of altimetry in orbit improvement. LOLA's five beams and 10-cm ranging precision along with the LR are expected to ultimately result in a radial RMS of order 0.5 m, including orbit and instrument errors, and horizontal errors should approach 10 m. Improvements due to the addition of the LR system are thus significant and will greatly improve knowledge of global and polar topography, footprint-scale surface slopes, and surface roughness. Knowledge of surface slopes improves from about 2° to 0.3°. The LR, however, does not contribute notably to knowledge of surface roughness within the footprint (5-m-scale). Radial and spatial improvements of altimetric footprints map directly into the global geodetic grid.

Table 3 Contributions to LRO orbit improvement and implications for LOLA data products

LOLA data products	S-band tracking alone	S-band + LOLA	S-band + LOLA + LR
Global topography			
Accuracy/ Resolution	R: 10 m; H: ~300 m	R: 10 m; H: ~200 m	R: 1 m; H: 50 m
Polar topography			
Accuracy/ Resolution	R: 10 m; H: ~ 300 m	R: 5 m; H: 200 m	R: 0.1 m; H: 25 m
Surface slopes			
Accuracy	2°	1.5°	0.3°
Resolution	300 m	200 m	25 m
Surface roughness			
Accuracy	35 cm	35 cm	35 cm
Resolution	~5 m	~5 m	~5 m

7 LR Archive

The deliverable product to the NASA Planetary Data System (PDS) are precision orbits. There is no explicit requirement to deliver the various radio tracking data sets, though the matter is currently under discussion.

While the LR uses LOLA as its data interface and the LOLA SOC as its pipeline, LR data are not part of the LOLA archive. The LR full-rate and normal point data will be processed and delivered into the ILRS, using CDDIS/ILRS services as its hosts. Any improved lunar gravity fields that result as a byproduct from the optical tracking of LRO will be archived on a best efforts basis with other planetary gravity fields in the Geoscience Node of the PDS.

8 Summary

Key objectives in the LRO mission include determination of a precise global topography model and a precise geodetic grid that will serve as a reference for LOLA, other LRO data sets, and, we anticipate, many other lunar data sets. Orbit determination is the limiting error source in the LOLA instrument's ability to measure lunar topography, as well as in the determination of the geodetic grid. The Laser Ranging system was added to the LRO spacecraft to enable the mission's precision orbit requirement to be achieved. The LR will allow LRO orbits to be determined at the meter scale radially and 50 m spatially, and these accuracies will approximately define the accuracy of LOLA and the resulting geodetic grid. LRO's LR experiment represents the first time that optical ranging will be used to track a planetary spacecraft, either experimentally or in an operational sense, and will demonstrate technology and analysis methods that will enable more sophisticated applications in future exploration of the planets.

Acknowledgements The Lunar Reconnaissance Orbiter Laser Ranging investigation is supported by the NASA Exploration Systems Directorate. We are grateful to the LRO Project, particularly Project Manager Craig Tooley, for support of the investigation.

Appendix: Acronym List

Acronym	Definition
CDDIS	Crustal Dynamics Data and Information System
CPF	Consolidated Prediction Format
CRD	Consolidate laser Ranging Data
fJ	femtoJoule = 10^{-15} Joule
FOB	Fiber Optic Bundle
FOV	Field of View
FWHM	Full Width at Half the Maximum amplitude
GEODYN	GSFC processing code for orbit determination
GHz	Gigahertz
GPS	Global Positioning System
GSFC	Goddard Space Flight Center
HGA	High-Gain Antenna
ILRS	International Laser Ranging Service
LHRS	Laser Hazard Reduction System
LOLA	Lunar Orbiter Laser Altimeter
LR	Laser Ranger
LRO	Lunar Reconnaissance Orbiter
LROC	Lunar Reconnaissance Orbiter Camera
LRT	Laser Ranging Telescope
MET	Mission Elapsed Time
NASA	National Aeronautics and Space Administration
Nd:YAG	Neodynium-doped, Yttrium, Aluminum, Garnet
NGSLR	Next Generation Satellite Laser Ranging
nm	nanometer = 10^{-9} m
ns	nanosecond = 10^{-9} s
OCXO	Oven-Controlled Crystal Oscillator
POD	Precision Orbit Determination
ps	picosecond = 10^{-12} s
RF	Radio Frequency
RGG	Range Gate Generator
SCLK	Spacecraft CLocK
SLR	Satellite Laser Ranging
SOC	Science Operations Center
STCF	Spacecraft Time Correction Factor
TOF	Time of Flight
USN	Universal Space Network
USO	UltraStable Oscillator
UTC	Coordinated Universal Time
VCL	Vegetation Canopy Lidar

References

G.W. Bush, A renewed spirit of discovery: The president's vision for U.S. space exploration, 9 pp., Executive Office of the President, Washington D.C., 2004

CDDIS website: http://cddis.nasa.gov/facility.html

J.F. Cavanaugh et al., The Mercury Laser Altimeter instrument for the MESSENGER mission. Space Sci. Rev. **131**, 451–480 (2007). doi:10.1007/s11214-007-9273-4

G. Chin et al., Lunar reconnaissance orbiter overview: The instrument suite and mission. Space Sci. Rev. **129**, 391–419 (2007). doi:10.10007/s11214-007-9153-y

ILRS website: http://ilrs.gsfc.nasa.gov

A.S. Konopliv et al., Recent gravity models as a result of the Lunar Prospector mission. Icarus **150**, 1–18 (2001)

A.S. Konopliv et al., Gravity field of the Moon from Lunar Prospector. Science **281**, 1476–1480 (1998)

F.G. Lemoine et al., A 70th degree and order lunar gravity model from Clementine and historical data. J. Geophys. Res. **102**, 16,339–16,359 (1997)

F.G. Lemoine et al., An improved solution of the gravity field of Mars (GMM-2B) from Mars Global Surveyor. J. Geophys. Res. **106**, 23,359–23,376 (2001)

LRO: Lunar Reconnaissance Orbiter Project Timing Specification, LRO NGIN 431-SPEC-000212, NASA/Goddard Space Flight Center, Greenbelt, MD, September 8, 2006

S.B. Luthcke et al., Spaceborne laser altimeter pointing bias calibration from range residual analysis. J. Spacecr. Rockets **37**, 374–384 (2000)

J. McGarry, T. Zagwodzki, SLR2000: The path towards completion, in *Proc. 15th Int. Workshop on Laser Ranging*, Canberra, Australia, 2006

T.W. Murphy Jr., et al.: The Apache Point Observatory lunar laser ranging operation, in *12th Int. Workshop on Laser Ranging*, Matera, Italy, 2000, 10 pp.

G.A. Neumann et al., The crossover analysis of MOLA altimetric data. J. Geophys. Res. **106**, 23,753–23,768 (2001)

C. Noll et al., Laser ranging archiving and infrastructure support through the ILRS data centers and web site, in *15th International Workshop on Laser Ranging*, Canberra, Australia, 2006

D.E. Pavlis et al., GEODYN operations manuals. Raytheon ITTS Contractor Report, Lanham, MD, 2001

M. Pearlman et al., The International Laser Ranging Service. *AOGS Advances in Geosciences: Solid Earth* (2008)

L. Ramos-Izquierdo et al., Optical system design and integration of the Lunar Orbiter Laser Altimeter (2008, manuscript in preparation)

R. Ricklefs, Consolidated laser prediction and data formats: Supporting new technology, in *Proc. 15th Int. Workshop on Laser Ranging*, Canberra, Australia, 2006

M.S. Robinson et al., The Lunar Reconnaissance Orbiter Camera (LROC). Space Sci. Rev. (2008)

D.D. Rowlands et al., The use of laser altimetry in the orbit and attitude determination of Mars Global Surveyor. Geophys. Res. Lett. **26**, 1191–1194 (1999)

D.D. Rowlands et al., A simulation study of multi-beam altimetry for Lunar Reconnaissance Orbiter and other planetary missions. J. Geodesy (2008). doi:10.1007/s00190-008-0285-y

D.E. Smith et al., The global topography of Mars and implications for surface evolution. Science **284**, 1495–1503 (1999)

D.E. Smith et al., Mars Orbiter Laser Altimeter (MOLA): Experiment summary after the first year of global mapping of Mars. J. Geophys. Res. **106**, 23,689–23,722 (2001)

D.E. Smith et al., Orbit determination of LRO at the Moon, in *7th Int. Laser Ranging Service Workshop*, Oct. 13–17, 2008, Poznan, Poland

D.E. Smith et al., The Lunar Orbiter Laser Altimeter (LOLA) investigation on the Lunar Reconnaissance Orbiter (LRO) mission. Space Sci. Rev. (2009, this issue)

M.T. Zuber et al., The Mars Observer Laser Altimeter investigation. J. Geophys. Res. **97**, 7781–7798 (1992)

Space Sci Rev (2010) 150: 81–124
DOI 10.1007/s11214-010-9634-2

Lunar Reconnaissance Orbiter Camera (LROC) Instrument Overview

M.S. Robinson · S.M. Brylow · M. Tschimmel · D. Humm · S.J. Lawrence · P.C. Thomas · B.W. Denevi · E. Bowman-Cisneros · J. Zerr · M.A. Ravine · M.A. Caplinger · F.T. Ghaemi · J.A. Schaffner · M.C. Malin · P. Mahanti · A. Bartels · J. Anderson · T.N. Tran · E.M. Eliason · A.S. McEwen · E. Turtle · B.L. Jolliff · H. Hiesinger

Received: 22 June 2009 / Accepted: 2 February 2010 / Published online: 11 March 2010

Abstract The Lunar Reconnaissance Orbiter Camera (LROC) Wide Angle Camera (WAC) and Narrow Angle Cameras (NACs) are on the NASA Lunar Reconnaissance Orbiter (LRO).

M.S. Robinson (✉) · M. Tschimmel · S.J. Lawrence · B.W. Denevi · E. Bowman-Cisneros · P. Mahanti · T.N. Tran
School of Earth and Space Exploration, Arizona State University, Tempe, AZ, USA
e-mail: mrobinson@ser.asu.edu

S.M. Brylow · J. Zerr · M.A. Ravine · M.A. Caplinger · J.A. Schaffner · M.C. Malin
Malin Space Science Systems, San Diego, CA, USA

D. Humm
Space Instrument Calibration Consulting, Annapolis, MD, USA

P.C. Thomas
Center for Radiophysics and Space Research, Cornell University, Ithaca, NY, USA

A. Bartels
National Aeronautics and Space Administration, Goddard Space Flight Center, Greenbelt, MD, USA

J. Anderson
Astrogeology Branch, United States Geological Survey, Flagstaff, AZ, USA

F.T. Ghaemi
Tony Ghaemi Optical Engineering, San Diego, CA, USA

E.M. Eliason · A.S. McEwen
Lunar and Planetary Lab, University of Arizona, Tucson, AZ, USA

E. Turtle
Applied Physics Lab, Johns Hopkins University, Laurel, MD, USA

B.L. Jolliff
Department of Earth and Planetary Sciences, Washington University, St. Louis, MO, USA

H. Hiesinger
Institut für Planetologie, Westfälische Wilhelms-Universität, Münster, Germany

The WAC is a 7-color push-frame camera (100 and 400 m/pixel visible and UV, respectively), while the two NACs are monochrome narrow-angle linescan imagers (0.5 m/pixel). The primary mission of LRO is to obtain measurements of the Moon that will enable future lunar human exploration. The overarching goals of the LROC investigation include landing site identification and certification, mapping of permanently polar shadowed and sunlit regions, meter-scale mapping of polar regions, global multispectral imaging, a global morphology base map, characterization of regolith properties, and determination of current impact hazards.

Keywords LRO · LROC · Instrument · Camera · Moon · Lunar · Calibration

1 Introduction

The observations, fieldwork, and sampling performed by the Apollo astronauts in lunar orbit and on the lunar surface yielded key insights into the origin, composition, and geologic evolution of the Moon and transformed humanity's knowledge of the Solar System. One of the key reasons that the Apollo missions were so successful was the intensive reconnaissance of the lunar surface performed using the five robotic Lunar Orbiter spacecraft, which provided synoptic photographic coverage of the entire Moon (Bowker and Hughes 1971), as well as high-resolution (1 m/pixel) observations specifically targeted at candidate Apollo landing sites. The Lunar Orbiter observations enabled the subsequent early Apollo landings and provided a valuable dataset for lunar geologic studies (El-Baz and Kosofsky 1970; Musgrove 1971). Existing lunar datasets (including Lunar Orbiter, Apollo, Clementine, and Lunar Prospector) are insufficient to effectively plan for future landings (human or robotic), especially in areas outside of the so-called "Apollo Zone" where meter-scale imagery and topographical data vital for landing site certification and operational planning is sparse or non-existent. Recent lunar missions (including JAXA's Kaguya mission, the CNSA Chang'E, and the ISRO's Chandrayaan-1) collected datasets of high scientific and exploration value, though none reveal lunar surface morphology at the scale needed to support the projected engineering requirements of NASA's next-generation human lunar exploration.

The Space Exploration Policy of the United States calls for a resumption of human lunar landings no later than 2020 under the aegis of Project Constellation, with the goal of eventually establishing a permanent outpost on the lunar surface (NASA 2004a). The Lunar Reconnaissance Orbiter (LRO) is designed to serve Project Constellation as Lunar Orbiter served Apollo, providing comprehensive remote sensing data that will enable the next generation of human lunar exploration and development. The Lunar Reconnaissance Orbiter Camera (LROC) is designed to address two of the LRO Level 1 measurement objectives: Landing site certification and determination of polar illumination (Chin et al. 2007; NASA 2004b). LROC data will help us redefine our understanding of the Moon's history and current state and guide lunar surface operations for decades to come.

LROC consists of three imaging subsystems (Figs. 1, 2, 5) with a supporting Sequence and Compressor System (SCS). The two Narrow Angle Cameras (NACs) are monochrome push-broom scanners, with an instantaneous field-of-view (IFOV) of 10-microradians (50-cm angular resolution from the nominal 50-km orbit (Fig. 2)). The multispectral Wide Angle Camera (WAC) is a push-frame imager with five visible wavelength filters (415 to 690 nm) with an IFOV of 1.498-milliradians (75-m angular size of pixel projected on the surface from a 50-km altitude) with the visible light filters, and IFOV of 9.672-milliradians

Fig. 1 The Lunar Reconnaissance Orbiter, with the positions of the two LROC Narrow Angle Cameras [NACs] and the Wide Angle Camera [WAC] highlighted (Image Credit: A. Bartels, NASA Goddard Space Flight Center)

(384-m angular size of pixel projected on the surface from a 50-km orbit) from two UV filters (320 and 360 nm). In addition to multispectral imaging, the WAC can operate in monochrome mode to provide a global large-incidence-angle (Sun low to the horizon) basemap and time-lapse movies of the illumination conditions at both poles (Fig. 5). The SCS transfers commands from the spacecraft to the camera to acquire image data, and receives and compresses data en route from the cameras to the spacecraft's solid state recorder. In this paper, we provide an overview of the LROC investigation, present the results of the preflight instrument calibration, and summarize LROC ground operations.

2 LROC Overview and Scientific Objectives

LRO was launched 18 June 2009 on an Atlas V 401 rocket from the Cape Canaveral Air Force Station Launch Complex 41. Following a four day Earth-Moon transit, the spacecraft will enter a 2-month commissioning phase, followed by a year-long nominal mission in a 50×50-km polar mapping orbit. Following the conclusion of the nominal mission, a multi-year extended mission in a fixed 27×216-km orbit is possible.

LROC has been specifically designed to address two of the primary LRO measurement requirements (NASA 2004b).

(1) Assessment of meter and smaller-scale features in order to facilitate safety analysis for potential lunar landing sites near polar resources and elsewhere on the Moon.

One of the two prime objectives of the LROC investigation is to provide the data necessary and sufficient to characterize potential landing sites. The LRO requirement is to assess the threat posed by "meter and smaller-scale features" (NASA 2004b). This assessment requires a 0.5-m pixel scale in order to ensure unambiguous two-pixel detection of meter-scale features. The NACs will detect blocks down to ~1 m horizontal scale with ≤ 0.5 m heights, and craters with diameters ≥ 2.5 m. In addition to high-resolution imaging, topographic data derived from both geometric and photometric stereo (photoclinometry) observations of potential landing sites at spatial scales of 2 to 5 m/pixel will provide quantitative information about the slopes in small craters that pose a risk for landing spacecraft. From both the Apollo and Surveyor experience, it is known that small craters, rather than blocks, may actually present a greater danger to landing vehicles because they are far more abundant. In fact, except for very young Copernican units, the lunar surface is saturated with craters smaller than 100 m in diameter (Gault 1970). Thus, a lunar landing site will contain many craters, presenting a sub-horizontal surface at the scale of any lander. Blocks, on the other hand, are much less common, especially in areas away from fresh craters (Cameron and Coyle 1971; Cintala et al. 1982; Cintala and McBride 1995; Lee et al. 1986; Moore 1971). The presence of small, shallow craters can be assessed from images with large (> 70°) incidence angles and meter-scale topography. LROC will provide essential morphology information and stereogrammetric and photometric stereo observations for high-priority exploration sites (NASA 2005; Taylor and Spudis 1990).

(2) Acquire multi-temporal synoptic 100 m/pixel imaging of the poles during every orbit to unambiguously identify regions of permanent shadow and permanent or near-permanent illumination.

The Moon's equator is tilted only 1.54° to the ecliptic, so seasonal lighting conditions change only a small amount during a year. Thus there are regions of permanent shadow and permanent or nearly permanent illumination in the vicinity of the lunar poles. Theory as well as radar and neutron measurements suggest that deposits of water-ice (a potentially valuable lunar resource) may occur in permanently shadowed regions near the poles (Feldman et al. 2000; Ingersoll et al. 1992; Nozette et al. 2001; Watson et al. 1961). Regions that are illuminated for extended periods are prime candidates for future lunar bases owing to relatively benign thermal conditions and constant solar power (Goddard 1920; Heiken et al. 1991). Clementine imaged the poles every 10 hours (250 to 500 m/pixel) allowing a synoptic reconstruction of lighting conditions over a period of two Earth months (i.e., two lunar days). At the time of the Clementine mission, the Moon was in southern winter (sub-solar point 1.5°N to 0.3°N), therefore the derived lighting maps more accurately show permanent shadows and near-permanent illumination in the northern and southern hemispheres, respectively. Clementine lighting maps revealed peaks near the South Pole that receive sunlight for almost 70% of a lunar day (Bussey et al. 1999; Noda et al. 2008). Unfortunately, the Moon was in a nearly identical configuration during the Apollo-era Lunar Orbiter polar missions, and so it is not possible to construct maps of regions of anomalously long illumination near the North Pole from any existing data.

In monochrome mode from the nominal 50-km orbit, the WAC FOV covers a swath ~104-km wide resulting in repeat coverage every orbit for the region between 88° and 90° at each pole. LROC will image 80° to 90° on the dayside and back to 80° on the night side on each orbit, enabling the creation of a year-long illumination movie with time steps at least every five hours from ±88–90°, and coverage to 85° with lesser but still sufficient time resolution over a full lunar year. WAC images will provide a calibrated, archived, and reduced dataset completely delimiting permanently shadowed regions and permanently (or near-permanently) illuminated terrain at both poles at a scale of 100 m.

In addition to these two primary objectives, LROC will return six other high-value datasets valuable both for planning future human exploration and supporting basic lunar science investigations.

(3) Meter-scale mapping of regions of permanent or near-permanent illumination of polar massifs.

While the combined NAC FOV will limit its temporal sampling of polar illumination conditions, it can provide meter-scale sampling of the boundaries of shadowed regions and complete mapping of illuminated terrain within 5° of the poles. During respective summers, when shadows are at a minimum, the NAC will acquire hundreds of images at 1-m/pixel to enable mosaics of each pole. The mosaics are built up by first imaging from 87.5° latitude to 0.8° past the pole on one orbit. For the next two orbits, the NACs then image from 85.5° to 88.8°. This three orbit cycle is repeated for 29 days until complete coverage of the map region is obtained. The summer mosaics will serve as a basemap of illuminated regions and will provide high-resolution details of the margins of permanently shadowed regions. During winter much of the polar regions are in shadow, so NAC imaging will concentrate on targets known to have extended periods of illumination. Two massifs near the South Pole collectively receive illumination for more than 95% of a winter day at a scale of $\sim$200 m/pixel (Bussey et al. 1999; Noda et al. 2008). Lighting conditions of these topographic peaks will be examined at 1 m/pixel over extended periods.

Identification and high-resolution mapping of highly illuminated sites is of critical importance for future human surface operations in the lunar polar region for both power and environmental reasons (Goddard 1920; Heiken et al. 1991; Watson et al. 1961). Since the persistently illuminated regions are by definition massifs, it is probable that their summit areas will receive the most illumination. Such regions could be as small as a few to tens of meters across, below the resolution of other lunar datasets. A 10×10 m region represents 1% of a WAC pixel, 4% of a SMART-1 AMIE pixel, 1 Kaguya terrain camera pixel, and 2×2 Chandrayaan Terrain camera pixels, but will be covered by 400 NAC pixels (20×20). The value of such small areas may actually be very high, providing mission planners with specific areas where installed solar arrays would provide continuous or nearly continuous power to nearby assets. High-priority landing and operational sites will be imaged at full resolution (0.5 m/pixel) allowing characterization of the suitability for landing and surface operations at these high value targets.

(4) Repeat observations of portions of potential landing sites and elsewhere for the derivation of high-resolution topography through stereogrammetric and photometric stereo analyses.

LROC will provide photogrammetric and photometric stereo data for high-priority candidate landing sites and select areas of high scientific interest. Most of the processing for topographic maps will need to be funded by programs other than LRO, but the LROC team will generate example geometric and photometric stereo-based products for select sites to demonstrate the utility of the methods and data reduction procedures. For photometric stereo, at least three nadir observations of the same scene, each under different illumination conditions will be acquired. For geometric stereo, the LRO will have to be rolled into an off-nadir geometry to collect cross-track images at appropriate lighting conditions, but these opportunities will be limited to at most three per 24-hour period due to constraints imposed by other LRO instruments.

(5) Global multispectral observations in seven wavelengths to characterize lunar resources, in particular ilmenite.

The WAC multispectral dataset will allow the science community to improve maps of some mineralogical components of the lunar crust, in conjunction with complementary datasets (Clementine, Kaguya, Chandrayaan-1, and SMART-1). The bandpasses of LROC WAC are particularly suited for the detection of ilmenite ($FeTiO_3$), which is identified as a valuable lunar resource. Ilmenite is notable for being an excellent source of oxygen, titanium metal (Williams et al. 1979; Zhao and Shadman 1993), and solar-wind-implanted volatiles (such as H and He), which are preferentially retained by ilmenite-rich soils (Cameron 1993). Volatile elements may be useful to facilitate the production of water and possibly fuels (H and ^{3}He). The accurate determination of titanium abundance in lunar surface deposits is therefore an import part of lunar resource assessment and development. No funding is provided to construct the global multi-spectral map.

(6) A global 100-m/pixel basemap with incidence angles (60°–80°) favorable for morphological interpretations.

Global coverage at scales of 100-m/pixel and large incidence angles (> 50°) is crucial for systematic geologic mapping and crater counting efforts. High precision crater counts of the lunar surface are especially important because the Moon is the only planetary body for which relative ages from crater counts can be correlated with absolute dating of rock samples collected by trained field observers (Stoffler et al. 2006, and references therein). The best existing lunar morphologic basemap is the recently completed Lunar Orbiter digital mosaic (Becker et al. 2004) with scales ranging from 60 to 600 m/pixel, although this dataset could eventually be superseded by the data collected by instruments aboard recent lunar missions (i.e., Chang'E, Kaguya). The current lack of globally uniform, high-resolution lunar imaging means that large parts of the Moon's surface (especially on the lunar farside) have not been systematically dated using crater statistics. Clementine observations were intentionally acquired at high Sun angles to enhance the analysis of the color data and are not suitable for morphologic mapping at non-polar latitudes (Nozette et al. 1994). LROC WAC will provide imaging to construct a 100-m/pixel global map with incidence angles between 55°–75° (except at in polar regions, where incidence angles are always higher) to enable the creation of a uniform global crater chronology dataset with consistent incidence angles.

(7) Sub-meter imaging of a variety of geologic units to characterize their physical properties, the variability of the regolith, and other key science questions.

The physical properties of the lunar regolith have been characterized in great detail from data and samples returned by Surveyor, Apollo, and Soviet missions. Meteorite bombardment over time has converted crystalline rock to a soil that is essentially uniform in terms of thickness, particle size distribution, bearing strength, cohesion, and many other physical properties (Carrier et al. 1991). This layer is known as the *regolith*, and it covers the entire lunar surface. Regolith typically varies in thickness between 1 and 10 m in the maria, and most likely greater than 10 m in depth in highlands regions (Cintala et al. 1982). All lunar surface operations will take place on or within the regolith, and all exploitable resources and construction materials will come from the regolith, which means that understanding the properties of the regolith is vital for successful lunar surface activities. Although much is known about the physical properties of lunar soils, LROC NAC imaging will provide the necessary data (0.5 m/pixel imaging) needed to apply proven techniques to estimate regolith thickness and other key parameters (Oberbeck and Quaide 1968, 1967; Quaide and

Oberbeck 1968). Among the science questions that will be addressed using LROC NAC images include the rates of regolith generation, local variability in regolith thickness, and the presence within the regolith of coherent blocks.

(8) Meter-scale coverage overlapping with Apollo-era panoramic images (1–2 m/pixel) to document the number of small impacts since 1971–1972. The current impact rate for bolides in the 0.5 to 10 m diameter range is not known to better than 1 or 2 orders of magnitude (Ivanov 2006). Elucidating the impact rate at these sizes enables engineering remediation measures for future surface operations and interplanetary travel.

Particles with masses 50 g to 1000 g (4 to 10 cm diameter at 2 g/cm^3) impact the Moon as often as 10 times/day (Duennebier et al. 1976). LROC NAC observations (0.5 m/pixel) of (1) very young mare surfaces and (2) regions photographed using the Apollo 15–17 Panoramic Camera will be used to place estimates on the number of small craters (1 to 100 m diameter) that have formed from 1971–2010. LROC NAC images will be collected from and compared to as many of the regions with comparable Apollo panoramic camera imaging (i.e., 1 m/pixel) as possible (Lawrence et al. 2008a). This comparison will provide the first direct measure of the current formation rate of 1 to 100 m craters and place an important constraint on the most energetic hypervelocity impact from which future lunar structures must be protected.

3 Instrument Design

Leveraging the extensive experience of LROC prime contractor Malin Space Science Systems (MSSS), the LROC instrument subsystems share heritage from the Mars Color Imager (MARCI) and Context Camera (CTX) aboard the Mars Reconnaissance Orbiter (MRO), the Mars Climate Orbiter (MCO), Mars Polar Lander (MPL), Mars Odyssey, and the Mars Phoenix Lander spacecraft (Malin et al. 2001, 2007; Bell et al. 2009).

3.1 Narrow Angle Cameras (NACs)

The two NACs (Fig. 2), designated NAC-Left (NAC-L) and NAC-Right (NAC-R), are designed to provide 0.5 m scale panchromatic images over a combined 5-km swath. Each NAC utilizes a 700-mm focal-length telescope that images onto a 5000-pixel charge coupled device (CCD) line-array, providing an IFOV of 10 µradians and a cross-track field-of-view of 2.85° each, which yields a total ground track swath-width for both cameras of 5-km at the nominal 50-km altitude. Each NAC internal buffer holds 256 MB of uncompressed data, enough for a full-swath image 25-km in length or a 2× binned image 104-km long. The NAC data are sampled to 12-bits, then companded (bit compression) to 8-bits. Prior to downlink an additional lossless first-difference Huffman compression is applied.

3.1.1 NAC Optics

The principal challenges of the NAC optical design were to provide near diffraction-limited performance at a half-meter ground sample distance over a relatively wide field, in the lunar orbit thermal environment and with low mass. While the starting point for the NAC design was the MRO CTX (Malin et al. 2007), that design was unable to meet these NAC requirements. The NAC resolution requirement was twice that of CTX (10 microradians per pixel rather than CTX's 20 microradians) but with the same field of view. The temperature range the instrument experiences in lunar orbit is considerably greater than what CTX sees at Mars, prompting us to seek a less thermally sensitive design than the CTX Maksutov telescope. Finally, cost minimization drove the need to maximize the reuse the CTX detector and electronics. Central to our strategy for addressing these requirements was to build two duplicate NACs with offset pointing to provide twice the field of view of a single copy of the instrument.

Using two instances of the instrument to get the full 5.7° swath requires an individual unit FOV of 2.85°. To limit the thermal sensitivity of NAC optics, we pursued optical designs without a large refractive front corrector group. The flight NAC telescope uses hyperbolic primary and secondary mirrors (Ritchey-Chretien, hereafter RC) with an all spherical triplet field corrector lens to provide an optical field of view of approximately 3° (large for an RC design). This f/3.59 telescope had a design MTF at Nyquist of greater than 56% over the entire field. The design also incorporates a 650 mm long sunshade to prevent illumination of the primary mirror by rays more than 30° off axis. The specifications of each of the two NAC telescopes are given in Table 2. To minimize both mass and thermal sensitivity, the metering structure is made of graphite-cyanate composite. Because of the composite resin is hygroscopic, there is a significant focus shift between fully dry, the case which pertains to lunar orbit, and "wet," the state the instruments were in the laboratory environment. Validation of focus for flight was by baking-out the cameras in at 65°C in vacuum while imaging a collimator stimulus. A final confirmation of focus was made by imaging the Moon immediately after removing each unit from thermal vac (Fig. 21). The NAC telescopes were built by LightWorks Optics of Tustin, CA in collaboration with Vanguard Composites of San Diego, CA.

3.1.2 NAC Electronics

The NAC electronics design is based on the Context Camera (CTX) on the Mars Reconnaissance Orbiter (MRO) mission, with modifications to increase the maximum pixel rate from 2 megapixels/second to 7.5 megapixels/second. The electronics are built around the Kodak KLI-5001G line array. This device is a 5064-element CCD with 7 micron pixels, and an average QE through the instrument bandpass of >60%. It is fabricated in an NMOS two-metal two-poly process by Kodak Image Sensor Solutions in Rochester, NY. Correlated Double Sampling (CDS) is incorporated on-chip. The device has dual readouts with a maximum readout rate of 12.5 MHz. Though the datasheet noise floor is fairly high (150 e^- at room temperature and 12.5 MHz operation) performance is better at operating temperature and slower clocking.

The NAC analog signal processing chain is considerably simplified by the on-chip CDS provided by the CCD. There are two duplicate analog signal chains, one handling the even pixels, the other handling the odd pixels. Although CDS is performed on the CCD die, the output signals still have a large DC pedestal, which varies device-to-device and (potentially) with temperature and radiation exposure. To mitigate excess background the design includes a black-level DAC that allows a constant black level to be removed prior to amplification. The black level was established during calibration and can be changed via ground command. The digitization range is set to 300 Ke^- maximum, yielding a scale factor of about 90 e^-/DN (90.5 NAC-L and 92.5 NAC-R), which is close to the best-expected noise performance of the CCD.

The NAC digital electronics provides clock generation, sampling of the CCD signal, conversion of the 12-bit samples to 8-bit encoded pixels, storage of the pixels, and finally readout of the pixels to the SCS. A dedicated field-programmable gate array (FPGA) rather than a general-purpose processor performs these functions. The specific functions implemented in the FPGA are:

Fig. 2 The LROC Narrow Angle Cameras side by side without their radiators. The black baffle prevents stray light from entering the optics (silver part of the tube). The focal plane array and associated electronics are in the triangular housing on the right end. The diameter of each tube is 27 cm

- Serial command interface from the SCS
- Serial data interface to the SCS
- DRAM interface
- CCD clock generation
- 12-to-8 bit pixel companding
- Commanding and DRAM readout

The NACs use 256 Mbit synchronous DRAM for image storage. Eight devices provide a total capacity of 256 MB (268,435,456 bytes). Each pixel value is mapped through an adjustable piecewise linear transfer function. Approximate square root encoding allows conversion of 12-bit per pixel data to 8-bit form with only minimal loss of information. Since the transfer function can be changed in flight, other forms of encoding (e.g., linear with offset) can be used as well (see Appendix A).

The SCS communicates with the NAC via a 3-signal (data, clock, and enable) synchronous serial LVDS interface, at a maximum clocking rate of 2 MHz. The command interface section of the FPGA is clocked by the interface clock and its outputs are synchronized to the system clock domain using dual flip-flops. The instrument communicates with the SCS via a 3-signal (data, clock, and enable) synchronous serial interface, at a maximum clocking rate of 30 MHz.

Only minimal instrument health monitoring is provided in the form of a thermistor at the focal plane (backside of CCD), at the field programmable gate array (FPGA) in the electronics box, and on the metering structure between the primary and secondary mirrors. Operating temperatures of the focal plane are expected to range from 0°C to 20°C over the course of the mission, mostly a function of local time-of-day.

3.1.3 NAC Orientation

The NACs are mounted on the spacecraft such that the CCDs are perpendicular to the spacecraft's X-axis. The NAC-L is off-pointed $\sim$2.85° from the NAC-R so that the footprints of

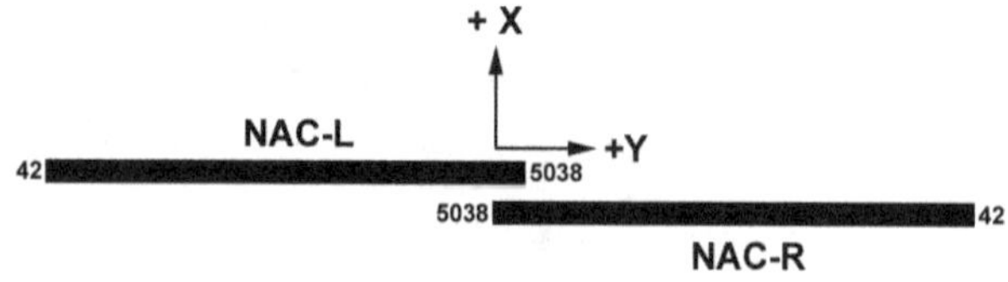

Fig. 3 Orientations of NAC CCDs as mounted on the spacecraft. Numbers show pixel addresses (zero based coordinates). The NACs overlap ~135 pixels in the crosstrack direction (Y) and are separated ~0.106° (185 pixels) downtrack (X)

the two images overlap only ~135 pixels. The NAC-R is also mounted 0.106° forward of the NAC-L. The NACS are mounted such that pixel 0 for the NAC-L is at the −Y (in spacecraft coordinates) end of its CCD and pixel 0 for the NAC-R is at the +Y end of its CCD (Fig. 3). This orientation requires that one of the NAC frames from a NAC-L and NAC-R paired observation must be transformed such that both images have the same ground orientation.

3.2 Wide Angle Camera (WAC)

The WAC (Fig. 4) is designed to provide global imaging at a scale of 100 m/pixel in seven color bands over a 105-km swath in monochrome mode and 57-km in color mode. The LROC WAC uses a 1000 by 1000 pixel CCD that has seven narrow-band interference filters bonded to the detector. Two of the filters are in the ultraviolet (UV) and five are in the visible. The WAC has two short-focal-length lenses, one UV and one visible, both of which image onto the same detector. From the nominal 50-km altitude orbit in monochrome imaging mode, the WAC will provide a nadir ground sample distance of 75 m/pixel (98 m/pixel at far field). In the UV system, the nadir resolution is 384 m/pixel. The seven-band color filter array provides different sections of the CCD with different bandpasses. Repeated imaging at a rate such that each of the narrow framelets of each color band overlap provides continuous coverage in any one color ("push-frame imaging"). The layout of the WAC filters is similar to the MRO MARCI instrument as described in Bell et al. (2009). WAC pixel values are digitized to 11-bits and subsequently companded to 8-bit values. After transmission to the SCS, the WAC data have a lossless first-difference Huffman compression that is applied prior to downlink.

3.2.1 WAC Optics

The WAC optics consist of four optical elements: the visible lens, the UV lens, the prism and the color filter array (CFA). The visible lens has a 6.0 mm focal length and a focal ratio of 5.5. It has six fused silica elements and is optimized for the wavelength range from 395 nm to 690 nm. It provides a 90° field of view over the full 1008 pixel width of the detector. The lens has a design MTF of greater than 60% at 56 line pairs per mm in all bands. The UV lens has a 4.7 mm focal length and a focal ratio of 5.1. It has five elements and is optimized for the wavelength range from 290 nm to 370 nm. It provides a 60° field of view over a 512 pixel width on the detector. Because signal is low in the UV bands, the UV system data is acquired by summing pixels 4 × 4. The optical design provides a 4 by 4-pixel ensquared energy of greater than 80% in both UV bands. The physical diameter of each of these lenses is greater than the format of the CCD detector. A prism is used to allow both lenses to image on the same CCD. This prism provides a straight-through path from the visible lens to the detector, and a periscope-type optical path for the UV lens. The latter

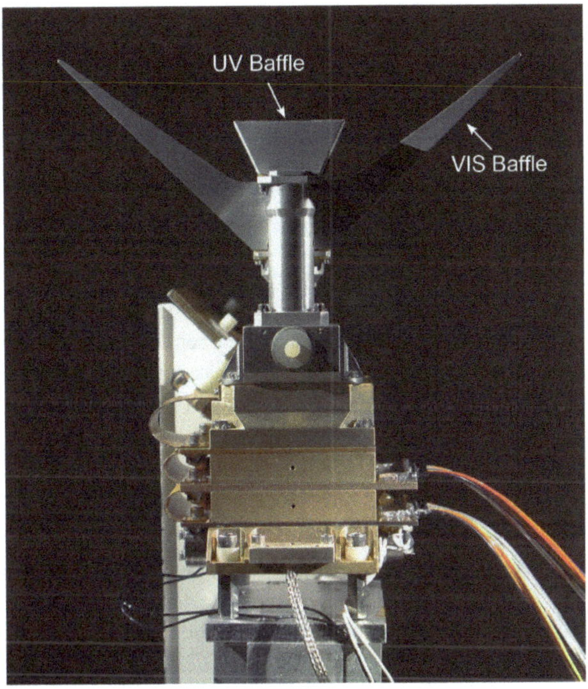

Fig. 4 LROC Wide Angle Camera. The width of the visible optic baffle is 15.95 m (Image credit: Mike Malin, Malin Space Science Systems, Inc.)

offsets the point at the UV image is formed laterally, placing it closer to the visible image (and therefore on the photoactive area of the detector). Each lens and the prism are integrated and aligned into a signal assembly with the WAC detector and electronics. The color filter assembly (CFA) is a 9.5 mm by 9.3 mm by 0.6 mm fused silica substrate with seven vacuum deposited interference filters laid down is the geometry appropriate for push-frame imaging. It is bonded to the surface of the CCD detector with optical cement. The final focusing of the system is done by lapping a spacer between the CCD package and the back of the prism housing. The WAC lens/prism assembly was built by LightWorks Optics and the CFA was manufactured by Barr Associates of Westford, MA.

3.2.2 WAC Electronics

The WAC electronics are designed around the Kodak KAI-1001 Charge-Coupled Device (CCD) detector. This detector has 1024 × 1024 9-μm pixels (1018 × 1008 photoactive, others masked for background/bias signal determination) and uses interline transfer to implement electronic shuttering. The KAI-1001's fill factor of 20% causes its quantum efficiency to be low, but the WAC optics are sufficiently "fast" to compensate and thus provide the required system sensitivity. To support selection of data through specific bands of the color filter array (CFA), the CCD's fast-dump feature can be used to read out only selected portions of the detector.

The output signal from the CCD is AC-coupled and then amplified. The amplified signal is digitized to 12-bits at 3 Mpixels/s. For each pixel, both reset and video levels are digitized and then subtracted in the digital domain to perform correlated double sampling (CDS), resulting in a net 11-bits of dynamic range. The digital electronics are responsible for clock generation, sampling of the CCD signal, conversion of the 11-bit samples to 8-bit encoded pixels, storage of the pixels, and readout of the pixels to the SCS. These functions are performed by a digital signal processor (DSP) with software emulation.

WAC instrument health monitoring includes real-time telemetry monitoring of voltage and temperature during S-band contacts, and trend analysis of voltage and temperature using stored telemetry downlinked during Ka-band contacts. Additional health monitoring is performed by statistical analysis of acquired images, especially WAC dark pixels.

3.2.3 WAC Monochrome Imaging

The highest-resolution existing morphologic basemap is the Lunar Orbiter digital mosaic (Becker et al. 2004) with resolutions ranging from 60 to 600 m/pixel (Bowker and Hughes 1971). The WAC will accordingly be operated in single-band mode (nominally the 605 nm filter) to acquire a global morphology basemap with incidence angles between 60° and 80° to provide an improved morphologic map of the Moon at a uniform resolution of 100 m/pixel.

3.2.4 WAC Multispectral Imaging

The WAC has two ultraviolet filters (320 and 360 nm), and five visible filters (415, 565, 605, 645, and 690 nm). These bandpass filters were selected to characterize spectral features for some common minerals found at the lunar surface and will complement Clementine UVVIS (415, 750, 900, 950, 1000 nm) and NIR datasets (1100, 1250, 1500, 2000, 2600, 2780 nm) (Nozette et al. 1994) as well as VIS-NIR data being collected by Chandrayaan-1 and Selene (Haruyama et al. 2008; Green et al. 2007).

The UV bandpasses are sensitive to the distribution of some opaque minerals such as ilmenite, which have a distinctive UV spectral signature consisting of a characteristic absorption between 450 and 500 nm (Fig. 5). This band results in a flat to upturned signal in the near-UV relative to other common lunar minerals (Wagner et al. 1987). The WAC visible bandpasses are also sensitive to large variations in the magnesium to iron ratio (magnesium number) of olivine, which exhibits maxima at 550 nm to 700 nm as the iron content increases (Fig. 6).

3.3 Sequence and Compressor System

The Sequence and Compressor System (SCS) interfaces the WAC and two NACs to the spacecraft avionics (Fig. 7). The SpaceWire link between the SCS and the spacecraft oper-

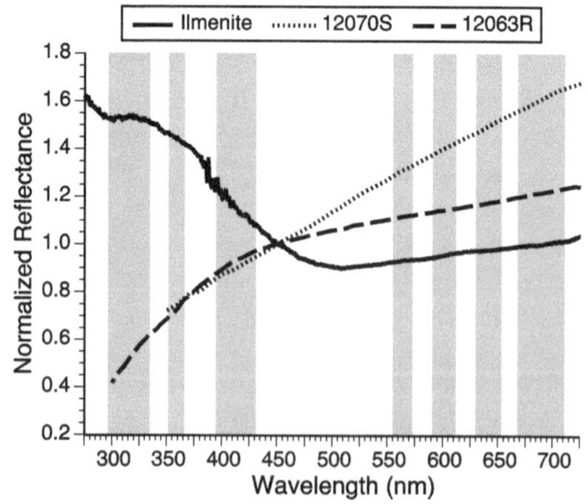

Fig. 5 Spectra of ilmenite (*solid line*) showing a broad minima at 500 nm. The WAC has the unique ability to acquire measurements in the UV (bandpasses at 320, 360 nm) that are well suited to discriminate ilmenite from other common lunar minerals. Low titanium rock (*dashed line* 12063R) and soil samples (*dotted line* 12070S) are plotted for reference (normalized at 450 nm). FWHM of the WAC bandpasses are shown as *gray bars*

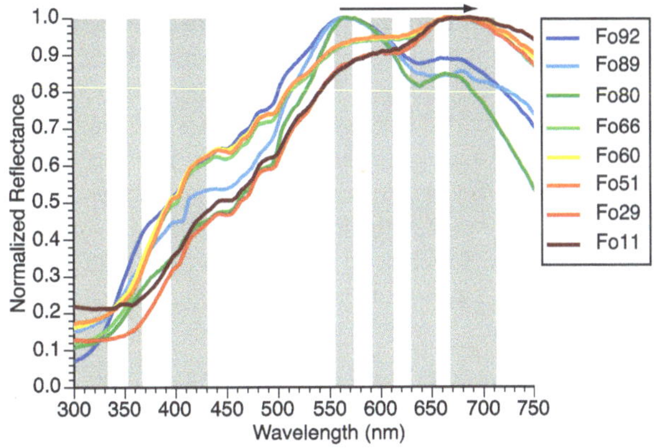

Fig. 6 Spectra of olivine as a function of Fo-number (i.e., Mg relative to Fe content). The *arrow* indicates the migration of the reflectance peak to longer wavelengths as the Fo number decreases. Spectra from RELAB public database, Brown University. LROC WAC bandpasses are shown for reference

Fig. 7 LROC Sequence and Compressor System (Image credit: Mike Malin, Malin Space Science Systems, Inc.)

ates at 2 Mbit/s while transferring commands to the SCS, and 40 Mbit/s carrying telemetry from the SCS to the spacecraft. Each NAC command/data interface is comprised of two synchronous serial 3-wire LVDS links; one transmits commands from the SCS to the NAC at 2 Mbit/s and the other receives NAC telemetry at 30 Mbit/s. The WAC command/data interface is comprised of a synchronous serial 4-wire RS-422 link that operates at 2.5 Mbit/s with a data signal in each direction and a common clock and frame signal generated by the SCS.

The SCS, WAC and NACs are powered directly from the spacecraft 28 V power bus, on separately switched circuits. The WAC heritage design includes an internal power switch and requires its DSP firmware to be loaded from an external source immediately after application of power. Both of these functions are provided by the SCS.

The core functionality of the SCS is implemented in a Xilinx Virtex-II FPGA. All interface, compression, and timing functions are implemented as logic peripherals to a MicroBlaze soft-processor core in that FPGA. NAC telemetry streams through the SCS in real-time, with minimal FIFO buffering and optional Huffman predictive compression performed in logic. To accommodate the interleaving of WAC and NAC data onto the single spacecraft interface, the SCS includes a double-buffer for WAC image frames. WAC data may also be compressed with a dedicated Huffman Compressor.

4 Instrument Calibration

The exploration and science goals of the LROC investigation require an accurate preflight calibration of the instrument system (i.e. optics, filters, detectors, and electronics), as well as continuing calibration activities in flight to ensure stability over the course of the mission. To those ends, the LROC team executed a comprehensive ground calibration sequence designed to accurately measure the response of the WAC and the NAC instruments to known, calibrated inputs. This section details the LROC instrument calibration results. The properties of the LROC cameras are summarized in Tables 2 and 3.

System-level testing of the LROC instruments was conducted using the facilities and equipment at MSSS (ambient laboratory calibrations), the University of California in San Diego (instrument thermal-vacuum tests) and the NASA Goddard Space Flight Center (LRO spacecraft thermal-vacuum tests). The calibration tests are listed in Table 1. Since the NAC design is based on the MRO/CTX camera and the WAC design on the MRO/MARCI camera, with the CCD detectors and electronics of the same type, the CTX/MARCI characterization tests described in Malin et al. (2005) and Bell et al. (2009) are both useful references.

Table 1 Calibration tests conducted for both NACs and the WAC. For the WAC, the test was done for each filter, if applicable

Calibration test	Description
Dark imaging	Determine dark current and bias; optimization of NAC bias level (DAC offset setting)
CCD performance validation and linearity	Measure linearity and photon transfer curve (gain, read noise, full-well)
Flat-field imaging	Measure system response non-uniformity on a pixel-by-pixel basis
System spectral response	Determine relative system throughput (in-band) and characterize out-of-band leaks
Absolute and relative radiometry	Determine conversion between measured DN and radiance units
Geometric calibration	Determine focal length, field of view and geometric distortion
Focus tests	Determine focus and MTF by imaging a bar target
Stray light test	Determine light scattering from out-of-field sources

Table 2 NAC system performance (where only one value is given it applies to both NACs)

Parameter	Performance NAC-L	Performance NAC-R
FOV	2.8502°	2.8412°
IFOV (nadir)	10.0042 µrad	9.9764 µrad
image scale at 50-km altitude	0.5 meter/pixel	
max. swath size at 50-km altitude	2.49 × 26.1-km	2.48 × 26.1-km
f/#	3.577	3.590
focal length	699.62 ± 0.08 mm	701.57 ± 0.09 mm
distortion coefficient *k*	0.0000181 ± 0.0000005	0.0000183 ± 0.0000005
optical center location	sample 2548 ± 8	sample 2568 ± 8
primary mirror diameter	198 mm	
system MTF (Nyquist)	0.23	
Gain	90.5 ± 2.6 e^-/DN	92.5 ± 1.5 e^-/DN
detector noise	101 ± 7 e^-	97 ± 1 e^-
detector full-well	334,000 ± 31,000 e^-	352,000 ± 4100 e^-
SNR (mare at 70° incidence angle)	> 42	> 42
Spectral response	400–760 nm	
radiometric accuracy*	1% relative, 10% absolute	
detector digitization	12 bit, encoded to 8 bit	
data link to SCS	LVDS, 30 Mbps	
temperature sensors	2 per electronics box, 2 per telescope	
lossless compression ratio	1.7:1	
Voltage	28 ± 7 V DC	
peak power	9.3 W	
orbit average power	6.4 W	
mass (both NACs)	16.4 kg	
volume (length x diameter)	118 cm × 27 cm (incl. radiator)	

*requires in-flight verification

4.1 NAC Calibration

4.1.1 Detector Characterization

The NAC utilizes a Kodak KLI-5001G linear array CCD that is read out with dual shift registers; channel A for the even pixels, and channel B for the odd pixels. The first 39 pixels are masked for dark signal monitoring, followed by 4 transition (non-imaging) pixels, then 4996 imaging pixels, then 4 transition pixels followed by 21 masked pixels.

The gain, read noise, and full-well of the NAC sensors were determined using the photon transfer method described in Janesick (2007). The setup for this experiment was a Quartz-Tungsten-Halogen (QTH) lamp illuminating a Spectralon® panel. At a given current setting, several images covering all exposure times between 0.34 ms (shortest possible) and 35.25 ms (longest possible) were acquired. After dark signal subtraction, the read-plus-shot noise and the shot noise were plotted versus the signal (the total noise also includes the fixed-pattern-noise). A NAC-L photon transfer curve is presented in Fig. 8, with the resulting derived parameters listed in Table 2. Precision was estimated by varying the endpoints of the linear

Table 3 WAC system performance

Parameter	Visible	UV
FOV (monochrome/color)	91.9°/61.4°	59.0°
IFOV (nadir)	1.498 mrad	7.672 mrad (4 × 4 binning)
image scale (nadir, 50-km altitude)	74.9 m/pixel	383.5 m/pixel
image frame width monochrome	104.6-km	–
image frame width 7-band color	59.6-km	56.8-km
image format monochrome	1024 samples × 14 lines	–
image format color (for each band)	704 samples × 14 lines	128 samples × 4 lines (binned)
f/#	5.052	5.65
focal length	6.013 mm	4.693 mm
system MTF (Nyquist)	0.37	
Gain	25.9 ± 0.7 e$^-$/DN	
detector noise	66 ± 4 e$^-$	
detector full-well	$46{,}100 \pm 3600$ e$^-$	
SNR (at 1000 DN)	> 150	
radiometric accuracy*	1% relative, 10% absolute	
detector digitization	11 bit, encoded to 8 bit	
data link to SCS	RS422, 2.5 Mbps	
temperature sensors	1	
lossless compression ratio	1.7:1	
Voltage	28 ± 7 V DC	
peak power	2.7 W	
orbit average power	2.6 W	
Mass	0.9 kg	
volume (width × length × height)	15.8 cm × 23.2 cm × 32.3 cm (incl. radiator)	

*requires in-flight verification

fit range, the number of pixels used for analysis, and the determination of the read noise. Above 200 DN both NACs have a highly linear response (Fig. 9).

4.1.2 NAC Calibration Pipeline

The image calibration pipeline for the NACs can be expressed with the following equation:

$$I_{cal}(x,\tau,T) = \frac{(I_{raw}(x,\tau,T) - m_I(\tau,T)) - (D(x,\tau,T) - m_D(\tau,T)) - S(x) - L(I_{off},x)}{F(x)\cdot\tau\cdot r},$$

where,

$I_{cal}(x,\tau,T)$ is the calibrated signal value of pixel x at exposure time τ and temperature T in radiance units of μW/(cm^2·sr·nm),
$I_{raw}(x,\tau,T)$ is the pixel's raw signal in DN,
τ the exposure time,
m is the mean DN value of the masked pixels,
$D(x,\tau,T)$ is the library dark image correction array,
$S(x)$ is the non-linearity offset array,

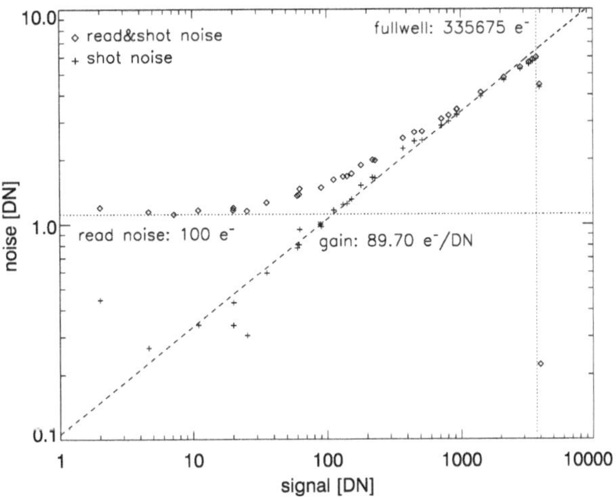

Fig. 8 Photon transfer curve for NAC-L. The linear fit of the shot noise between signal levels of 600 DN and 3000 DN yields a gain of 89.7 e⁻/DN. This value is within the uncertainty range of 2.6 e⁻/DN (average of 90.5 e⁻/DN), which was estimated by varying the endpoints of the linear fit range, the number of pixels used for analysis, and the determination of the read noise

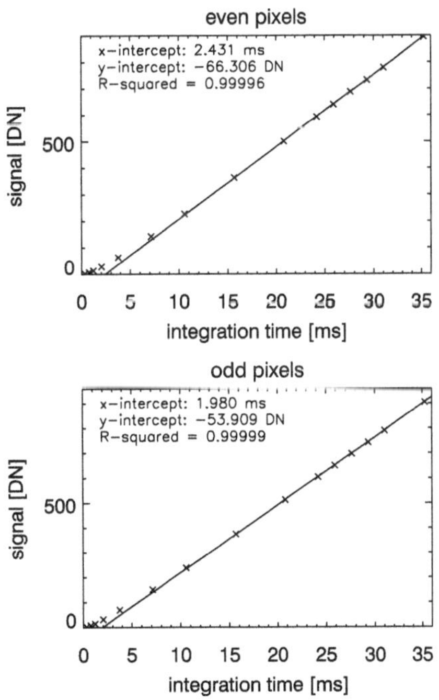

Fig. 9 NAC-R linearity fit for pixel-averaged DN values above 600 DN (dark corrected). Signal response is non-linear for exposure times below 10 ms. The averages of all even illuminated pixels yield a y-intercept of about −66.3 DN (*top*). The corresponding linear fit for the odd pixels yields a y-intercept of −53.9 DN (*bottom*)

$L(I_{off}, x)$, with $I_{off} = I_{raw} - m_I - (D - m_D) - S$, is the logistic function value for low-signal non-linearity correction (only if $I(x, \tau)_{raw} < 600$ DN),

$F(x)$ is the sensitivity non-uniformity array (flat-field), and

r is the spectral responsivity for conversion of DN into radiance units.

The individual steps are discussed in detail in the following sections (Fig. 10).

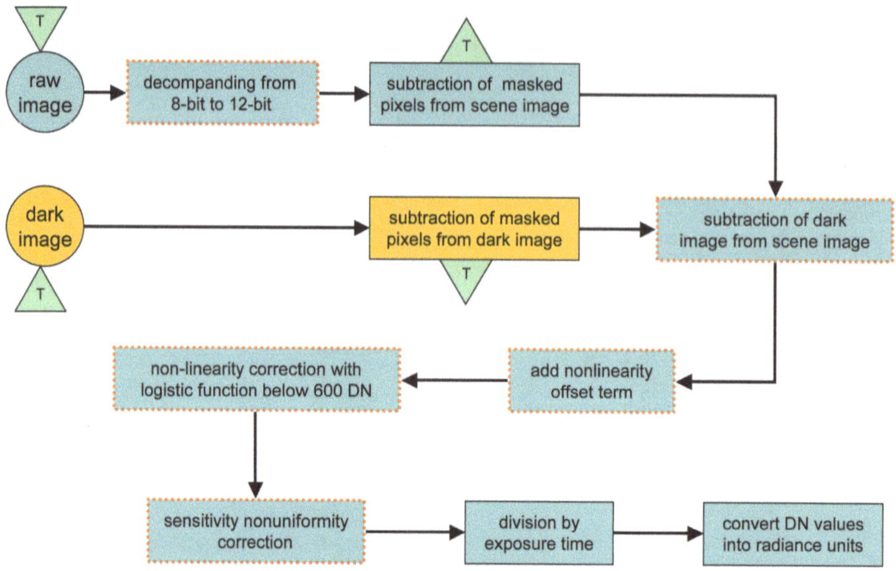

Fig. 10 Schematic overview of the NAC calibration pipeline. Depending on the electronics temperature "T," images have a different bias level, which is subtracted with the masked pixels (both for the raw scene image and the library dark image). The library dark image subtraction then corrects the pixel-to-pixel variation of the dark current. The non-linearity correction is a two-step process applied to DNs below 600: additive offset term and a linearization process using a logistic function. The final steps are sensitivity non-uniformity correction (flat-fielding), division by exposure time, and the conversion into radiance units. Steps applied on a pixel-by-pixel basis are indicated with orange dotted border

4.1.3 Interleaving and Conversion to 12 bits

For each NAC the even and odd pixels are read out independently during image acquisition. This strategy enables line readout and transfer to the buffer in the required 337.6 microseconds that corresponds to one pixel of down-track ground motion (nominal 50-km orbit). The even and odd pixels are stored in separate buffers and thus raw data files must be interleaved to form a coherent image. The 12-bit pixel values are companded to 8-bit values using one of six on-board schemes, three schemes are optimized for different signal levels, one is general purpose, while two others are used for calibration sequences (Fig. 11). To enable the fast pixel readout each bit-compression scheme consists of a piecewise linear function composed of up to 5 segments. After ground reception, these 8-bit digital values are converted back to 12-bit with the corresponding decompression tables. The linear functions of each segment are listed for each table in Appendix A.

4.1.4 Bias Level and Dark Images

The offset level of the digital-to-analog converter (DAC) determines the bias level for digitization, i.e. the coarse DN value of a dark image (no incident photons) at shortest exposure. A finer adjustment of the bias level is set with an additional offset parameter for each even and odd channel. Using dark images acquired during thermal vacuum tests, the dependence of the DAC offset setting as a function of focal plane temperature was determined. Thus, in flight, the bias level will be optimized for each individual image based on predicted detector temperature and exposure time (*cf.* Fig. 12).

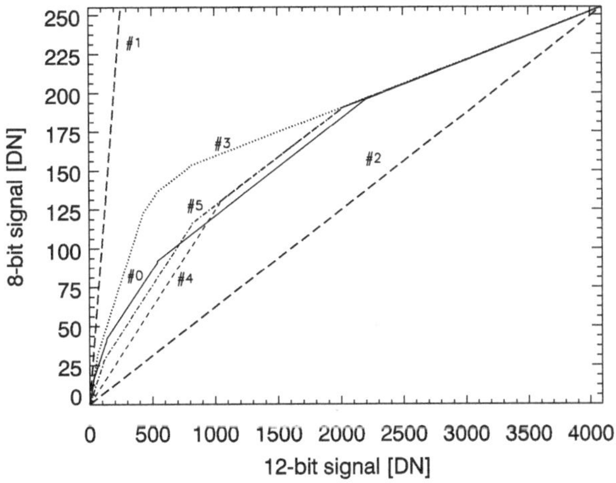

Fig. 11 Six NAC bit-compression schemes (12-bit to 8-bit). Scheme #0 approximates a square root function, scheme #1 is used for dark imaging, scheme #2 bins the input into 16-bit bins, and the remaining are options optimized for different illumination conditions (dim scene, scheme #3; bright scene, scheme #4; cap quantization error, scheme #5)

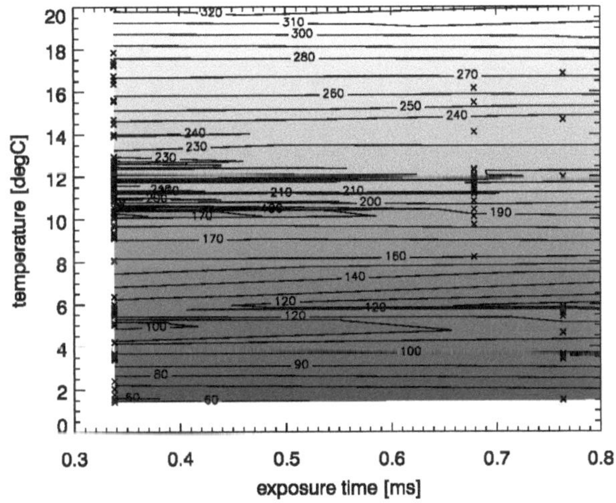

Fig. 12 The dependence of the dark signal (shown as contour lines in DN; for NAC-L even pixels) on exposure time [ms] and temperature [°C]. These data are used to optimize the bias level for each image

In the NAC calibration pipeline (Sect. 4.2.2), the background level, which is the sum of bias and dark current, is removed by subtracting the average of the masked even (odd) pixels on the edges from each illuminated even (odd) pixel. This step is done line-by-line in order to compensate for increase in electronics temperature over a full 52,224-line image that results in an increase of the dark current of up to 3 DN.

Subtraction of the masked pixels corrects the overall average value of the background level, but not pixel-to-pixel variations. This high frequency term is characterized from a library of dark images acquired in thermal vacuum tests over a range of operating temperatures (and eventually with flight data). The average background level is removed from the library dark image in the same manner as the surface image, by subtracting the masked pixel value. After the masked-pixel subtraction the library dark image is averaged over lines and converted into a one-line correction array. Because the background-subtraction is done for both the scene and the library dark image it is not necessary to have a library dark image with the same DAC and offset value. The library dark image subtraction only corrects for the pixel-to-pixel variations of the dark current.

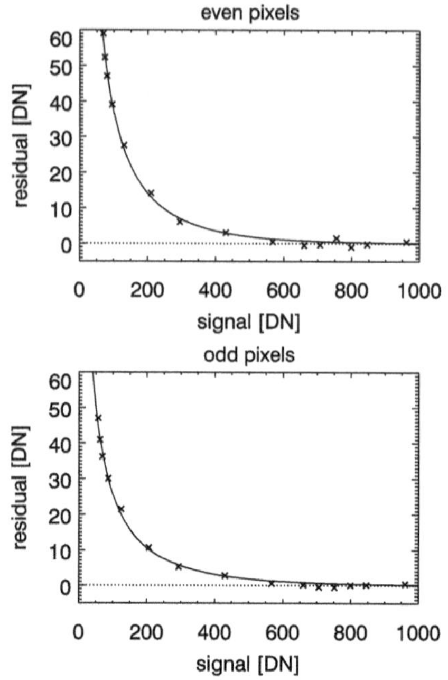

Fig. 13 The residual DN values (measured signal values minus linear fit values) plotted against the offset-corrected signal values follow a logistic function of the form $1/(a \cdot b^x + c)$. Here the results from averaged NAC-R images are shown. The resulting even parameters of the logistic function are $a = 0.03359405$, $b = 1.00561273$ and $c = -0.03180369$ (*top*). The odd parameters are $a = 0.05827176$, $b = 1.00466108$ and $c = -0.05361603$ (*bottom*). If the measured signal is below 600 DN the value calculated from the logistic function is subtracted from the offset-corrected signal in order to linearize the signals

4.1.5 Non-linearity Correction

NAC response at low signal levels (below 600 DN) is non-linear (Fig. 9). The required linearity correction is a two-step process. First a non-linearity offset term is calculated for each pixel (4996 array) as the y-intercept of the linear fit of the background-corrected data between 600 DN and 2000 DN. This offset term is independent of the DAC setting, i.e. the non-linearity depends on the magnitude of the incoming signal.

The second step is performed only for pixel values below 600 DN (background corrected). The difference $L(x)$ between the actual measured signal (after the non-linearity offset correction) and the expected signal from the linear fit function follows a logistic function of the form:

$$L(I_{off}, x) = \frac{1}{a(x) \cdot b(x)^{I_{off}(x)} + c(x)}$$

where $I_{off}(x)$ is the background and non-linearity offset corrected signal in DN (Fig. 13). After applying the non-linearity offset term to all values, those below 600 DN are linearized by simply subtracting the linearity residual $L(x)$ from the input $I_{off}(x)$.

4.1.6 Sensitivity Non-uniformity Correction

Laboratory flat-field images were acquired by evenly illuminating a Spectralon panel with Britek photo studio lamps powered by regular AC current. The setup is basically the same as shown in Fig. 18, just without the black plate. Several thousand integrations were acquired with various exposure times. Small cyclic brightness variations caused by the AC current are eliminated by line averaging. The line-averaged flat-field images are then normalized to the mean value of the central 500 pixels providing a sensitivity non-uniformity correction array. For illustration the non-uniformity arrays of both NACs are presented in Fig. 14.

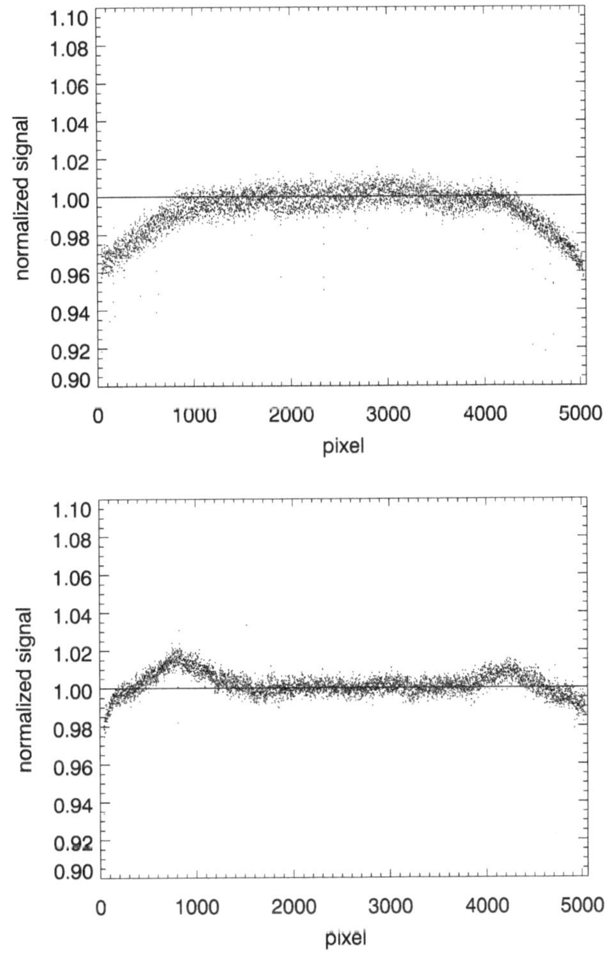

Fig. 14 (*Top*) NAC-L non-uniformity array. (*Bottom*) NAC-R non-uniformity array. Both are the average of 2560 lines and normalized to the mean of the central 500 pixels

4.1.7 Spectroradiometric Calibration

The experimental setup to measure the spectral responsivity $r(\lambda)$ [(DN/s)/(μW/(cm2·sr·nm))] of each NAC consisted of a monochromator with a Xenon lamp. The exit slit of the monochromator was effectively a point source a few meters distant from the NAC. The source was offset in the aperture so that the secondary mirror would not block it. The resulting out of focus image had a spatial FWHM, identical at the different wavelengths, of approximately 2000 pixels. Both the spectral FWHM of the monochromator and the wavelength step-size were set to 10 nm. At each wavelength step the power output of the monochromator was recorded, using an Oriel model 71639 calibrated silicon photodiode fed by a beam-splitter mounted after the exit slit of the monochromator. In addition, before each NAC calibration the monochromator spectral irradiance at the aperture of the NAC was sampled at the brightest spot of the beam by a calibrated spectroradiometer at each wavelength. These power measurements represent only a relative responsivity $r_M(\lambda)$. Absolute calibration was obtained by analyzing flat-field and background corrected images (see Sect. 4.1.6) dividing the image's count rate S [DN/s] by the integral over the relative responsivity multiplied with

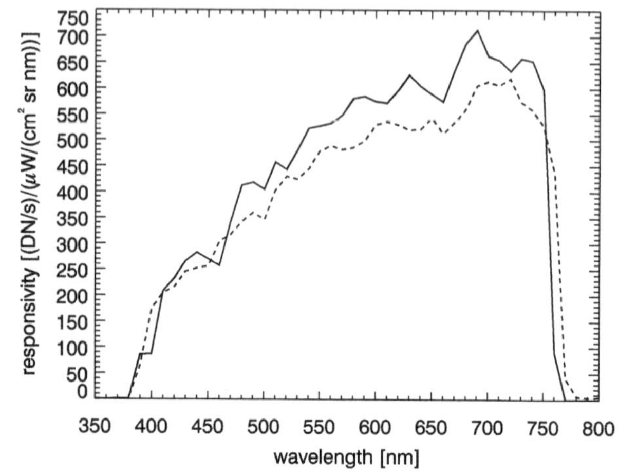

Fig. 15 Spectral responsivity of NAC-L (*solid line*) and NAC-R (*dashed line*). The high-frequency variation reflects noise in the measurement, but have no significant effect on the derived calibration

the spectral radiance of the source $I'_S(\lambda)$ [μW/(cm^2·sr·nm)]. This spectral radiance was measured with a spectroradiometer. However, it turned out that the spectroradiometer measurements had a systematic error that required a cross-calibration with a calibrated integrating sphere in order to obtain the necessary scaling factor $\gamma = 2.3 \pm 0.1$. Thus $I_S(\lambda) = \gamma \cdot I'_S(\lambda)$. In other words, for the absolute calibration the relative responsivity $r_M(\lambda)$ was scaled by a factor equal to the ratio of the actual count rate S observing the laboratory source to the count rate predicted if the absolute responsivity were equal to the relative responsivity. This approach can be expressed as follows:

$$r(\lambda) = r_M(\lambda) \cdot \frac{S}{\int_{\lambda'} r_M(\lambda') \cdot I_S(\lambda') d\lambda'}.$$

The spectral responsivity characterization (Fig. 15) allows the calibrated DN values to be converted to radiance. For a given observation the measured count rate S_L [DN/ms] constitutes the spectral radiance $I_L(\lambda)$ of the lunar scene weighted by the NAC responsivity, i.e.:

$$S_L = \int_\lambda r(\lambda) \cdot I_L(\lambda) d\lambda.$$

So the weighted mean radiance measured by the NAC is:

$$\overline{I} = \frac{\int_\lambda r(\lambda) \cdot I_L(\lambda) d\lambda}{\int_\lambda r(\lambda) d\lambda} = \frac{S_L}{\int_\lambda r(\lambda) d\lambda} = \frac{S_L}{r_{NAC}}.$$

The radiometric conversion parameters are $r_{NAC\text{-}L} = 180.56$ (DN/ms)/(μW/(cm^2·sr·nm)) and $r_{NAC\text{-}R} = 166.83$ (DN/ms)/(μW/(cm^2·sr·nm)).

Radiance can be further converted into the reflectivity I/F (Minnaert 1961), with I the observed radiance depending on the observation geometry and F the solar radiance coming from a normally illuminated Lambertian surface. The term I/F is also called radiance factor by Hapke (1993).

With Φ_E the solar irradiance at a distance of 1 AU the measured reflectivity of a lunar scene is:

$$I/F = \frac{I_L \cdot \pi \cdot d^2}{\Phi_E}$$

with d [AU] the distance Sun–Moon. With the formulas above the weighted mean of I/F as measured by the NAC is:

$$\overline{I/F} = \frac{\int_\lambda r \cdot I/F \cdot \Phi_E d\lambda}{\int_\lambda r \cdot \Phi_E d\lambda} = \frac{S_L \cdot \pi \cdot d^2}{\int_\lambda r \cdot \Phi_E d\lambda} = \frac{S_L \cdot d^2}{\varphi_{NAC}}.$$

Using the solar irradiance Φ_E at 1 AU of (Allen 1976) the conversion parameters are $\varphi_{NAC\text{-}L} = 9308.5$ (DN/ms)/AU^2 and $\varphi_{NAC\text{-}R} = 8504.1$ (DN/ms)/AU^2.

Finally, the signal-to-noise ratio (SNR) is calculated from assumed radiances for mare and highlands materials under various incidence angles. The signal is the expected mean radiance for those conditions, and the noise is composed of read noise, shot noise and quantization noise. The minimum SNR is estimated with a 70° solar incidence angle on mare material and presented in Table 2.

4.1.8 Geometric Calibration

Geometric calibration was performed for both NACs by rotating each instrument on a high-precision Ultradex rotary stage and imaging a bar pattern. In this experiment the bar pattern is at the focus of a collimator of identical optical design to the NAC. The bar target is tilted with respect to the collimator focal plane, so the bars are in focus only at one place in the image and progressively go out of focus toward the edges of the image. The Ultradex stage rotations are only in steps of one degree, accurate to better than one arc second, and thus direct collection of sample versus angle can provide only three data points across the detector. Because distortion of over ten pixels is expected for these optics, finer sampling is required. Thus the basic data set obtained was a series of measures of the number of samples that subtend one degree at different locations across the detector. These observed values can then be compared to those predicted by camera models across the detector based on pixel spacing, focal length, and distortion coefficients. In the analysis the desired measurements are the pixel locations (samples) of identifiable locations in the bar patterns. We chose the left sides of up to five bars to the left of the fiducial and one bar to the right. The edge was determined using existing software for measuring limb coordinates of objects that models a Gaussian decrease of signal from a sharp edge. For bars close to the fiducial the out-of-focus character gave a spread somewhat greater than the software was nominally adjusted to, and as is noted below, some data had uncertainties of more than a half pixel and were thus excluded. This did not substantially impact the spread of useful data across the detector. This relatively narrow-angle device does not require complex functions for distortion description. The basic geometry is shown in Fig. 16. Here $xc/fl = \tan(\alpha)$, where xc is the mm coordinate of the undistorted image, and fl is focal length in mm, and α is the angle from the optical center.

A simple quadratic distortion correction is given by Davies et al. (1994):

$$xd = xc \cdot (1 + k \cdot r^2),$$

where xd is the distorted position (the actual measured position), k the distortion coefficient, and r is the distance from the optical center (the difference in samples from the optical

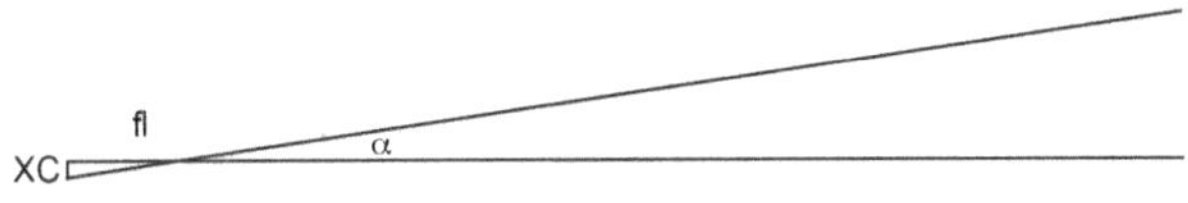

Fig. 16 Idealized camera geometry. Here $xc/fl = \tan(\alpha)$, where xc is the mm coordinate of the undistorted image, and fl is focal length in mm, and α is the angle from the optical center

center). In this one-dimensional case, offset of the true optical center from the line array cannot be measured, and would largely affect the value of the distortion coefficient. The undistorted coordinates thus are:

$$xc = \frac{xd}{1 + k \cdot r^2}.$$

For any combination of focal length, optical center position, and distortion coefficient we can calculate left and right sample coordinates for 1 degree changes in α across the entire field of view. We choose to measure this difference as right (greater) sample – left (lesser) sample, and thus are able to tabulate this difference for samples up to 5000 samples minus the one degree difference in samples, or roughly from sample 1 to sample ~3200. Our fit is made by comparing the observed one degree difference in samples versus the predicted value for the (left, low-pixel) samples tabulated. Residuals are the predicted minus observed number of samples, and the best fit is taken at minimum root mean square of the residuals.

It is assumed that the detector lies in the plane of rotation of the stage. If the detector is tilted relative to this rotation, a given angular rotation will result in a larger change in position on the detector, which would result in a larger calculated focal length. Review of the mechanical arrangement indicates possible misalignment of $0.88 \pm 0.1°$. Error induced across the field of view from this would be ~0.8 pixels. We choose to leave the numbers uncorrected.

The fitting result for NAC-L is displayed in Fig. 17. The resulting distortion coefficients and optical center position for both cameras are included in Table 2.

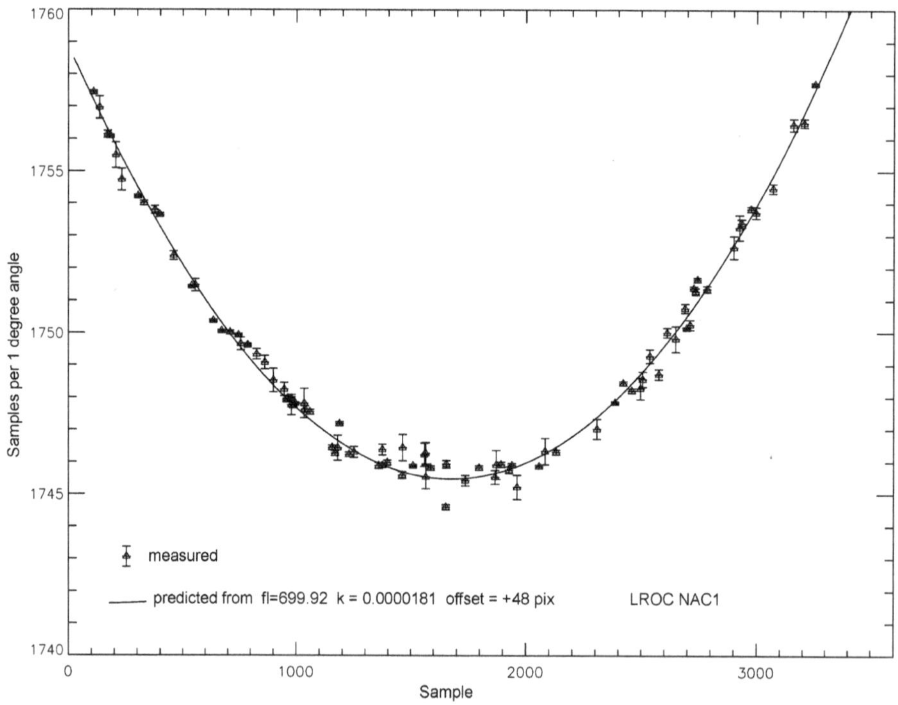

Fig. 17 Fitting results for NAC-L. Offset is distance of optical center from sample 2500

4.1.9 Stray Light Measurements

To search for stray light effects, the beam of the QTH calibration lamp was aimed directly into the NAC's aperture at various angles (up to about 60 degrees in the crosstrack direction only). A photo of this setup is shown in Fig. 18. No light leaks and no large-angle stray-light problems were identified. However, two out-of-field low-level stray light features were detected affecting the pixels centered at samples 1000 and 4000. These features were further quantified with two setups: (a) an integrating sphere illuminated with a QTH lamp was moved perpendicularly to the camera's line of sight, with small increments centered around an angle of about 5°; (b) the flat-field images described in Sect. 4.2.5 were compared with flat-field images where increasingly large parts of the Spectralon panel were covered with black panel (a photo of this setup is shown in Fig. 19). Both the NAC-L and NAC-R stray light features have magnitudes of 1–3% of a uniform flat field wide enough to include the entire out-of-field source of the stray light (Fig. 20). The test data acquired is sufficient to fully characterize and develop a mitigation strategy if necessary.

Additionally, with no ambient illumination a high intensity focused flashlight beam was held adjacent to all structural seams and fasteners for both the telescope and electronics housing (both NACs) while long integration images were acquired. This test revealed no light leaks.

Fig. 18 NAC stray light test setup. The quartz tungsten halogen lamp in the back-left corner was moved perpendicularly to the line of sight between the NAC (right foreground) and the black plate on the wall while rotating the lamp such that it pointed at the telescope aperture

Fig. 19 Setup for characterization of the NAC out-of-field stray light. The NAC (foreground) viewing a Spectralon panel (*white*) illuminated by studio lamps. In the experiment a black plate (*left*) is moved with one inch (2.54 cm) increments across the spectralon panel

4.1.10 Focus Test

The NAC carbon-fiber structure absorbs water under ambient conditions resulting in expansion that in turn affects focus. Thus accurate measures of focus must be performed after baking out the telescopes in a vacuum environment. Focus verification was performed during environmental testing by imaging a tilted bar target mounted at the focus of a collimator built with optics identical to that of the NACs. Measured MTFs were above requirement (0.2 at Nyquist) for both NACs across the full range of telescope operating temperatures. Shortly after removing each NAC from the vacuum chamber the Moon was imaged as a qualitative focus check (Fig. 21). Comparison of the moon image with an equivalent CTX image, and analysis of the limb to space edge, indicates both NACs to be in focus. A quantitative measure of focus from the lunar images is not useful due to less than optimal seeing conditions and hand scanning.

4.2 WAC Calibration

4.2.1 Detector Characterization

The detector gain, read noise, and full-well were calculated using the photon transfer method (Janesick 2007). The setup for this experiment consisted of a Xenon lamp illuminating a 10″ (25.4 cm) diameter integrating sphere with a 4″ (10.16 cm) diameter exit portal. Three sets of linearity images were taken, for each set the exposure times ranged from zero ms to sufficiently long exposures to saturate all filters. After background subtraction the total noise, read-plus-shot noise, and the shot noise are plotted versus the signal (Fig. 22). The measured gain, read noise and full well are listed in Table 3. Errors were estimated by varying the input parameters (boundary of the linear fit and the determination of the read noise). The

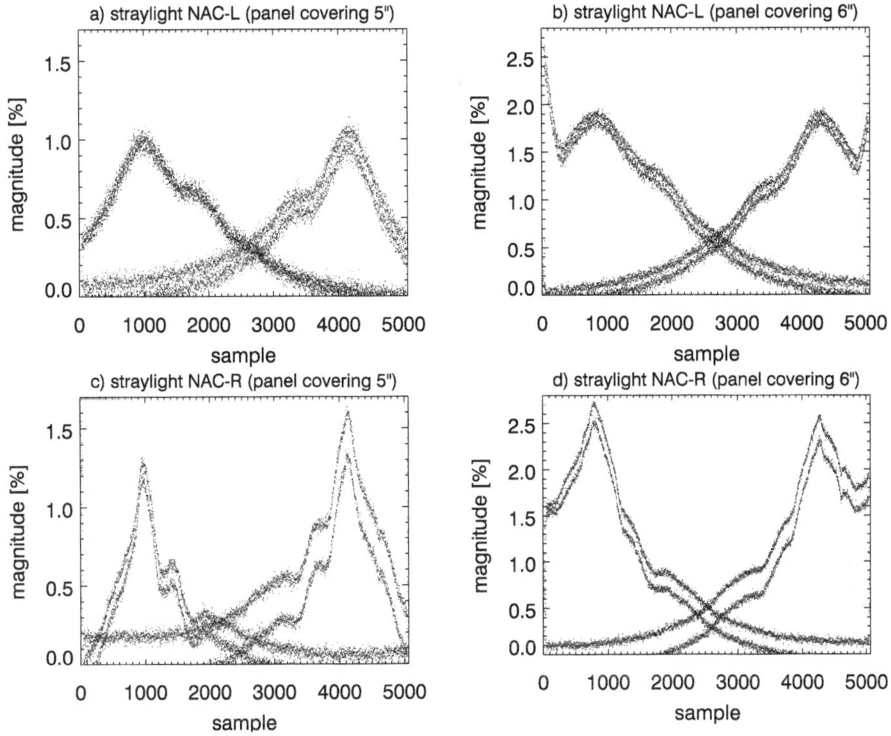

Fig. 20 The magnitude of the NAC stray light features was determined by dividing normal flat-field images with those where parts of the Spectralon panel were covered with a black panel. Here the unit is this ratio minus 1, given in percent. **a** and **b**: NAC-L results with the black panel covering 5″ (12.7 cm) and 6″ (15.2 cm) of the Spectralon panel. **c** and **d**: NAC-R results with the black panel covering 5″ (12.7 cm) and 6″ (15.2 cm) of the Spectralon panel. The two distinct populations represent even and odd pixels

WAC CCD has a highly linear response between 5% and 90% of the saturation value (see Fig. 23). Around 1800 DN the sensor behaves in a non-linear manner.

4.2.2 WAC Calibration Pipeline

The calibration to radiance or I/F is a standard procedure described by:

$$I_{cal}(x, y, f, \tau, T) = \frac{I_{raw}(x, y, f, \tau, T) - D(x, y, f, \tau, T) + N(x, y, f)}{F(x, y, f) \cdot \tau \cdot r_f},$$

where

$I_{cal}(x, y, f, \tau, T)$ is the calibrated signal of the pixel at position (x, y) in radiance units of $\mu W/(cm^2 \cdot sr \cdot nm)$,

$I_{raw}(x, y, f, \tau, T)$ is the pixel's raw signal in DN,

f is the filter,

τ is the exposure time,

T is the temperature,

$I_{raw}(x, y, \tau, T)$ is the raw signal in DN,

$D(x, y, f, \tau, T)$ is the dark image,

Fig. 21 NAC-L image of the Moon obtained immediately after removing instrument from thermal vacuum testing. Slight deviations from sphericity are due to scanning the NAC by hand, Moon is about 940 pixels in diameter

$N(x, y, f)$ is non-linearity correction (term above 1600 dn)
$F(x, y, f)$ is the sensitivity non-uniformity correction matrix (flat-field), and
r_f is a filter's responsivity coefficient to convert DN to radiance

The individual steps are discussed in detail in the following sections. The flowchart provided as Fig. 24 provides a graphical illustration of the WAC calibration pipeline. The WAC data are first decompressed. The subsequent conversion from 8-bit to 11-bit data follows the inverse of the conversion look-up-table presented in Fig. 25.

4.2.3 Background Characterization

For the WAC, unlike the NAC, it is in principle possible with the WAC to remove the electronic bias by taking dark images with 0 ms exposure. Yet unlike the NAC electronics it is not possible to change the bias level by simply uploading a parameter on a per image basis, but rather the bias is a fixed value in the flight software and can only be changed with a flight software modification. In the calibration pipeline (Sect. 4.2.2) the total background, which includes the bias, is removed from the scene image in one step. To characterize the dark signal dependence on exposure time and temperature, a comprehensive set of images was acquired during thermal vacuum tests over a range of temperatures (−40°C to +25°C) spanning expected flight conditions (Fig. 26).

4.2.4 Linearity Correction

Above 1800 DN (background corrected) the WAC response is non-linear (Fig. 23). For background corrected values between 1800 and 1950 a linearity correction is applied by

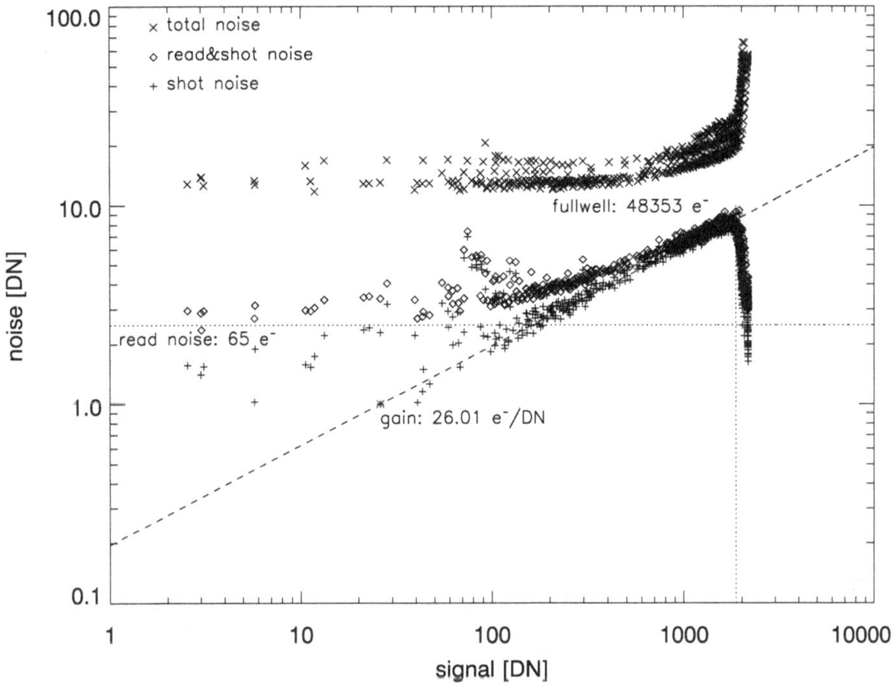

Fig. 22 WAC Photon transfer curve. The linear fit of the shot noise between signal levels of 500 DN and 1600 DN yields a gain of 26.0 e⁻/DN. This value is within the uncertainty range of 0.7 e⁻/DN (around an average of 25.9 e⁻/DN), which was estimated by varying the endpoints of the linear fit range, the number of pixels used for analysis, and the determination of the read noise

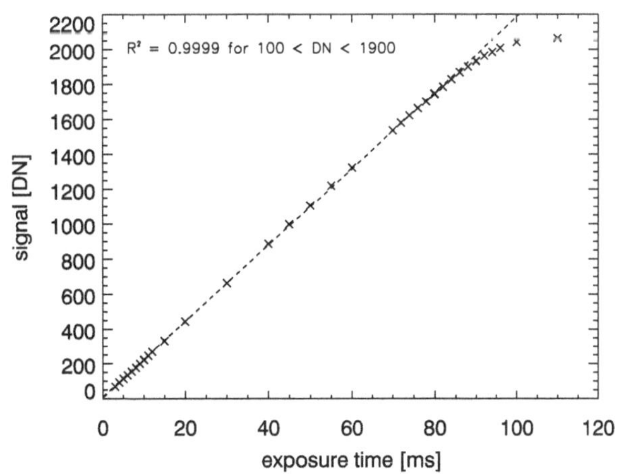

Fig. 23 WAC linearity measurements for the 645 nm band. The system response is highly linear until ~1800 DN (background corrected)

elevating a particular DN to match corresponding line fitted from the linear range. Above 1950 the data are considered saturated and thus not useful for quantitative analysis.

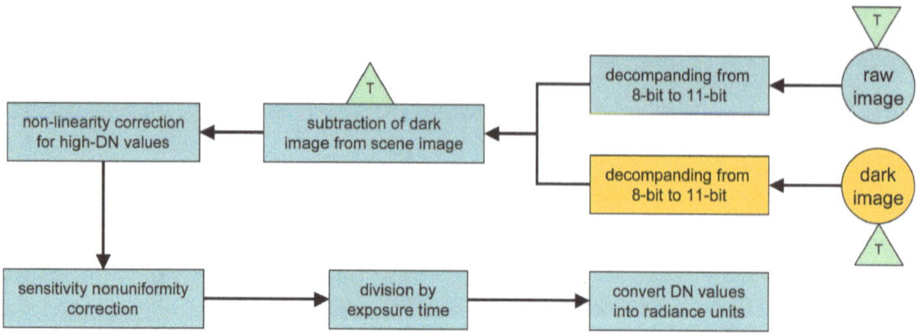

Fig. 24 Schematic overview of the WAC calibration pipeline. Depending on the electronics temperature "T," images have a different background level (bias plus dark current), which is subtracted with the corresponding dark image. The non-linearity for high-DN values is corrected, as well as the sensitivity non-uniformity (flat-fielding). The final steps are the division by exposure time and the conversion into radiance units

Fig. 25 Graphic representation of the WAC companding scheme (11-bit to 8-bit)

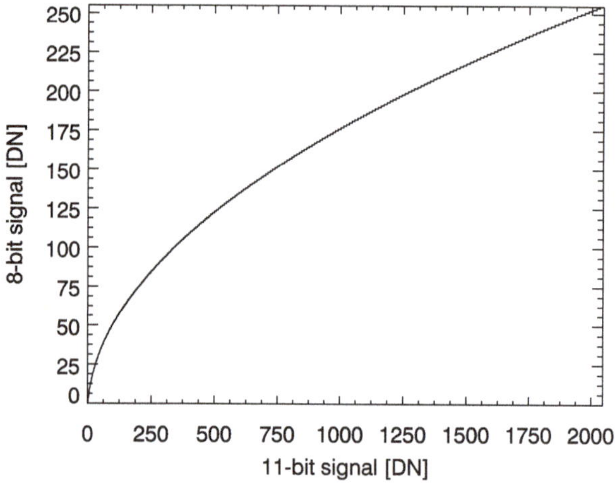

Fig. 26 WAC dark signal (contour lines) raw DN as function of exposure time [ms] and temperature [°C]

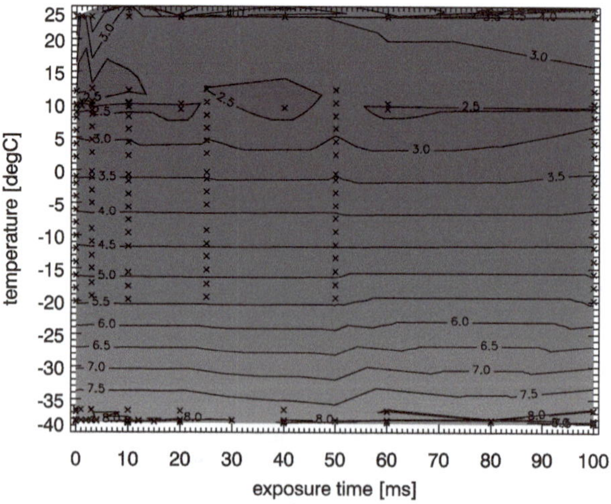

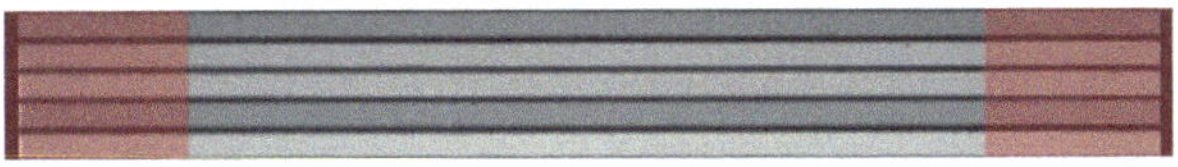

Fig. 27 Non-uniformity matrix of the 5 visible filters (415 nm band to 690 nm band, top to bottom). The *red shading* is not part of the system response but rather delimits portions of the CCD readout in multispectral mode. Monochrome images are acquired with the 645 nm filter (4th band from top)

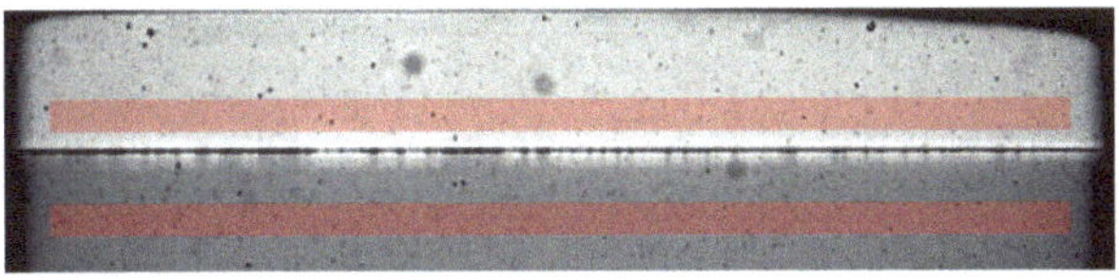

Fig. 28 Flat-field image of the two UV filters (*top*: 320 nm; *bottom*: 360 nm). The areas shaded in *red* are used for the actual imaging; note that 4 × 4 binning has not been applied to the data shown here

4.2.5 Sensitivity Non-uniformity Correction

The wide angle FOV (90°) of the WAC was much larger than the aperture of the available integrating sphere thus the sensitivity non-uniformity was characterized through a series of images. The WAC was rotated through 98 degrees such that every three degrees an image of the sphere aperture was acquired. The central portion of these images were mosaicked to construct the final sensitivity non-uniformity matrices (Figs. 28, 29). This method is identical to that described in Bell et al. (2009).

4.2.6 Spectroradiometric Calibration

The full system spectral response of the WAC was first characterized with a monochromator illuminated with a calibrated Xenon light source. Two tests were performed with the monochromator illuminating discrete spots on each filter. First, out of band leaks for each filter were tested by stepping through the full spectral sensitivity of the detector (250 nm to 1100 nm) at 10 nm steps with 100× overexposures. Out of band leaks were shown to be less half a percent in all visible bands (at a single point on each filter). To test if any of the visible filters suffered from spatially variable out of band leaks a series of images (5 second exposures) were taken through narrow band filters (420 nm ± 5 nm, 550 nm ± 5 nm, 600 nm ± 5 nm, 650 nm ± 5 nm, 750 nm ± 5 nm) placed in front of the Xenon lamp illuminating the integrating sphere. These images indicated no out of band leaks, though we note this test did not test the full spectral sensitivity range of the CCD. Therefore it is concluded that all filters have a maximum out-of-band leak of less than 0.5%.

Characterizing the system sensitivity profile for each filter was next accomplished by stepping the monochromator through 2-nm increments across a given filter's full spectral width. The projected image of the monochromator spot exhibited Gaussian brightness contours with the peak area being on the order of the size of a pixel. As the monochromator was adjusted in wavelength (2-nm steps) the location and shape of the spot changed while drifting across the sensitive area of the center pixel. Thus accurate in-band spectral response proved to be problematic due to the unfavorable pixel fill factor (20%) of the KAI-1001.

Final characterization of the band sensitivity was ultimately accomplished by calculation from the vendor supplied measurements of the individual components (optics with coatings, filters, CCD quantum efficiency). The estimated WAC filter bandpasses are summarized in

Table 4 WAC filter bandpasses and responsivity coefficient (RC)

Band name	λ_{eff} [nm]	FWHM [nm]	RC [(W/m^3/sr)/(DN/s)]
320 nm	321	32	865
360 nm	360	15	918
415 nm	415	36	3727
565 nm	566	20	3310
605 nm	604	20	3377
645 nm	643	23	3273
690 nm	689	39	2612

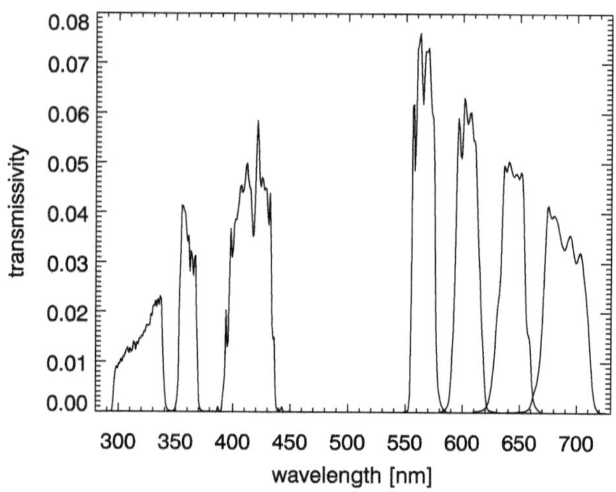

Fig. 29 Estimated spectral system throughput of the seven WAC bandpasses

Table 4 and their spectral transmissivities are displayed in Fig. 29. Images of a color standard were acquired by tilting the WAC through a series of angles to build up a complete scan. These data were background corrected and flat-fielded to check for low frequency flat field residuals, pinholes, and undetected out of band leaks (Fig. 30), no problems were detected.

4.2.7 Absolute Calibration

With the knowledge of the WAC system spectral response and the radiometric source calibration the system spectral responsivity in units of (DN/s)/(μW/(cm^2·sr·nm)) is calculated for each filter (Table 4). For the independent determination of each filter's signal-to-noise ratio (SNR) the dark-subtracted images of the photon transfer curve analysis (Sect. 4.2.1) were analyzed. Due to the fixed-pattern-noise correction with flat-field images the remaining noise for the calculation is the read-plus-shot noise. Thus the SNR is calculated as the signal divided by the read-plus-shot noise in a selected "clean" area of each filter. Above 1000 DN the SNR is ∼150 for the UV and visible bands (Fig. 31).

4.2.8 Geometric Calibration

Geometric calibration was performed by mounting the WAC on the Ultradex rotary stage for "azimuthal" control in 1-degree steps, with elevation separately provided by another rotary

Fig. 30 WAC RGB composite (RGB: 689, 643, 415 nm) of standard Munsell Color chart and geologic samples

stage with continuous variability. The one degree rotation steps provided by the Ultradex stage are accurate to less than 1-arcsecond; the elevation rotations are not as accurate; they are checked in the analysis. The target is a collimated spot image of selectable size. Illumination is by a QTH lamp with a collimating lens illuminating the pinhole for the visible images, a Xenon lamp with a collimating lens illuminating the pinhole for the UV images. The resulting images, one each for every azimuth and elevation of the collimator, provide spots spread over a few pixels width. Existing centroiding software provided the lines and sample of spot centers, from which the detector mm coordinates were recorded. The goal is to provide a function relation between line and sample in the detector and viewing angles from the camera.

Due to its wide field-of-view the WAC exhibits more distortion than the NAC, thus an additional distortion term is required:

$$xd = xc \cdot (1 + k_1 \cdot r^2 + k_2 \cdot r^3),$$

$$yd = yc \cdot (1 + k_1 \cdot r^2 + k_2 \cdot r^3).$$

For the visible detector, $k_1 = -0.0099$, and $k_2 = -0.00050$. For UV detector, $k_1 = -0.024$, and $k_2 = -0.0070$. The resulting focal length and FOV solutions are given in Table 3.

4.2.9 Stray Light Measurements

A discrete collimated point source was rastered in and out of the WAC Visible FOV to check for glints and scattered light (Fig. 32). The collimated beam (pinhole 0.1 cm) formed a circle of light ~3 pixels in diameter (FWHM) at unsaturated exposure levels. For each position 100× and 600× over-exposed images were acquired to check for low level effects. Measurable effects were only found on one side of the detector (Fig. 33). During lunar imaging the first 20 to 40 columns will suffer from scattered light contamination approximately 5 to 10% of scene brightness during monochrome full field imaging. No scattered light was detected in the multispectral readout zone.

In the darkened calibration facility a high intensity focused flashlight beam was held adjacent to all structural seams and fasteners for both the optics and electronics housings while long integration (1000 ms) images were acquired. This test did not reveal any light leaks.

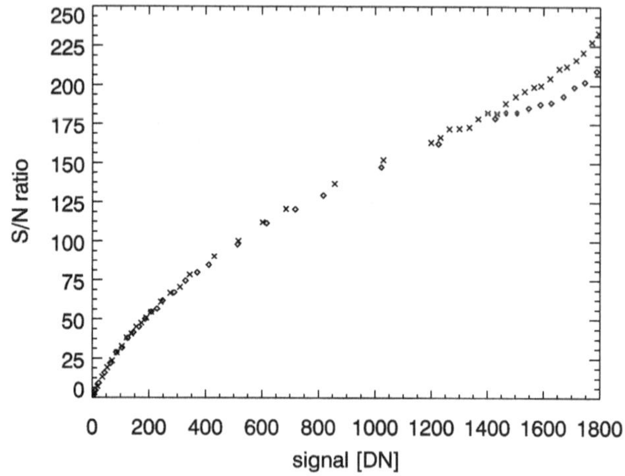

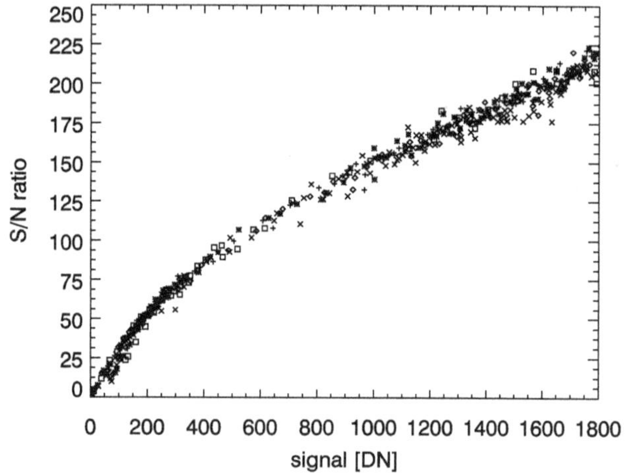

Fig. 31 (*Top*) The SNR for the two UV bands: 320 nm (*crosses*)/360 nm (*diamonds*). (*Bottom*) SNR estimates for the five visible bands: 415 nm (*crosses*)/565 nm (*diamonds*)/605 nm (*plus-signs*)/645 nm (*asterisks*)/690 nm (*squares*)

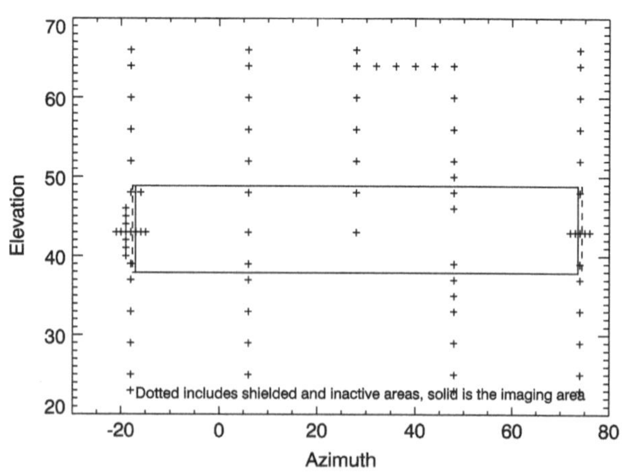

Fig. 32 Schematic showing approximate locations of collimated light beam in WAC 1 (not the flight unit) stray light testing, similar test results were obtained from the flight unit

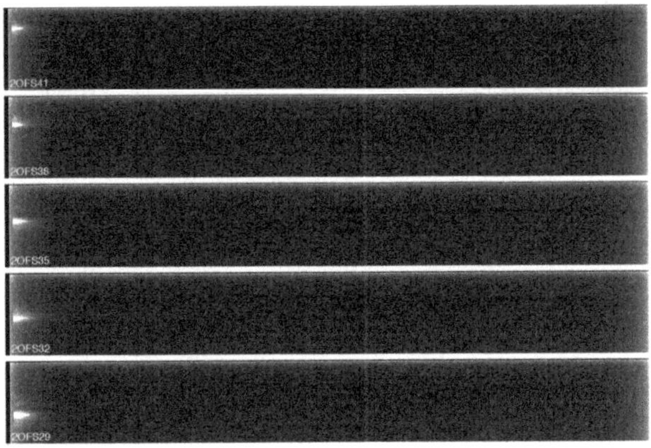

Fig. 33 WAC 1 (not the flight article) frames showing glare artifact when collimated source was moved vertically outside field of view, similar results were obtained with the flight WAC. Exposure time was ~600× that of normal exposure. At normal exposures levels the glare peaks at about 20 DN above background and affects the first 20 imaging columns

5 LROC Operations

The LROC Science Operations Center (SOC) is located at the School of Earth and Space Exploration at Arizona State University in Tempe, AZ. All LROC uplink, downlink, processing, and archival activities take place at this state-of-the-art facility. The LROC SOC supports LROC operations through all phases of the LRO mission. In this section, we provide a brief overview of LROC Operations.

5.1 Data Volume

The SOC accommodates LROC's ~400 Gbits of data downlink per day with lossless compression, which will produce a total of almost 30 terabytes (TB) of raw data over the nominal mission. Production of higher-level data products will eventually produce a total of > 100 TB for archive in the Planetary Data System (PDS).

5.2 Software

To reduce schedule risk and cost, LROC is leveraging existing software to the fullest practical extent. Brief summaries of the software packages deployed by the LROC Operations Team to support mission goals are provided below.

5.2.1 REACT

The LROC Science Team populates a planning database with NAC targeting requests using the Rapid Environmental Assessment Composition Tools (REACT) planning tool. REACT was developed by Applied Coherent Technologies (ACT) as a generic spacecraft remote-sensing target planning tool and is also used operationally by other spacecraft instrument teams including Chandrayaan-1's Moon Mineralogy Mapper, MESSENGER, and the Mars Reconnaissance Orbiter's Compact Reconnaissance Imaging Spectrometer. REACT is a GIS-based application that allows the user to specify Regions of Interest (ROIs) using various lunar global basemaps. Target-request ROIs entered into the database through REACT are defined as points, lines, or polygons and contain specific acquisition parameters such as lighting (high sun or low sun), percent coverage, repeat observations, and resolution. ROIs specified by the Science and Operations Teams are uploaded to a central database on the ACT server. The targeting strategy is discussed in detail in Sect. 5.3.1.

5.2.2 JMoon

JMoon (JMars 2010) is a derivative of the Java Mission-Planning and Analysis for Remote Sensing (JMARS) software, the observation and planning tool that was developed to support the Mars Odyssey THEMIS experiment. JMoon uses the spacecraft's predict ephemeris to determine the LROC observing opportunities that lie under LRO's orbit track, allowing analysts to interactively select and position LROC NAC targets from the target request database created by the LROC Science Team. The LROC Operations Team uses JMoon to convert the ROIs created by the LROC Science Team in REACT into the command loads sent to the instrument in flight.

5.2.3 AutoTarget

AutoTarget is a tool developed by the LROC Operations Team designed to facilitate the automated generation of LROC command loads. AutoTarget automatically creates LROC command loads using the LRO predict ephemeris to locate targets in the LROC target request database or by randomly placing NAC targets where no target requests are available. AutoTarget is used for specific LROC targeting functions, including random observations along a complete orbit track, planning observations for the global or polar mapping campaigns, and pairing complementary stereo or mosaic observations with existing observations.

5.2.4 CommandGen

Once the LROC Operations Team has completed an observational plan, the software tool CommandGen (short for **Command Gen**erator) collects the details of all the planned observations for the given orbit window and compiles them into command load files. These CommandGen products are then sent to the LRO Mission Operations Center (MOC) at the NASA Goddard Space Flight Center for integration with the LRO command load and subsequent uplink to the spacecraft.

5.2.5 LROC Database

The LROC SOC uses PostgreSQL 8.3 to store data accumulated and created during LROC operations. PostgreSQL was chosen due to its extensive and mature feature set, including robust transaction logging, foreign key relationships, point-in-time recovery, and its demonstrated ability to handle large volumes of data under heavy load. The database will support LROC operations both by providing an archive of the data produced during operations and also by providing data storage and retrieval services (session management) for all aspects of LROC data processing. The LROC Database will facilitate creation of PDS deliveries by pre-packaging all pertinent information prior to the PDS delivery process. The LROC database natively supports geographic objects and functions, enabling the LROC SOC to store and produce OpenGIS compliant data products.

5.3 Concept of Operations

Here, we provide an overview of the concept of operations for the LROC Science and Operations Teams during the LRO nominal mission. These operations include science targeting, operational target definition, operational target selection, command sequence generation, transfer of downlinked image data, data processing, special product generation, and data archiving.

5.3.1 LROC Science Targeting Request Process and Strategy

LROC NAC science target requests are generated by the LROC Science and Operations teams interactively using the REACT software package (Sect. 5.2.1) and are organized in several ways, the most important of which is target priority. The priorities are as follows:

(1) Exploration targets requested by NASA's Project Constellation
(2) Targets that fulfill other LROC Level 1 Requirements (detailed in Sect. 2)
(3) Targets that support LROC Science Team and LRO mission science objectives
(4) Science requests from outside the LRO team
(5) Public target requests

During the first year of the LRO mission, the Priority 1 targets consist of sites specified by the NASA Constellation Project. These sites were selected on the basis of scientific, resource utilization, and operational merits as well as a desire to span the range of lunar terrain and geologic features, including polar sites, mare and cryptomare surfaces, pyroclastic deposits and other volcanic landforms, impact basins and crater terrains, locations with representative rock-type exposures, and unusual features such as radar anomalies, magnetic anomalies, albedo swirls, sites of possible outgassing, and compositional anomalies that have been observed from orbit. The Priority 1 targets are derived primarily from three reference target sets (1) The ten design reference sites listed by the 2005 Exploration Systems Architecture Study, (2) the six sites evaluated by the 1991 *Site Selection Strategy for a Lunar Outpost* workshop for the First Lunar Outpost study, and (3) the 59 sites identified by the 1988 workshop on *Geoscience and a Lunar Base: A Comprehensive Plan for Lunar Exploration* (Taylor and Spudis 1990; NASA 2005, 1990).

The LROC Science Team is also generating a comprehensive set of NAC high-resolution image acquisition target requests that achieve LROC measurement objectives, science objectives proposed by the LROC participating scientists, and targets relating to features of generally known geologic interest. These targets include, but are not limited to, spectral anomalies (Wood and Head 1975), Copernican craters (Wilhelms et al. 1987), Apollo Panoramic photo areas (Masursky et al. 1978; Lawrence et al. 2008b), crater central peaks and peak rings, sites of lunar pyroclastic volcanism (Gaddis et al. 1985, 1995, 2000, 2003), lava flow margins (Gifford and El-Baz 1981), possible lava tubes (Coombs and Hawke 1992), mare surfaces for crater counting (Hiesinger et al. 2003), volcanic domes (Head and McCord 1978), geologic units in the Apennine Mountains and Imbrium backslope (Wilhelms 1980; Wilhelms et al. 1987), tectonic structures, dark haloed impact craters and cryptomaria (Blewett et al. 1995; Giguere et al. 2003, 2007; Hawke et al. 2004, 2005, 2007, 2008; Lawrence et al. 2008b), concentric craters (Wood 1978), sites of observed impact flashes (Ortiz et al. 2000, 2002, 2006; Yanagisawa and Kisaichi 2002; Cudnik et al. 2003), and highland scarps.

5.3.2 Operational Target Definition

The LROC SOC uses the software tools described in Sect. 5.2 to plan observations and generate command loads for the NAC and WAC on a daily basis. Based on the most recent predicted LRO spacecraft ephemeris (which will be received and updated daily from the LRO MOC) and the NAC Target Request Database for upcoming orbits, an operations plan for the next 72 hours of normal operations will be generated and submitted to the LRO MOC each day. The NACs can be commanded individually or together, and in normal operating mode can acquire images of any length in 1024-line increments up to the full length of

52,224 lines with a maximum of 15 NAC image-pairs of the illuminated surface possible per orbit, for a theoretical daily total of 180 NAC pairs (or 360 images). Each LROC operational plan will therefore include NAC requests for at least 1000 images. Before NAC images are written to the spacecraft's solid-state data recorder, they are temporarily stored in a 256 MB buffer. It takes approximately 15 seconds to acquire a full NAC image pair and another 220 seconds to read that image pair through the SCS to the LRO solid-state recorder. It is not possible to obtain another NAC image until the buffer is cleared, so after the collection of each NAC pair there is an 11° down-track latitude zone where additional imaging cannot be acquired. In addition to NAC commands for target requests, each 3-day plan will also include commands for the WAC. The WAC is not buffer limited and will almost always be imaging when over the daylight side of the Moon.

5.3.3 Spacecraft Slews

The daily target plans include on average three off-nadir spacecraft roll (up to 20°) observations per 24-hour period and any number of slews 2° or less, barring no operational conflicts. More than three slews greater than 20° per day are not permitted in order to minimize disruption to the other LRO science instruments. Off-nadir slews are necessary to fulfill stereo and mosaic requests. Geometric stereo requires two images of the same terrain with a 10–50° angle between the two observations with respect to the target and incidence angles as similar as possible (< 20° difference). Photometric stereo requires three or more images of the same terrain, with each observation in the set differing by at least 20° in incidence and azimuth angles from one another. Instrument slew maneuvers for star calibrations, limb views, 90° yaw offset, or slews > 20° are specially planned in advance due to the operational complexity and disruption to normal mapping for the payload suite.

5.4 In-flight Calibration

Comprehensive in-flight calibration sequences are planned for the operational phase of the LRO mission. Details for both the NAC and WAC subsystems are provided below.

5.4.1 LROC NAC Calibration Imaging

The NACs will acquire images on the lunar farside during non-illuminated periods in order to characterize the dark current behavior of the CCDs. This operation is not possible while the spacecraft is on the night-time nearside because of Earth-shine. During the initial cruise to lunar orbit (0–3 days after the launch), the NACs will image open space while the Moon is: (1) far from the edge of the field of view (2) near the edge of the FOV and (3) within the FOV. This series of observations will be used to characterize stray light within the NAC instruments.

5.4.2 LROC WAC Calibration Imaging

Calibration images will be obtained with the WAC during the cruise phase by imaging the Earth and Moon, both inside and outside of the instrument's imaging field of view. The imaging of point-sources such as stars and nearby planets will also be performed. In the 50-km nominal mission orbit, image calibration will typically include the acquisition of images on the non-illuminated far side of the Moon for dark reference calculation and the regular imaging of select stars and planets.

Table 5 LROC data measurement sets

Measurement Set	Estimated Storage Allocation (GB)	Pixel Scale (m/pixel)	Product Description
EDR/CDR	60,000	0.5 to 100	Level 0, Level 1A (EDR,CDR)
SPECIAL DATA PRODUCTS TO SATISFY LROC MEASUREMENT OBJECTIVES			
Landing Site Certification	500	0.5	50 Hi-priority uncontrolled site mosaics
Synoptic Polar Illumination	179	100	2 Polar movies, Illumination Maps
Meter Scale Polar Illumination	293	1	2 Summer uncontrolled mosaics
High-resolution topography	25	1–10 (vertical)	Select DEMs, Co-registered sets
Global Multispectral	639	100	Select uncontrolled mosaics
Global Black-and-white Basemap	27	100	Uncontrolled mosaics
Regolith Investigation	128	0.5	Catalog, Mosaics
Recent Small-impact rate	1280	1	Catalog, Sample co-registration sets
Total: 63,017 GB			

5.5 LROC Downlink Operations

Spacecraft downlink will occur during S-band (256 kbps, for housekeeping) and Ka-band (100 Mbps, for image data) contacts. The MOC will provide the LROC SOC with access to real-time housekeeping telemetry in order to allow monitoring of instrument operations and health. In general, there will be an average of four Ka-band contacts per day and at minimum one 30-minute S-band contact per orbit. Up to 15 NAC pairs/orbit can be accommodated given sufficient downlink bandwidth, for a total of approximately 180 NAC pairs per day.

The instrument data files returned from the spacecraft first arrive at the MOC and are then sent to the SOC. As the raw LROC data files arrive at the SOC, they are validated, queued for image processing and reduction, processed, and archived with the PDS. An enterprise-grade storage array at the SOC will provide 200 TB for production-file storage. LROC image processing will be done using Linux servers running ISIS software to generate PDS data products (EDRs, CDRs, and RDRs see Sect. 5.5.1.1). Image processing is pipelined using the Rector software, which has been self-developed by the LROC Operations Team. Rector automates the data processing steps by controlling product migration through a series of scripts across networked machines. Failures in the data processing pipeline will be recorded for resolution. There are no manual steps in the pipeline (except when failures arise) in order to mitigate time-intensive tasks associated with the processing of the large volume of LROC data received daily during the nominal mission.

5.5.1 Data Products

5.5.1.1 EDRs and CDRs LROC's key data products are the documented, verified PDS-format Experiment Data Records and Calibrated Data Records. These products form the basis of all Reduced Data Records (RDR) produced in order to extract scientific information from various measurement data sets. RDRs might include: image cubes of multispectral data, mosaics of multiple images, DEMs, perspective view maps, and various color composites. All RDRs are produced on a best-effort basis, using the United States Geological Survey's ISIS image processing software (USGS 2010) in a geometrically uncontrolled format to support the LROC measurement goals. The LROC RDRs that will be generated to fulfill the LROC proposed measurement objectives are detailed in Table 5.

Acknowledgements We thank the many contractors who helped bring LROC to realization. This paper was improved by comments from two anonymous reviewers.

Appendix A

NAC bit-compression tables (from 12-bit to 8-bit data):

NAC images are companded using a piecewise linear transfer function with up to five segments. The companding scheme can be described as linear functions with corresponding segments defined by 12-bit DN ranges. Note that the y-values (8-bit) are always truncated.

Code 0: NAC nominal table (square-root-like)

	linear function	12-bit DN range (y)	8-bit DN range (x)
Segment 1	$y = 1/2 \cdot x + 0$	0–31	0–15
Segment 2	$y = 1/4 \cdot x + 8$	32–135	16–41
Segment 3	$y = 1/8 \cdot x + 25$	136–542	42–92
Segment 4	$y = 1/16 \cdot x + 59$	543–2206	92–196
Segment 5	$y = 1/32 \cdot x + 128$	2207–4095	196–255

Code 1: NAC lin1 (0 DN to 255 DN mapped one-to-one)

	linear function	12-bit DN range (y)	8-bit DN range (x)
Segment 1	$y = x$	0–255	0–255
Segment 2	$y = x - 256$	256–510	0–254
Segment 3	$y = 1/32 \cdot x$	511–4095	15–127

Code 2: NAC lin16 (12 bit to 8 bit linear)

	linear function	12-bit DN range (y)	8-bit DN range (x)
Segment 1	$y = 1/16 \cdot x$	0–4094	0–255
Segment 2	$y = 1/32 \cdot x$	4095–4095	127–127

Code 3: NAC low signal table (optimized for DN < 500)

	linear function	12-bit DN range (y)	8-bit DN range (x)
Segment 1	$y = 1/2 \cdot x + 0$	0–63	0–31
Segment 2	$y = 1/4 \cdot x + 16$	64–423	32–121
Segment 3	$y = 1/8 \cdot x + 69$	424–535	122–135
Segment 4	$y = 1/16 \cdot x + 103$	536–799	136–152
Segment 5	$y = 1/32 \cdot x + 128$	800–4095	153–255

Code 4: NAC high signal table (optimized for 500 < DN < 2000)

	linear function	12-bit DN range (y)	8-bit DN range (x)
Segment 1	$y = 1/8 \cdot x + 0$	0–1039	0–129
Segment 2	$y = 1/16 \cdot x + 65$	1040–1999	130–189
Segment 3	$y = 1/32 \cdot x + 128$	2000–4095	190–255

Code 5: NAC cap Nq/Ne table (minimize quantization noise for low DN)

	linear function	12-bit DN range (y)	8-bit DN range (x)
Segment 1	$y = 1/4 \cdot x + 0$	0–111	0–27
Segment 2	$y = 1/8 \cdot x + 14$	112–815	28–115
Segment 3	$y = 1/16 \cdot x + 65$	816–1999	116–189
Segment 4	$y = 1/32 \cdot x + 128$	2000–4095	190–255

Simply inverting the companding equations allows the 8-bit value to be decompanded back to the original 12-bit DN. However there is an ambiguity when inverting, the 8-bit value could have been any of the 12-bit values within the particular bin. Simply inverted equation returns the lowest 12-bit DN within the bin. An alternate method is to use a lookup table and for a particular implementation the analyst can choose the lowest, middle, or highest value within a bin (or any value that meets the particular requirements).

References

C.W. Allen, *Astrophysical Quantities*, 3rd edn. (The Athlone Press, London, 1976)

T. Becker, L. Weller, L. Gaddis, D. Soltesz, D. Cook, A. Bennett, D. Galuszka, B. Redding, J. Richie, Progress on reviving Lunar Orbiter: scanning, archiving, and cartographic processing at USGS, in *Lunar Planet. Sci. Conf. XXXV. Abstract 1791* (Lunar and Planetary Institute, Houston, 2004), CD-ROM

J.F. Bell III, M.J. Wolff, M.C. Malin, W.M. Calvin, B.A. Cantor, M.A. Caplinger, R.T. Clancy, K.S. Edgett, L.J. Edwards, J. Fahle, F. Ghaemi, R.M. Haberle, A. Hale, P.B. James, S.W. Lee, T. McConnochie, E. Noe Dobrea, M.A. Ravine, D. Schaeffer, K.D. Supulver, P.C. Thomas, Mars Reconnaissance Orbiter Mars Color Imager (MARCI): instrument description, calibration, and performance. J. Geophys. Res. **114**, E08S92 (2009). doi:10.1029/2008JE003315

D.T. Blewett, B.R. Hawke, P.G. Lucey, G.J. Taylor, R. Jaumann, P.D. Spudis, Remote sensing and geologic studies of the Schiller-Schickard region of the Moon. J. Geophys. Res. **100**, 16959–16978 (1995)

D.E. Bowker, J.K. Hughes, *Lunar Orbiter Photographic Atlas of the Moon*. NASA Special Publication, vol. 206 (1971)

D.B.J. Bussey, P.D. Spudis, M.S. Robinson, Illumination conditions at the lunar South Pole. Geophys. Res. Lett. **26**, 1187–1190 (1999)

E.N. Cameron, *Evaluation of the Regolith of Mare Tranquillitatis as a Source of Volatile Elements* (University of Wisconsin-Madison, Madison, 1993)

W.S. Cameron, G.J. Coyle, An analysis of the distribution of boulders in the vicinity of small lunar craters. Earth Moon Planets **3**, 159–188 (1971)

W.D. Carrier, G.R. Olhoeft, W. Mendell, Physical properties of the lunar surface, in *Lunar Sourcebook*, ed. by G.H. Heiken, D.T. Vaniman, B.M. French (Cambridge University Press, Cambridge, 1991), pp. 475–594

G. Chin, S. Brylow, M. Foote, J. Garvin, J. Kasper, J. Keller, M. Litvak, I. Mitrofanov, D. Paige, K. Raney, M. Robinson, A. Sanin, D. Smith, H. Spence, P. Spudis, S.A. Stern, M. Zuber, Lunar Reconnaissance Orbiter overview: the instrument suite and mission. Space Sci. Rev. **129**, 391–419 (2007)

M.J. Cintala, J.B. Garvin, S.J. Wetzel, The distribution of blocks around a fresh lunar mare crater, in *Lunar Planet. Sci. Conf. XIII* (Lunar and Planetary Institute, Houston, 1982), pp. 100–101

M.J. Cintala, K. McBride, Block distribution on the lunar surface: a comparison between measurements obtained from surface and orbital photography. NASA, Johnson Space Center, Houston, TX, 1995

C.R. Coombs, B.R. Hawke, A search for intact lava tubes on the Moon: Possible lunar base habitats, in Lunar and Planetary Institute, Houston, TX (1992), pp. 219–229

B.M. Cudnik, D.W. Palmer, D.M. Palmer, A. Cook, R. Venable, P.S. Gural, The observation and characterization of lunar meteoroid impact phenomena. Earth Moon Planets **93**, 97–106 (2003)

M.E. Davies, T.R. Colvin, M.J.S. Belton, J. Veverka, P.C. Thomas, The direction of the north pole and the control network of asteroid 951 Gaspra. Icarus **107**, 18–22 (1994)

F.K. Duennebier, Y. Nakamura, G.V. Latham, H.J. Dorman, Meteoroid storms detected on the Moon. Science **192**, 1000–1002 (1976)

F. El-Baz, L.J. Kosofsky, *The Moon as Viewed by Lunar Orbiter*. NASA SP (United States Government Printing Office (USGPO), Washington, 1970)

W.C. Feldman, D.J. Lawrence, R.C. Elphic, B.L. Barraclough, S. Maurice, I. Getenay, Polar hydrogen deposits on the Moon. J. Geophys. Res. **105**, 4175–4195 (2000)

L.R. Gaddis, C.M. Pieters, B. Ray Hawke, Remote sensing of lunar pyroclastic mantling deposits. Icarus **61**, 461–489 (1985)

L.R. Gaddis, A.S. McEwen, T.L. Becker, Compositional variations on the Moon: recalibration of Galileo solid-state imaging data for the Orientale region and farside. J. Geophys. Res. **100**, 26345–26356 (1995)

L.R. Gaddis, B.R. Hawke, M.S. Robinson, C. Coombs, Compositional analyses of small lunar pyroclastic deposits using Clementine multispectral data. J. Geophys. Res. **105**, 4245–4262 (2000)

L. Gaddis, M.I. Staid, A. Tyburczy, B.R. Hawke, N.E. Petro, Compositional analyses of lunar pyroclastic deposits. Icarus **161**, 262–280 (2003)

D.E. Gault, Saturation and equilibrium conditions for impact cratering on the lunar surface: criteria and implications. Radio Sci. **5**, 273–291 (1970)

A.W. Gifford, F. El-Baz, Thicknesses of lunar mare flow fronts. Moon Planets **24**, 391–398 (1981)

T.A. Giguere, B.R. Hawke, D.T. Blewett, D.B.J. Bussey, P.G. Lucey, G.A. Smith, P.D. Spudis, G.J. Taylor, Remote sensing studies of the Lomonosov-Fleming region of the Moon. J. Geophys. Res. **108**, 5118 (2003)

T.A. Giguere, B.R. Hawke, D.T. Blewett, B.A. Campbell, J.J. Gillis-Davis, P.G. Lucey, C.A. Peterson, C. Runyon, G.J. Taylor, *Cryptomare and Pyroclastic Deposits in the Gassendi Region of the Moon* (2007), p. 1132

R. Goddard, in *The Papers of Robert H. Goddard*, ed. by E.C. Goddard, G.E. Pendray (McGraw-Hill, New York, 1920), pp. 413–430

R.O. Green, C. Pieters, P. Mouroulis, G. Sellars, M. Eastwood, S. Geier, J. Shea, The Moon mineralogy mapper: characteristics and early laboratory calibration results, in *Lunar Planet. Sci. Conf. XXXVIII*, Abstract 2354 (Lunar and Planetary Institute, Houston, 2007), CD-ROM

B. Hapke, *Theory of Reflectance and Emittance Spectroscopy* (Cambridge University Press, Cambridge, 1993)

J. Haruyama, T. Matsunaga, M. Ohtake, T. Morota, C. Honda, Y. Yokota, M. Torii, Y. Ogawa, LISM Working Group, Global lunar surface mapping experiment using the Lunar Imager/Spectrometer on Selene. Earth Planets Space **60**, 243–256 (2008)

B.R. Hawke, J.J. Gillis, T.A. Giguere, D.T. Blewett, D.J. Lawrence, P.G. Lucey, G.A. Smith, P.D. Spudis, G.J. Taylor, *Remote Sensing and Geologic Studies of the Balmer Region of the Moon* (2004), p. 1190

B.R. Hawke, J.J. Gillis, T.A. Giguere, D.T. Blewett, D.J. Lawrence, P.G. Lucey, G.A. Smith, P.D. Spudis, G.J. Taylor, Remote sensing and geologic studies of the Balmer-Kapteyn region of the Moon. J. Geophys. Res. **110**, E06004 (2005)

B.R. Hawke, T.A. Giguere, D.T. Blewett, B.A. Campbell, J.J. Gillis-Davis, J.J. Hagerty, D.J. Lawrence, P.G. Lucey, C.A. Peterson, G.A. Smith, P.D. Spudis, G.J. Taylor, *Remote Sensing Studies of the Schiller-Schickard Region of the Moon: Final Results* (2007), p. 1474

B.R. Hawke, T.A. Giguere, D.T. Blewett, J.M. Boyce, J.J. Gillis-Davis, J.J. Hagerty, P.G. Lucey, C.A. Peterson, G.A. Smith, P.D. Spudis, G.J. Taylor, *Light Plains and Cryptomare Deposits on the Northeastern Portion of the Lunar Nearside* (2008), p. 1512

J.W. Head, T.B. McCord, Imbrian-age highland volcanism on the Moon: the Gruithuisen and Mairan Domes. Science **199**, 1433–1436 (1978)
G.H. Heiken, D.T. Vaniman, B.M. French (eds.), *Lunar Sourcebook* (Cambridge University Press, Cambridge, 1991)
H. Hiesinger, J.W. Head, U. Wolf, R. Jaumann, G. Neukem, Ages and stratigraphy of mare basalts in Oceanus Procellarum, Mare Nubium, Mare Cognitum, and Mare Insularum. J. Geophys. Res. **108**, 5065 (2003)
A.P. Ingersoll, T. Svitek, B.C. Murray, Stability of polar frosts in spherical bowl-shaped craters on the Moon, Mercury, and Mars. Icarus **100**, 40–47 (1992)
B.A. Ivanov, Earth/Moon impact rate comparisons: Searching constraints for lunar secondary/primary cratering proportion. Icarus **183**, 504–507 (2006)
J.R. Janesick, *Photon Transfer*. SPIE Publications (2007)
JMars (2010). http://jmars.asu.edu/node/2055
S.J. Lawrence, M.S. Robinson, M. Broxton, J.D. Stopar, W. Close, J. Grunsfeld, R. Ingram, L. Jefferson, S. Locke, R. Mitchell, T. Scarsella, M. White, M.A. Hager, T.R. Watters, E. Bowman-Cisneros, J. Danton, J. Garvin, The Apollo digital image archive: new research and data products. LPI Contrib. **1415**, 2066 (2008a)
S.J. Lawrence, B.R. Hawke, J.J. Gillis-Davis, G.J. Taylor, D.J. Lawrence, J.T. Cahill, J.J. Hagerty, P.G. Lucey, G.A. Smith, K. Keil, Composition and origin of the Dewar geochemical anomaly. J. Geophys. Res. (Planets) **113**, 02001 (2008b)
S.W. Lee, P. Thomas, J. Veverka, Phobos, Deimos, and the Moon: size and distribution of crater ejecta blocks. Icarus **68**, 77–86 (1986)
M.C. Malin, J.F. Bell III, W. Calvin, R.T. Clancy, R.M. Haberle, P.B. James, S.W. Lee, P.C. Thomas, M.A. Caplinger, Mars Color Imager (MARCI) on the Mars Climate Orbiter. J. Geophys. Res. **106**, 17651–17672 (2001)
M.C. Malin, J.F. Bell III, W.M. Calvin, B.A. Cantor, M.A. Caplinger, R.T. Clancy, T. Ghaemi, R.M. Haberle, P.B. James, S.W. Lee, T. McConnochie, E.N. Dobrea, M.A. Ravine, P.C. Thomas, M.J. Wolff, Mars Reconnaissance Orbiter Mars Color Imager (MARCI) and Context Camera (CTX) pre-flight calibration report [MSSS-CTX-DOC-2002]. Malin Space Science Systems, San Diego, CA, 2005
M.C. Malin, J.F. Bell, B.A. Cantor, M.A. Caplinger, W.M. Calvin, R.T. Clancy, K.S. Edgett, L. Edwards, R.M. Haberle, P.B. James, S.W. Lee, M.A. Ravine, P.C. Thomas, M.J. Wolff, Context camera investigation on board the Mars Reconnaissance Orbiter. J. Geophys. Res. **112**, E05S04 (2007)
H. Masursky, G.W. Colton, F. El-Baz (eds.), *Apollo over the Moon: A View From Orbit*. NASA SP, vol. 362 (United States Government Printing Office, Washington, 1978)
M. Minnaert, Photometry of the Moon, in *Planets and Satellites*, ed. by G.P. Kuiper, B.M. Middlehurst (University of Chicago Press, Chicago, 1961), p. 601
H.J. Moore, Large blocks around lunar craters, in *Analysis of Apollo 10 Photography and Visual Observations* (National Aeronautics and Space Administration, Washington, 1971), pp. 26–27
R.G. Musgrove, *Lunar Photographs from Apollos 8, 10, and 11*. NASA Special Publication, vol. 246 (1971)
NASA, *A Site Selection Strategy for a Lunar Outpost: Science and Operational Parameters* (Lunar and Mars Exporation Program Office, Johnson Space Center, Houston, 1990)
NASA, *The Vision for Space Exploration [NP-2004-01-334-HQ]* (NASA, Washington, 2004a)
NASA, Announcement of opportunity Lunar Reconnaissance Orbiter measurement investigations. NNH04ZSS003O, 2004b
NASA, Exploration systems architecture study. NASA-TM-2005-214062, 2005
H. Noda, H. Araki, S. Goossens, Y. Ishihara, K. Matsumoto, S. Tazawa, N. Kawano, S. Sasaki, Illumination conditions at the lunar polar regions by KAGUYA(SELENE) laser altimeter. Geophys. Res. Lett. **35**, L24203 (2008)
S. Nozette, P. Rustan, L.P. Pleasance, D.M. Horan, P. Regeon, E.M. Shoemaker, P.D. Spudis, C.H. Acton, D.N. Baker, J.E. Blamont, B.J. Buratti, M.P. Corson, M.E. Davies, T. Duxbury, E.M. Eliason, B.M. Jakosky, J.F. Kordas, I.T. Lewis, C.L. Lichtenberg, P.G. Lucey, E. Malaret, M.A. Massie, J.H. Resnick, C.J. Rollins, H.S. Park, A.S. McEwen, R.E. Priest, C. Pieters, R.A. Reisse, M.S. Robinson, R.A. Simpson, D.E. Smith, T.C. Sorensen, R.W. Vorder Breugge, M.T. Zuber, The Clementine mission to the Moon: scientific overview. Science **266**, 1835–1839 (1994)
S. Nozette, P.D. Spudis, M.S. Robinson, D.B.J. Bussey, C. Lichenstein, R. Bonner, Integration of lunar polar remote-sensing data sets: evidence for ice at the lunar south pole. J. Geophys. Res. **106**, 23253–23266 (2001)
V.R. Oberbeck, W.L. Quaide, Estimated thickness of a fragmental surface layer of Oceanus Procellarum. J. Geophys. Res. **72**, 4697–4704 (1967)
V.R. Oberbeck, W.L. Quaide, Genetic implications of Lunar regolith thickness variations. Icarus **9**, 446–465 (1968)

J.L. Ortiz, P.V. Sada, L.R. Bellot Rubio, F.J. Aceituno, J. Aceituno, P.J. Gutierrez, U. Thiele, Optical detection of meteoroidal impacts on the Moon. Nature **405**, 921–923 (2000)

J.L. Ortiz, J.A. Quesada, J. Aceituno, F.J. Aceituno, L.R.B. Rubio, Observation and interpretation of Leonid impact flashes on the Moon in 2001. Astrophys. J. **576**, 567–573 (2002)

J. Ortiz, F. Aceituno, J. Quesada, J. Aceituno, M. Fernández, P. Santos-Sanz, J. Trigo-Rodríguez, J. Llorca, F. Martín-Torres, P. Montañés-Rodríguez, E. Pallé, Detection of sporadic impact flashes on the Moon: Implications for the luminous efficiency of hypervelocity impacts and derived terrestrial impact rates. Icarus **184**, 319–326 (2006)

W.L. Quaide, V.R. Oberbeck, Thickness determinations of the lunar surface layer from lunar impact craters. J. Geophys. Res. **73**, 5247–5270 (1968)

D. Stoffler, G. Ryder, B.A. Ivanov, N.A. Artemieva, M.J. Cintala, R.A.F. Grieve, Cratering history and lunar chronology. Rev. Min. Geochem. **60**, 519–596 (2006)

G.J. Taylor, P.D. Spudis, Geoscience and a Lunar Base: A Comprehensive Plan for Lunar Exploration, 1990

USGS, 2010. http://isis.astrogeology.usgs.gov/UserDocs/index.html

J.K. Wagner, B.W. Hapke, E.N. Wells, Atlas of reflectance spectra of terrestrial, lunar, and meteoritic powders and frosts from 92 to 1800 nm. Icarus **69**, 14–28 (1987)

K. Watson, B.C. Murray, H. Brown, On the possible presence of ice on the Moon. J. Geophys. Res. **66**, 1598–1600 (1961)

D.E. Wilhelms, *Stratigraphy of Part of the Lunar Near Side*. United States Geological Survey Professional Paper, vol. 1046-A (United States Government Printing Office, Washington, 1980)

D.E. Wilhelms, J.F. McCauley, N.J. Trask, *The Geologic History of the Moon*. U.S. Geol. Surv. Prof. Pap., vol. 1348 (United States Government Printing Office, Washington, 1987)

R.J. Williams, D.S. McKay, D. Giles, T.E. Bunch, Mining and beneficiation of lunar ores, in *Space Resources and Space Settlements*, ed. by G.K. O'Neill, J. Billingham, W. Gilbreath, B. O'Leary (NASA, 1979), pp. 275–188

C.A. Wood, *Lunar Concentric Craters* (1978), pp. 1264–1266

C.A. Wood, J.W. Head, Geologic setting and provenance of spectrally distinct pre-mare material of possible volcanic origin, in *LPI Contributions* (Lunar Science Institute, Houston, 1975), p. 189

M. Yanagisawa, N. Kisaichi, Lightcurves of 1999 Leonid impact flashes on the Moon. Icarus **159**, 31–38 (2002)

Y. Zhao, F. Shadman, Production of oxygen from lunar ilmenite, in *Resources of Near-Earth Space*, ed. by J. Lewis, M.S. Matthews, M.L. Guerrieri (University of Arizona Press, Tucson, 1993), pp. 149–178

Space Sci Rev (2010) 150: 125–160
DOI 10.1007/s11214-009-9529-2

The Lunar Reconnaissance Orbiter Diviner Lunar Radiometer Experiment

D.A. Paige · M.C. Foote · B.T. Greenhagen · J.T. Schofield · S. Calcutt · A.R. Vasavada · D.J. Preston · F.W. Taylor · C.C. Allen · K.J. Snook · B.M. Jakosky · B.C. Murray · L.A. Soderblom · B. Jau · S. Loring · J. Bulharowski · N.E. Bowles · I.R. Thomas · M.T. Sullivan · C. Avis · E.M. De Jong · W. Hartford · D.J. McCleese

Received: 24 January 2009 / Accepted: 7 May 2009 / Published online: 26 June 2009

Abstract The Diviner Lunar Radiometer Experiment on NASA's Lunar Reconnaissance Orbiter will be the first instrument to systematically map the global thermal state of the Moon and its diurnal and seasonal variability. Diviner will measure reflected solar and emitted infrared radiation in nine spectral channels with wavelengths ranging from 0.3 to 400 microns. The resulting measurements will enable characterization of the lunar thermal environment, mapping surface properties such as thermal inertia, rock abundance and silicate mineralogy, and determination of the locations and temperatures of volatile cold traps in the lunar polar regions.

D.A. Paige (✉) · B.T. Greenhagen · M.T. Sullivan
Dept. of Earth and Space Sciences, University of California, Los Angeles, USA
e-mail: dap@moon.ucla.edu

M.C. Foote · J.T. Schofield · A.R. Vasavada · D.J. Preston · B. Jau · S. Loring · J. Bulharowski · C. Avis · E.M. De Jong · W. Hartford · D.J. McCleese
Jet Propulsion Laboratory, California Institute of Technology, Pasadena, USA

S. Calcutt · F.W. Taylor · N.E. Bowles · I.R. Thomas
Dept. of Physics, University of Oxford, Oxford, UK

C.C. Allen
Lyndon B. Johnson Space Center, Houston, USA

K.J. Snook
NASA Headquarters, Washington DC, USA

B.M. Jakosky
Dept. of Geological Sciences, University of Colorado, Boulder, USA

B.C. Murray
Div. of Geological and Planetary Sciences, California Institute of Technology, Pasadena, USA

L.A. Soderblom
US Geologic Survey, Flagstaff, USA

Keywords Moon · Lunar · Diviner · Thermal · Radiometer · Mapping · Temperature · Infrared · Mineralogy · Petrology · LRO · Reconnaissance Orbiter

1 Introduction

The Moon's surface thermal environment is among the most extreme of any planetary body in the solar system. Lunar temperatures at the subsolar point approach 400 K, whereas temperatures in permanent shadow may be lower than 40 K. The surface temperature of the Moon also represents a fundamental boundary condition that governs the thermal state of the Moon's regolith and interior, and the behavior of near-surface volatiles. The lunar thermal environment is also a significant challenge for human and robotic exploration systems, which must be designed to handle its extremes.

This paper describes the Diviner Lunar Radiometer Experiment, which is one of seven instruments aboard NASA's Lunar Reconnaissance Orbiter (LRO) mission. Diviner will be the first experiment to systematically map the global thermal state of the Moon and its diurnal and seasonal variability. In the sections below, we will describe the goals, instrumentation, observational strategy, and anticipated data products of the Diviner investigation.

1.1 The Lunar Thermal Environment

Observations and models show that the Moon has three distinct thermal environments: daytime, nighttime and polar. The Moon's daytime thermal environment is controlled by the flux of solar radiation. Because of the low thermal conductivity of the lunar soil, and the length of the lunar day, the temperatures of illuminated surfaces on the Moon are always close to radiative equilibrium. As illustrated in Fig. 1, the latitudinal and diurnal variation in lunar daytime temperatures is controlled mostly by the angular distance to the subsolar

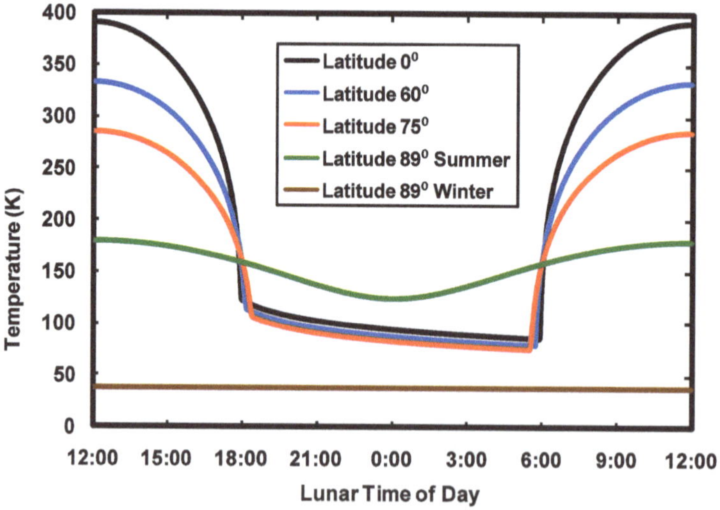

Fig. 1 Model calculations of lunar surface temperature variations as a function of local time and latitude. (Vasavada et al. 1999). Local time is expressed in lunar hours which correspond to 1/24 of a lunar month. At 89° latitude, diurnal temperature variations are shown at summer and winter solstices

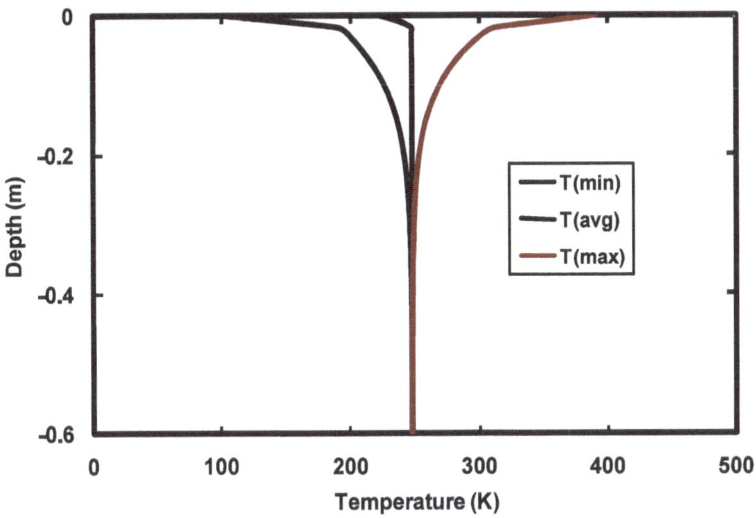

Fig. 2 Model calculations of diurnal minimum, maximum and average temperatures at the lunar equator as a function of depth (Vasavada et al. 1999)

point. The lunar nighttime environment is characterized by extreme cold. With no oceans or appreciable atmosphere to buffer surface temperatures, the only heat source to balance the loss of thermal radiation to space during the long lunar night is the sensible heat stored in the lunar subsurface during the day. The low thermal conductivity of the uppermost layers of the lunar regolith impedes the flow of heat from the subsurface to the surface, and as a consequence, lunar nighttime temperatures hover near 100 K for the duration of the 14-day lunar night. As illustrated in Fig. 2, the depth of penetration of the diurnal temperature wave on the Moon is ~30 cm.

The Moon also experiences seasonal insolation variations due to the combined effects of the 5.14° obliquity of the Moon's orbital plane relative to the ecliptic, and the 6.68° obliquity of the Moon's spin axis relative to the Moon's orbital plane. The net effect is that the latitude of the subsolar point undergoes a seasonal variation with an amplitude of ~1.54° and a period of ~346 days, which is less than a full Earth year due to the precession of the Moon's orbital plane. As a consequence, the polar regions experience very little direct solar illumination, and the illumination levels are strongly influenced by the effects of topography, as well as the lunar seasons. As illustrated in Fig. 1, a hypothetical horizontal surface at 89° latitude would have temperatures ranging from 128 to 180 K during peak continuous illumination conditions at summer solstice, and temperatures near 38 K during un-illuminated polar night conditions at winter solstice. The depth of penetration of seasonal temperature waves is roughly square root of the ratio of the annual to diurnal insolation periods, or ~120 cm at the lunar poles.

Superposed on the large-scale trends due to latitude, time of day, and season, the surface temperature of the Moon also exhibits variations due to topographic relief (slopes, roughness, shadows, and scattered solar and infrared radiation), spatial variations in solar reflectance and infrared emissivity, and spatial variations in the thermal properties of lunar surface materials (shown in Fig. 3). Because of the low thermal conductivity of the bulk of the lunar regolith, the lack of an appreciable atmosphere, and the effects of slopes and shadowing, the surface temperature of the Moon can exhibit extreme spatial variations all the

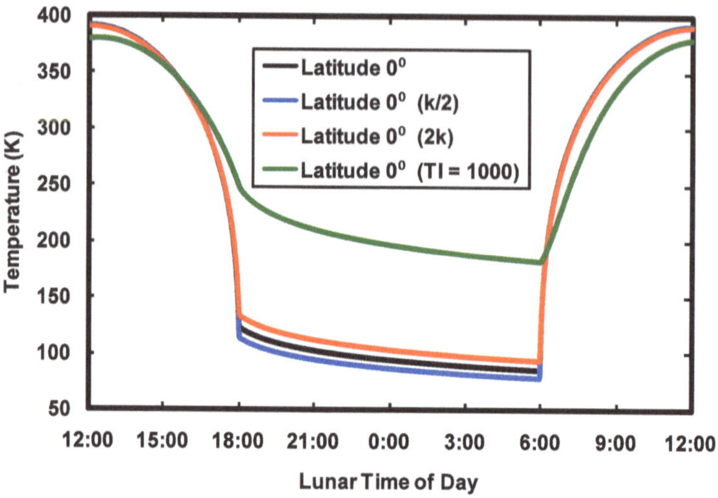

Fig. 3 Model calculations of the effects of regolith thermal properties and rocks on diurnal temperature variations at the equator (Vasavada et al. 1999). Shown are the effects of halving or doubling the thermal conductivity of the regolith, and those anticipated for a semi-infinite rock surface with a thermal inertia of $1000\ \mathrm{J\,m^{-2}\,k^{-1}\,s^{-1/2}}$

way down to the ~10 cm length scale of diurnal thermal skin depth. This fact significantly complicates the interpretation of lunar thermal observations and thermal model results, and the application of the information they provide to the design of practical lunar exploration systems.

Of special interest is the thermal behavior in high latitude regions that do not receive direct solar illumination. Thermal models that include the effects of multiply scattered solar and infrared radiation show that daily maximum surface temperatures in the coldest parts of larger permanently shadowed polar craters range from 40 to 90 K, depending on latitude and crater geometry (Vasavada et al. 1999). Even lower temperatures are possible in smaller-scale topographic features within permanently shadowed regions that are shaded from first-bounce indirect sunlight and/or infrared emission from warmer surfaces (Hodges 1980). The surface and subsurface thermal behavior in these coldest regions places fundamental constraints on the stability of cold-trapped water ice and other volatiles that may be cold-trapped at the lunar poles (Watson et al. 1961).

1.2 Past Ground-Based and Spacecraft Measurements

The first radiometric measurements of the temperature of the lunar surface were made by placing two uncooled thermopiles side by side at the focus of a reflecting telescope, one receiving radiation from the Moon and the other from space. Refinements of this original technique led to measurements of diurnal surface temperature variations over most of a lunation period, and during lunar eclipses (Petit and Nicholson 1930). Thermal modeling results showed that observed nighttime and eclipse temperatures were consistent with a ~2 cm-thick lunar surface layer composed of fine-grained low thermal conductivity regolith overlying material of higher thermal conductivity below (Jaeger and Harper 1950). More sophisticated ground-based infrared observations revealed the existence of spatial variations in nighttime brightness temperatures associated with bright rayed craters, which suggested

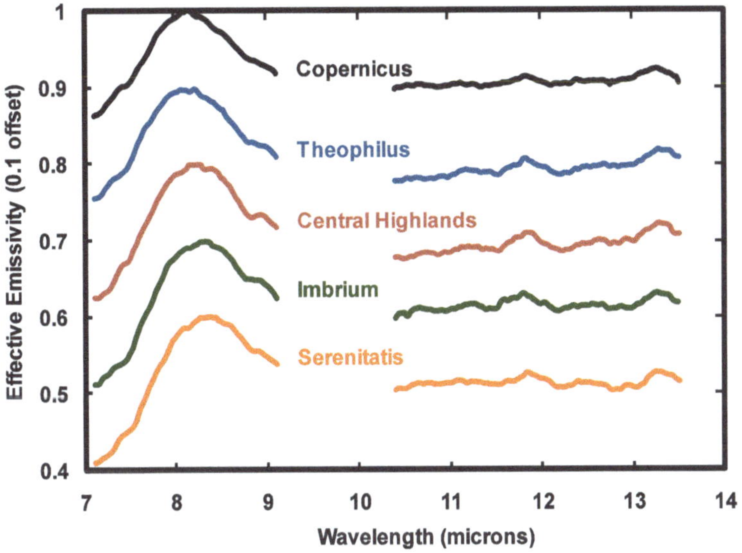

Fig. 4 Balloon-borne observations showing spatial variations in the lunar infrared emission spectrum from Murcray et al. (1970). The prominent emissivity maximum near 8.2 is the Christiansen feature (Salisbury et al. 1970)

an uneven distribution of regolith thermal properties and/or the presence of rocks (Murray and Wildey 1964). Balloon-borne telescopic measurements revealed the presence of a distinct peak in the Moon's thermal emission spectrum near 8 μm which was interpreted to be the Christiansen feature, an emissivity maximum associated with Si–O stretching vibrations (Murcray et al. 1970; Salisbury et al. 1970) (see Fig. 4).

Spacecraft missions to the Moon enabled more detailed infrared observations, *in situ* surface and subsurface temperature measurements, and determinations of the thermal and spectroscopic properties of lunar samples. Measurements and samples acquired at the Surveyor and Apollo landing sites confirmed the diurnal temperature behavior, and the general physical and spectral properties of lunar soils deduced from ground based measurements (Stimpson and Lucas 1970; Cremers and Birkebak 1971; Keihm and Langseth 1973; Logan et al. 1973; Salisbury et al. 1973).

The Apollo 17 Infrared Scanning Radiometer (ISR) experiment acquired thermal maps of approximately 25% of the lunar surface at a resolution of better than 10 km (Mendell and Low 1974). The instrument recorded brightness temperatures in a single uncooled thermopile bolometer channel with a spectral sensitivity of 1.2 to 70 μm that ranged from 80 to 400 K with ±2 K absolute precision. The data confirmed the presence of a large population of nighttime positive thermal anomalies or hot spots associated with the ejecta of small and large fresh impact craters (see Fig. 5). The accepted explanation for this phenomenon is that impacts excavate and expose blocky material with higher thermal inertia than the fine-grained lunar regolith. After exposure to the lunar environment over time, this blocky material is processed by smaller impacts and covered with fresh ejecta that gradually reduces its thermal inertia (Mendell 1976).

The Clementine Long Wavelength Infrared (LWIR) Camera mapped 0.4% of the Moon's surface from polar orbit with resolutions ranging from 200 m/pixel near the poles to 55 m/pixel at the equator (Lawson et al. 2000). The instrument employed a mechanically

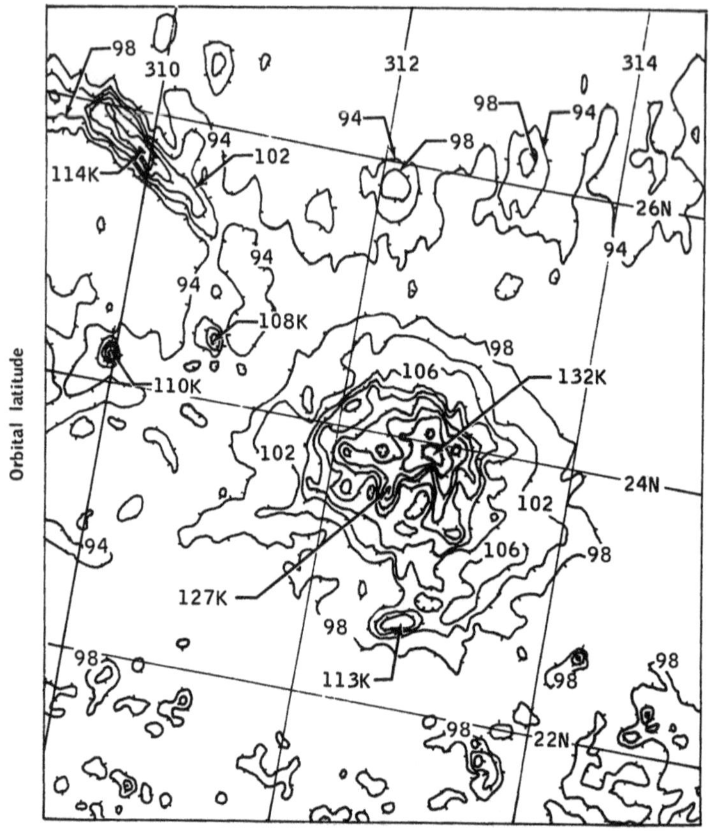

Fig. 5 Pre-dawn thermal map of the Aristarchus crater region by the Apollo 17 ISR (Mendell 1976). The contour interval is 4 K. The +32 K positive thermal anomaly associated with the interior of the crater has been attributed to the presence of exposed high thermal inertia blocks that were excavated during the relatively recent impact event that formed the crater

cooled 128 × 128 pixel HgCdTe detector array and had a single thermal infrared channel centered at 8.75 μm. Because the instrument's minimum detectable temperature was in the neighborhood of 200 K, useful data were only acquired in directly illuminated regions. Since the lunar surface is in radiative equilibrium during the lunar day, and the rate of thermal emission is linearly proportional to the absorbed solar flux, most of the LWIR thermal images are indistinguishable from broad-band visible reflectance images of the Moon (Fig. 6).

Our current understanding of the lunar thermal environment is incomplete in a number of respects. While we have a good understanding of the bulk thermal properties of typical lunar regolith, and the large-scale diurnal thermal behavior at low and mid latitudes, we have much to learn regarding global thermal spatial variability, and its implications for lunar geological history. We also have good evidence for observable variations in the infrared spectral emissivity of the lunar surface, but the wavelength and spatial coverage of presently available observations has been insufficient to correlate with other compositionally-related datasets. We also have relatively little direct information regarding the thermal behavior of the Moon

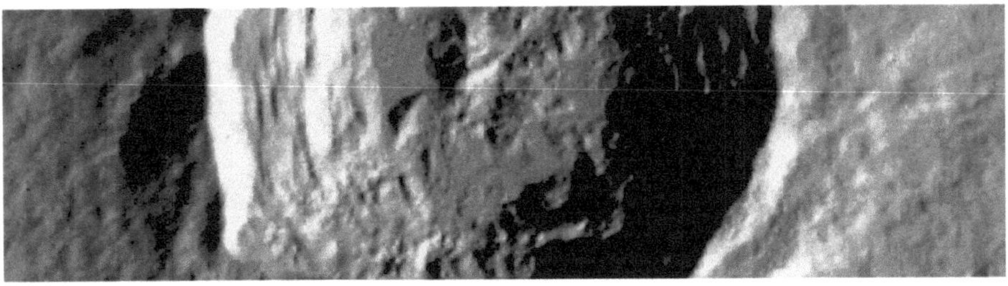

Fig. 6 Clementine LWIR daytime thermal image of Anaxagoras Crater (75°N, 10°W; diameter 52 km) at ~200 m spatial resolution (Lawson et al. 1997), which is similar to the mapping resolution that will be achieved by Diviner

at smaller spatial scales—particularly in the polar regions where scattered solar and infrared radiation are the dominant heat sources. As described in the next section, the Diviner investigation will provide a significant new source of lunar thermal data that will provide a basis for an improved understanding of many aspects of the lunar thermal environment, as well as the implications for the Moon's geologic and volatile history.

2 Investigation Description

2.1 History

The Diviner Lunar Radiometer was conceived and proposed in response to NASA's LRO Measurement Investigations announcement of opportunity (AO) in 2004. The LRO AO solicited investigations that could provide temperature mapping of the Moon's polar shadowed regions, search for near surface water ice in the Moon's polar cold traps, and identify safe landing sites for future landed missions. Accomplishing these measurements would require a thermal mapping instrument with a low minimum detectable temperature, a small and well-defined field of view, and stable absolute calibration. At this same time, an instrument that possessed these same characteristics, the Mars Climate Sounder (MCS), was being integrated and tested for the Mars Reconnaissance Orbiter (MRO) mission at NASA's Jet Propulsion Laboratory (JPL). MCS was developed primarily as an atmospheric sounder, designed to scan the limb of Mars to obtain vertical profiles of temperature, dust and water vapor (McCleese et al. 2007). Like the earliest ground based infrared observers and the Apollo ISR, MCS employed un-cooled thermopile detectors that were coupled to a set of nine spectral filters that spanned a wavelength range from 0.3 to 50 μm. After some initial study, it became clear that a follow-on MCS instrument could meet and exceed the LRO measurement requirements with relatively minor changes to the MCS filter set, thermal design, software and electronics. Diviner, so named because of its potential ability to locate resources, was competitively selected as part of the original LRO payload in late 2004, designed and developed at the Jet Propulsion Laboratory, and delivered to the Goddard Space Flight Center for spacecraft integration in early 2008.

2.2 Investigation Goals

The goals of the Diviner investigation are driven by the exploration and science goals of the LRO mission. They can be summarized as follows:

1. Characterize the Moon's surface thermal environments:
   a) Daytime
   b) Nighttime
   c) Polar
2. Map properties of the lunar surface:
   a) Bulk thermal properties
   b) Rock abundance
   c) Composition
3. Characterize polar cold traps:
   a) Map cold trap locations
   b) Determine their temperatures and thermophysical properties
   c) Assess potential lunar volatile resources

2.3 Investigation Approach

Diviner is a nine-channel pushbroom radiometer designed to map emitted thermal radiation and reflected solar radiation from the surface of the Moon. Diviner will employ the following approaches to accomplishing the goals of the investigation:

2.3.1 Characterizing the Moon's Thermal Environments

Diviner will spend the majority of its time mapping the thermal emission from the lunar surface from low polar orbit. The goal will be to accumulate as consistent and complete a dataset as possible given the parameters of the LRO orbit and the duration of the mission, in order to characterize temperature variations as a function of latitude, longitude, time of day and season. Diviner will have the capability to make accurate radiometric measurements for the warmest and coldest surfaces on the planet, as well as simultaneous measurements of broadband solar reflectance. The Diviner dataset should be sufficiently complete to allow confident prediction of lunar surface temperatures in daytime, nighttime and polar thermal environments. Characterization of surface temperature variations at a given location over time will also place strong constraints on the thermal state of the lunar subsurface, which is of interest for a range of scientific problems and lunar exploration activities. Two immediate applications for this information would include constraining subsurface thermal conditions in suspected volatile cold traps at the lunar poles, and the definition of sites for future lunar bases and permanent habitats.

2.3.2 Mapping Lunar Surface Properties

Diviner radiometric temperature and solar reflectance measurements will be compared with the results of models to infer the bulk thermal properties of lunar surface materials and aspects of their composition. The combination of Diviner data and thermal models will enable mapping of the thermal inertia of the uppermost centimeters of the lunar soil, as well as the areal coverage of exposed rocks. While Diviner's spatial resolution will not be sufficient to resolve even the largest lunar ejecta blocks, the wavelength range of Diviner's infrared channels will enable determination of whether a given footprint includes surface materials emitting at differing temperatures, and estimation of the population of emitting temperatures within a given footprint. By analyzing patterns of spectral emission from the lunar surface over a complete diurnal cycle, it will be possible to map thermal inertia, rock abundance, surface roughness and infrared spectral emissivity at every location where sufficient

Diviner coverage is obtained. In addition to new information regarding physical properties, Diviner's spectral emissivity measurements will provide new information regarding spatial variations in the composition of lunar surface materials that can be correlated with other lunar compositional datasets. The generation of Diviner's surface properties mapping products will benefit from the infusion of independently acquired topographic information at or below the spatial scale of the Diviner footprint. Such data, which can be acquired from orbital laser altimetry and stereo imagery investigations, will provide information regarding slopes and shadowing. The interpretation of Diviner's surface properties mapping products will be significantly enhanced through comparison with other mapped datasets relating to the composition and physical properties of the lunar surface.

2.3.3 Investigating Polar Cold Traps

The polar regions of the Moon are largely unexplored, and a number of significant questions remain regarding the possible existence of cold-trapped volatiles in the permanently shaded regions. The LRO payload includes multiple instruments that will contribute to a better understanding of these regions. Diviner's primary contribution will be to provide detailed thermal maps of permanently shadowed regions and adjacent areas. These maps can be used to determine the locations and surface temperatures of polar cold traps, and their diurnal and seasonal temperature variability that in turn, will place first-order constraints on the composition of volatile material that they may contain. Of particular significance is the potential correlation between the abundance of hydrogen in the uppermost lunar regolith and temperature. The Lunar Prospector (LP) Neutron Spectrometer data have strongly suggested the presence of enhanced hydrogen abundances in both the north and south polar regions. These data have been interpreted to be consistent with the presence of roughly 1.5% water-equivalent hydrogen in the uppermost ~50 cm of the lunar regolith if concentrated in the floors of craters unresolved by the spectrometer's footprint (Feldman et al. 2001). The spatial resolution of the LP neutron data makes it difficult to uniquely establish the spatial distribution and concentration of the detected hydrogen. The LRO LEND instrument (Mitrofanov 2009) will acquire a higher-resolution dataset that should enable better localization of lunar polar neutron anomalies. Correlating compositional constraints provided by neutron datasets with thermal constraints provided by Diviner will make it possible to assess whether the neutron hydrogen signatures are localized within cold-trap locations, and correlate neutron hydrogen signature with cold-trap surface and subsurface temperatures. Enhanced hydrogen abundance in cold-trap locations would be strong evidence for the presence of cold-trapped water ice that would have important implications for our understanding of the volatile history of the Moon, and for future robotic and human exploration (Duke et al. 2006).

3 Instrument Description

The LRO Diviner radiometer (Fig. 7) is largely identical to the MRO MCS instrument described previously (McCleese et al. 2007). Key instrument specifications are listed in Table 1. Two identical three-mirror off-axis telescopes are co-boresighted and mounted within an optical bench assembly. At the telescope focal planes are nine 21-element thermopile detector arrays, each with a separate spectral filter. The instrument will predominantly point in the nadir direction, operating as a multi-spectral pushbroom mapper. Two actuators, each allowing 270° of rotation, provide pointing in azimuth and elevation. Diviner is mounted to

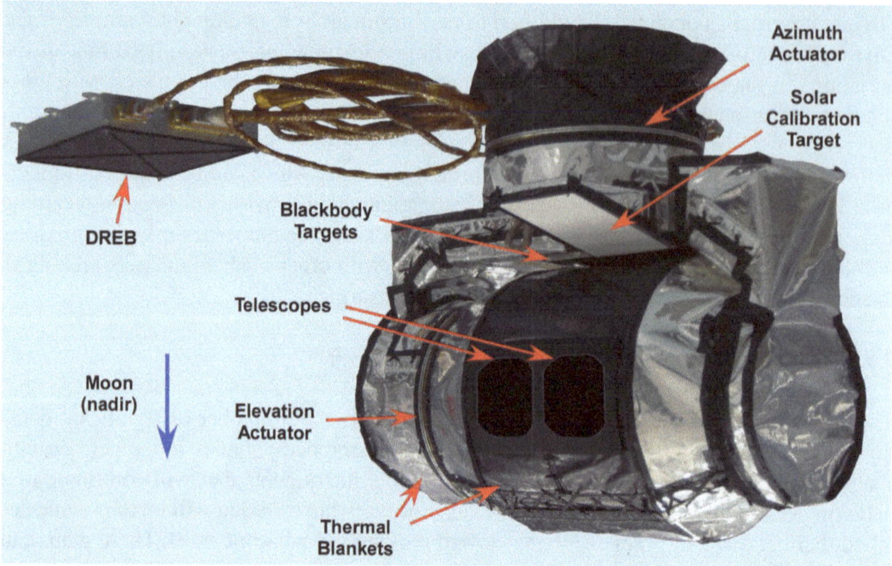

Fig. 7 Diviner instrument with major components labeled

the nadir panel of the LRO spacecraft, allowing it to view the Moon directly below. When the telescopes are pointed up into the instrument base they view a corrugated blackbody calibration target; when pointed horizontally they view space. The solar calibration target allows the solar channels to be calibrated using diffuse sunlight. The Diviner Remote Electronics Box (DREB) provides a Mil Spec 1553 interface to the spacecraft. The DREB also contains additional computing power compared to the MCS heritage design to allow the faster frame time required for pushbroom mapping.

Table 2 lists the nine spectral bands, their spectral passbands and measurement functions. In the text and figures they are often referred to by the channel names listed in the third column. Channels 1 and 2 measure reflected solar radiation from the lunar surface and utilize glass filters. These channels are referred to as the high- and reduced-sensitivity solar channels. Channels 3–5 use interference filters on zinc sulfide substrates provided by the University of Reading, UK. These channels, which have relatively narrow passbands near 8 μm, have sufficient signal-to-noise ratio to allow accurate spectral location of the emissivity maximum (Christiansen feature) for most known lunar materials at temperatures above 250 K (Greenhagen and Paige, 2006, 2009). Knowledge of the Christiansen feature spectral location provides compositional information. These channels are referred to as the 8 μm channels. Channels 6–9 divide the thermally emitted radiation roughly into four octaves, ranging approximately from 12.5 to 400 μm. Channel 6 uses an interference filter on a germanium substrate provided by the University of Reading. Channels 7–9 use stacks of hot-pressed copper mesh filters on polypropylene substrates provided by Cardiff University. Channels 6–9 are referred to as the thermal channels.

Channels 1–6 have a KCl window for long-wave blocking. Channel 7 has an interference filter and channels 8 and 9 have a quartz absorbing filter, both for short-wavelength blocking. All channels have a field-limiting aperture in front of the filter stack, running the length of the detector array and limiting the in-track field of view. The apertures for channels 2 and 6 are narrower than those over the other channels to avoid saturation at the highest signal

Table 1 Diviner instrument specifications

Parameter	Property
Instrument type	Infrared and solar radiometer
Spectral range	0.35 to 400 μm in nine spectral channels
Telescopes	Two identical three-mirror, off-axis, f/1.7 telescopes with 4 cm apertures
Detectors	Nine 21-element linear arrays of uncooled thermopile detectors Pixel size 240 μm × 480 μm
Fields of view	Detector Geometric IFOV: 6.7 mrad in-track 3.4 mrad cross track 320 m on ground in track for 50 km altitude 160 m on ground cross track for 50 km altitude Swath Width (Center to center of extreme pixels): 67 mrad; 3.4 km on ground for 50 km altitude
Instrument articulation	Two-axis azimuth/elevation, range 270°, resolution 0.1°
Operating modes	Single operation mode, 0.128 s signal integration period
Observation strategy	Primarily nadir pushbroom mapping

levels. These aperture widths are reduced to roughly 16% of the full width for channel 2 and 24% of the full width for channel 6. These apertures reduce the in-track fields of view from the values listed in Table 1.

The channel layout in the two telescopes is shown in Fig. 8. The telescope A focal plane contains the six shorter wavelength channels. The longer wavelength mesh filters used in telescope B require more space for mounting, and hence only three filter channels fit on the telescope B focal plane. The channel colors used in Fig. 8 are the same as used in other figures in this paper.

Figure 9 illustrates where the Diviner spectral bands lie with respect to Planck blackbody emission from scenes at temperatures from 30 to 400 K.

The internal calibration blackbody target (Fig. 10) consists of a grooved aluminum block coated with Martin Black. A blackened baffle assembly reduces stray light and divides the blackbody into two regions corresponding to the two Diviner telescopes. The baffle dimensions are slightly larger than the telescope aperture dimensions. The blackbody assembly is thermally isolated by ceramic spacers. Platinum resistance thermometers embedded in the aluminum block and a heater mounted on the backside of the target, are used for temperature control and measurement.

The solar calibration target (SCT) serves as Diviner's in-flight photometric calibration reference. The SCT is the last of five uniquely constructed targets originally created for Mars Observer Pressure Modulator Infra-Red Radiometer (PMIRR) (McCleese et al. 1986, 1992) and also used on MCS (McCleese et al. 2007). The SCT consists of a 6061 T6 aluminum base covered by an arc-sprayed coating of 99.999% pure aluminum that was grit blasted to remove specular facets (Ono 1999).

Diviner takes advantage of a great deal of heritage from the MCS instrument. Spare focal-plane assemblies from MCS were used. The vast majority of Diviner mechanical parts

Table 2 Diviner spectral channel passbands and measurement functions

Channel number	Channel type	Channel name	Passband μm	Measurement function
1	Solar	High Sensitivity Solar	0.35–2.8	Reflected solar radiation, high sensitivity
2	Solar	Reduced Sensitivity Solar	0.35–2.8	Reflected solar radiation, reduced sensitivity
3	8 μm	7.8 μm	7.55–8.05	Christiansen feature
4	8 μm	8.25 μm	8.10–8.40	Christiansen feature
5	8 μm	8.55 μm	8.38–8.68	Christiansen feature
6	Thermal	13–23 μm	13–23	Surface temperature (most sensitive channel for >178 K)
7	Thermal	25–41 μm	25–41	Surface temperature (most sensitive channel for 69–178 K)
8	Thermal	50–100 μm	50–100	Surface temperature (most sensitive channel for 43–69 K)
9	Thermal	100–400 μm	100–400	Surface temperature (most sensitive channel for <43 K)

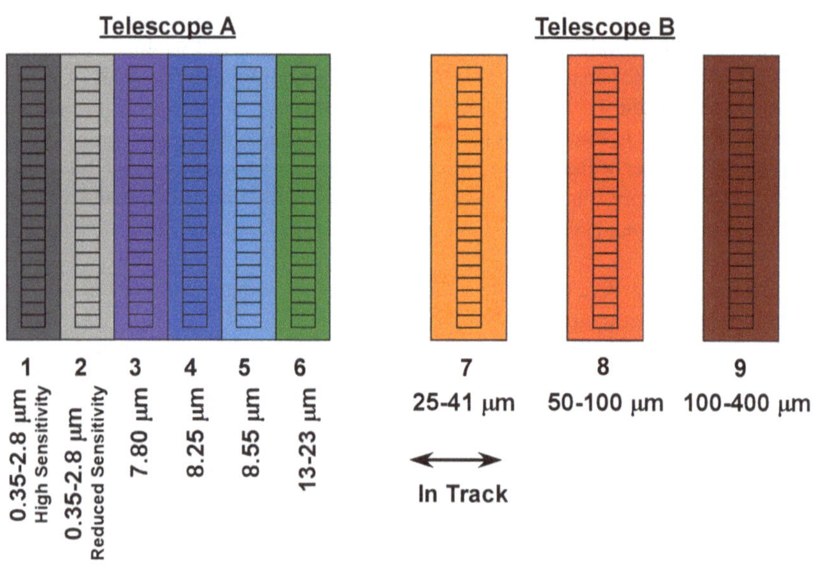

Fig. 8 Diviner focal-plane layout. The two telescopes are co-boresighted but are separated in this figure for clarity

were build-to-print from MCS designs. A large fraction of the MCS electronics and software design was also reused for Diviner. However, several key changes, listed below, were made to enable Diviner to accomplish its lunar science goals.

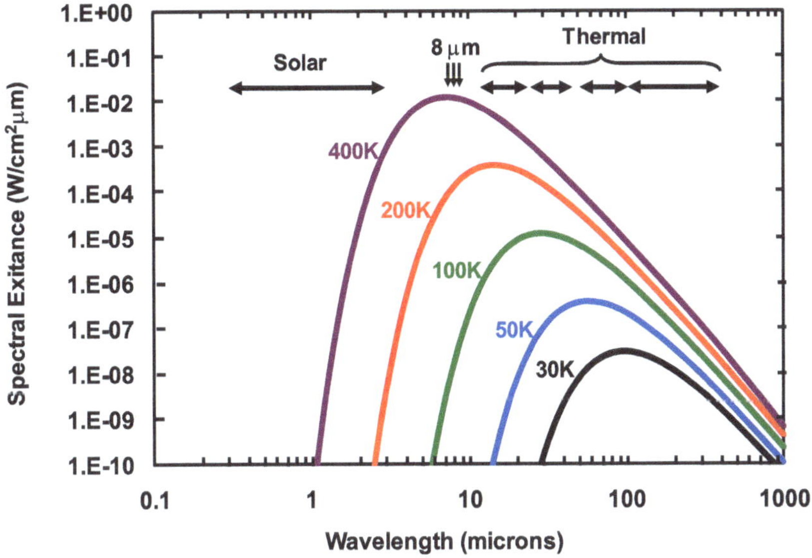

Fig. 9 The locations of the Diviner spectral channels and representative *blackbody curves*

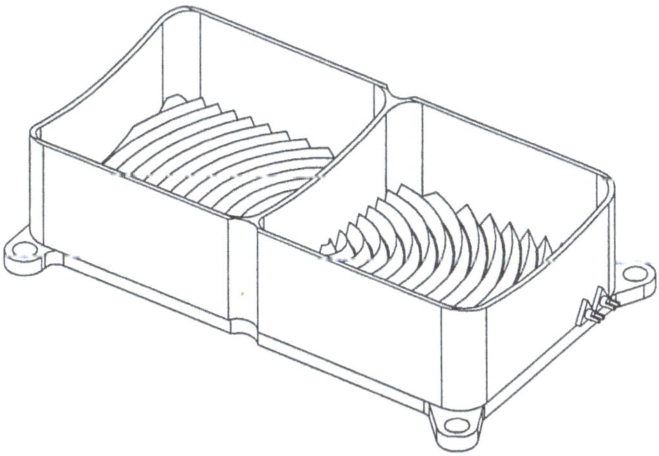

Fig. 10 Diviner internal calibration blackbody target

1. Diviner utilizes a different set of spectral passbands. The MCS bands were tuned to Mars atmospheric spectral features, while Diviner's bands are designed for measurements of lunar surface temperature, albedo, and the Christiansen feature.
2. Diviner requires a faster integration time to allow pushbroom mapping of the lunar surface. MCS was able to stare at the Mars limb for two second periods, while Diviner must integrate for only 128 ms to achieve the required spatial resolution and to minimize image smearing due to spacecraft motion. This change increases the data volume by roughly a factor of 16 and required numerous changes to the Field-Programmable Gate Array (FPGA) and software programs.

3. The harsher thermal environment in lunar orbit compared to Mars dictated a modified thermal blanket and heater design. MCS has multilayer blanketing for insulation from the environment. The Diviner blanket is thinner, and some internal parts are painted black to allow more heat radiation to space. Additionally, an outer layer of silver Teflon was used over much of Diviner to reflect sunlight.
4. The DREB was added to provide a 1553 interface to the spacecraft (MCS used a low-voltage differential signaling (LVDS) interface) and to provide additional computing power needed due to the sixteen-fold increase in data rate of Diviner compared to MCS.

An electrical block diagram is shown in Fig. 11. The LRO spacecraft provides three power circuits, one for instrument power, one for DREB electronics and supplemental heater power, and one for survival heater power. LRO also monitors five thermistors distributed through the instrument and DREB. A one-second time tick is supplied by the spacecraft through an RS422 interface. All other communication to and from the spacecraft occurs through the 1553 interface located in the DREB. The DREB and instrument communicate through an RS422 interface. The RS422 interface allows commands and time ticks to be sent from the DREB to the instrument, telemetry to be sent from the instrument to the DREB, and resets to be initiated by either the instrument or DREB. The instrument electronics provide science data collection, command and data handling, housekeeping (temperature and voltage monitoring), heater control, and actuator control and position sensing. The DREB provides additional science data processing as well as command and data handling.

3.1 Observation Strategy

Diviner employs a nearly continuous nadir viewing strategy. Minimizing gaps in nadir viewing enables accumulation of maximum spatial/temporal coverage for mapping. Although Diviner will spend much of its operational life in nadir staring, its MCS heritage provides a high degree of operational flexibility. The independent elevation and azimuth actuators are each capable of a 270° range of motion divided into 2673 steps (0.101° per step). Diviner can slew the elevation and azimuth actuators at rates of 25.1 and 23.9 deg/sec, respectively. However, due to electronic crosstalk, Diviner cannot collect useful science data while slewing.

The pointing direction of Diviner is controlled by three types of tables through which the instrument software repeats continuously. The Orbit Schedule Table (OST) has twelve entries, each representing one orbit. The OST directs which Event Schedule Table (EST) will be used for each orbit. After the twelfth orbit, the OST repeats from the first entry. The EST controls activities over the course of a single orbit. The EST calls individual Scan Sequence Tables (SST) based on the time since the last equator crossing, which because of LRO's near polar orbital inclination, is effectively latitude. Each orbit, the LRO spacecraft generates and sends to Diviner a command containing the descending orbit equator-crossing time. The descending orbit will vary between day and night sides during the mission. An SST describes a frequently used scan pattern and includes multiple entries each of which addresses the following: (1) which actuator to move, (2) which step to move to, and (3) how long to stay at a given step. When an SST is completed, the first entry will be repeated continuously until the EST directs a new SST be used. The SSTs must be "sun safe" because Diviner's detectors and filters will be near-instantaneously destroyed if they cross the Sun.

Diviner routinely points at a number of locations, which are defined by specific actuator positions and ranges. These locations, highlighted in Fig. 12, include stowed, nadir push-broom, internal blackbody calibration target, space view, and solar calibration target. The instrument stowed position, located at azimuth 0° and elevation 0°, is the position where

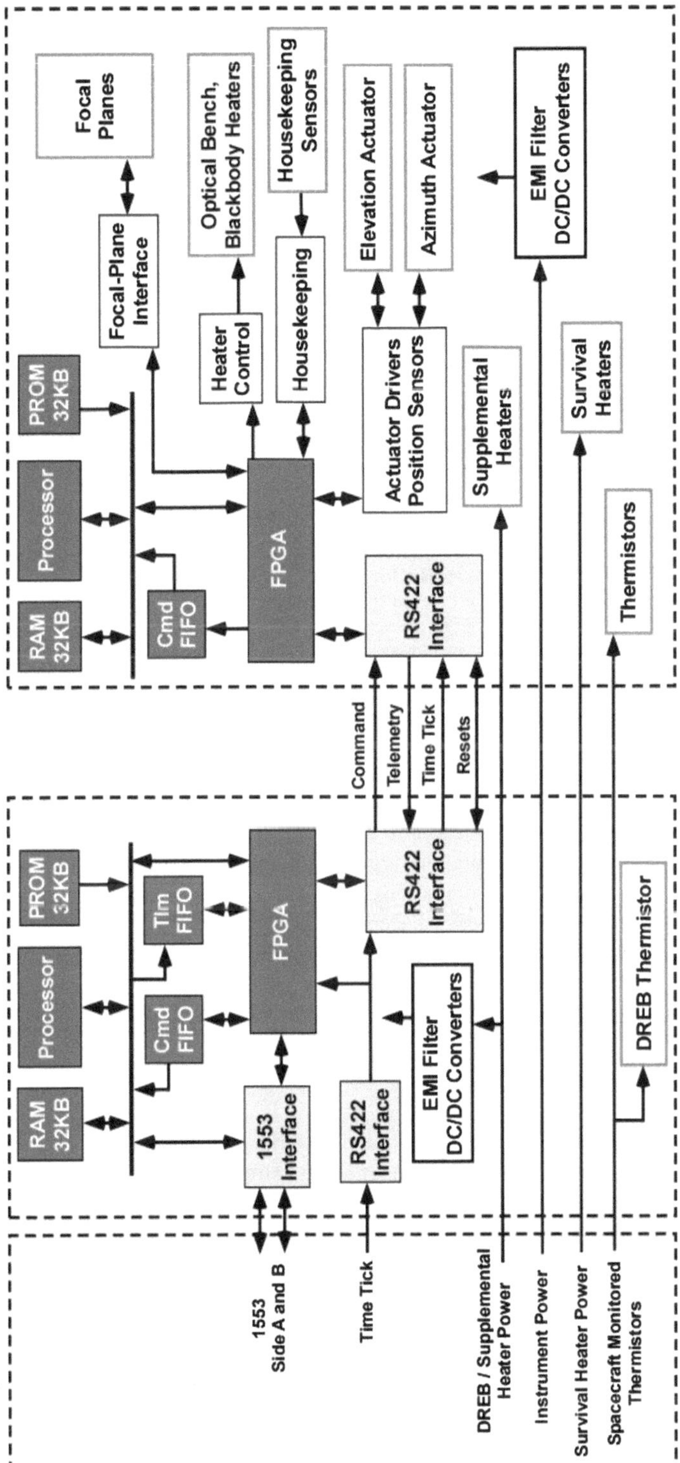

Fig. 11 Electrical block diagram of Diviner instrument and DREB

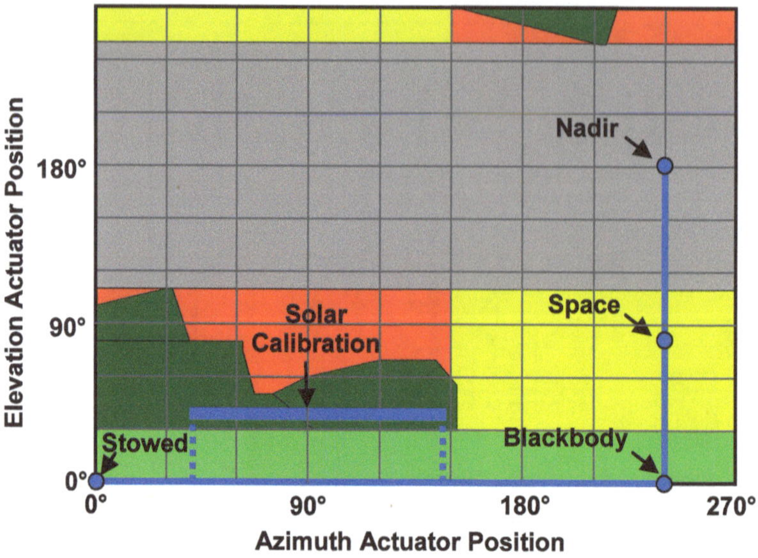

Fig. 12 The field of regard of the Diviner B telescope. The gray box represents the Moon for a 100 km circular orbit. The *dark green areas* are blocked by the spacecraft and instruments. The *light green area* is blocked by Diviner's yoke. Space regions are shown as both *red* and *yellow*: *red* is the region of space that could include the Sun, while yellow should normally never contain the Sun. Diviner will typically move along the *blue line* between common observation locations, nadir pointing, space views, blackbody calibration target, and stowed/safe position. The solar calibration target observations are made by the A telescope along the thick *blue line* between 40 and 150° azimuth

the instrument starts after power on and resets and is the ultimate destination of all safing routines. The nadir pushbroom position is at azimuth 240° and elevation 180°. The yoke-mounted internal blackbody calibration target can be observed at any azimuth with elevation 0°. Nominally, blackbody observations are made at azimuth 240° in conjunction with nadir pushbroom mapping. Views of space, necessary for both in-flight radiometric and photometric calibrations are typically made at azimuth 240° and elevation 80°. The solar calibration target can be observed at any azimuth with elevation 37°; however, the azimuth angle must be such that the target is illuminated.

3.2 In-flight Calibrations

Diviner performs routine two-point in-flight radiometric and photometric calibrations to calculate signal gain and offset values to correct for the effects of thermal and electronic drift. Both calibrations use space as the zero reference. The in-flight radiometric calibration uses the internal blackbody calibration target and the in-flight photometric calibration uses the solar calibration target mounted in front of telescope A (Channels 1–6) as the second reference. The solar calibration target was characterized pre-flight as described in a later section. The in-flight radiometric calibration sequence, which occurs approximately 10 times per orbit, is performed by pointing to space for 10 seconds, then pointing to the internal blackbody for 10 seconds, and then returning to space for 10 seconds. The total time required for the radiometric calibration is approximately 50 to 60 seconds depending on starting position. The in-flight photometric calibration sequence involves pointing to space for 10 seconds, then moving (in a sun-safe manner) to the solar calibration target position for ten seconds,

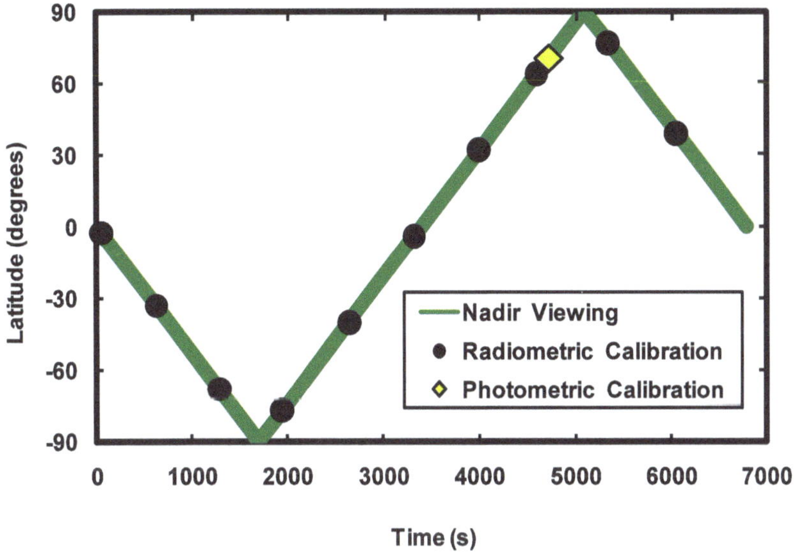

Fig. 13 Diviner observation and calibration activities during a typical LRO orbit. The timing of in-flight calibrations will be staggered to prevent latitudinal gores

then pointing to space for another 10 seconds. The total time required for the photometric calibration is approximately 75 to 85 seconds. The photometric calibration sequence is typically followed or preceded by a radiometric calibration and a duplicate space view is not performed. Values of gain and offset obtained through these calibrations are interpolated using a simple polynomial fit to obtain gain and offset values for each data sample.

There are several operational considerations regarding the timing and execution of the in-flight calibrations. First, the in-flight radiometric and photometric calibration sequences are deviations from the nominal nadir staring and can cause latitudinal gaps in the spatial and temporal coverage (see Fig. 13, for example). Therefore, multiple ESTs will be employed to stagger the timing of the calibration sequences such that consecutive orbits do not have these gaps at the same latitudes. Weekly EST updates will incrementally change the latitudinal positions of the staggered calibrations. Second, the in-flight photometric calibration is ideally performed with the Sun low over the horizon to minimize additional radiation from sunlight reflected off the lunar surface and onto the solar calibration target. The latitude at which with these sun angles occur will vary throughout the mission (e.g. midnight/noon vs. terminator orbit). Finally, the azimuth position of the in-flight photometric calibration will be chosen to minimize sunlight reflected off Diviner itself and onto the solar calibration target. Bi-weekly updates will change the azimuth pointing of the photometric calibration.

3.3 Spacecraft and Payload Considerations

Diviner is programmed to automatically respond to a number of scenarios including large spacecraft rolls, Lunar Reconnaissance Camera (LROC) high resolution imaging, and spacecraft/instrument anomalies. Targeted narrow-angle LROC images may require the LRO spacecraft to roll up to 20° off-nadir three times per day. During rolls, Diviner will execute its standard nadir viewing observation sequence to maintain its boresight relative to the other LRO instruments along the spacecraft $+Z$ axis. In-flight calibrations would continue even though some rolls will affect space and solar calibration target views. Testing may prove that Diviner's movements cause vibrations in the spacecraft sufficient to degrade LROC narrow angle images. For this reason, Diviner has the capability to "freeze", that is, to remain in its nadir position or to move there in a sun-safe manner. When frozen, Diviner will ignore scheduled calibration sequences while LROC narrow angle images are being acquired. Lastly, if there is an anomaly in the spacecraft or instrument, Diviner will execute a safing routine, abandoning its observation sequence and immediately moving to the stowed position in a sun-safe manner.

3.4 LRO Orbit and Lunar Seasons

The LRO orbit plane has an inclination of approximately 90° relative to the lunar equator, and is nearly fixed in inertial space. As a consequence, the LRO sub-spacecraft longitude during the ascending or descending legs of the orbit will cycle between 0° and 360° once each month. Furthermore, the LRO sub-spacecraft local time during each orbit leg will cycle between 0 and 24 hours once each Earth year. Therefore, for nadir viewing geometry, Diviner will have an opportunity to observe a given point on the planet at a given local time twice each year, once on the ascending leg of the orbit and once on the descending leg.

The lunar spin axis is tilted at an angle of ~1.54 degrees relative to the ecliptic. As a consequence, the lunar arctic circles are located at latitudes of ±88.46° and insolation conditions at these highest lunar latitudes will depend on both local time and season. Because the Moon's orbital plane precesses with a period of ~18.6 years, the length of the Draconic Year (the period of the lunar seasonal cycle) is approximately 346.62 days, which is 18.13 days shorter than the 365.25 day Sidereal Year (the period of the orbit of the Earth and Moon around the Sun). Because the period of the Moon's rotation is a non-negligible fraction of its annual period, diurnal and seasonal insolation does not repeat exactly from one year to the next. For instance, the longitude that experiences maximum insolation (noon at summer solstice) will change each year. In order to observe maximum insolation conditions at the poles, the local time of the LRO Orbit will be roughly aligned with noon when solar declination is at its maximum. The degree to which this orbital alignment will be achieved depends on the exact launch date.

3.5 Spatial and Temporal Coverage

The percentage of the area of the Moon that Diviner will sample can be estimated using a statistical coverage model that assumes nadir pointing, average orbital altitudes and randomized ground track longitudes. At any given latitude, the fractional coverage, including random overlap, is $F = 1 - e^{-sn/c}$, where F is the fractional coverage, s is the swath width, n is the number of latitude crossings by the orbit track (subject to various constraints), and c is the circumference of the latitude circle. Figures 14 and 15 show calculated Diviner fractional coverage for nighttime observations (defined here to be between 7 pm and 5 am local time) and noontime observations (defined here to be 10 am to 2 pm). The calculations assume that Diviner will spend 85% of its time in nadir pushbroom mapping mode, with remaining fraction devoted to Diviner blackbody and solar calibration and off-nadir rolls by the LRO spacecraft. The LRO mapping orbit is assumed to be an ellipse with perilune altitude of 36 km over the south pole and an apolune altitude of 64 km over the north pole. The LRO extended mission orbit is assumed to be an ellipse with a perilune altitude of 30 km over the south pole and an apolune altitude of 206 km over the north pole. As Diviner's

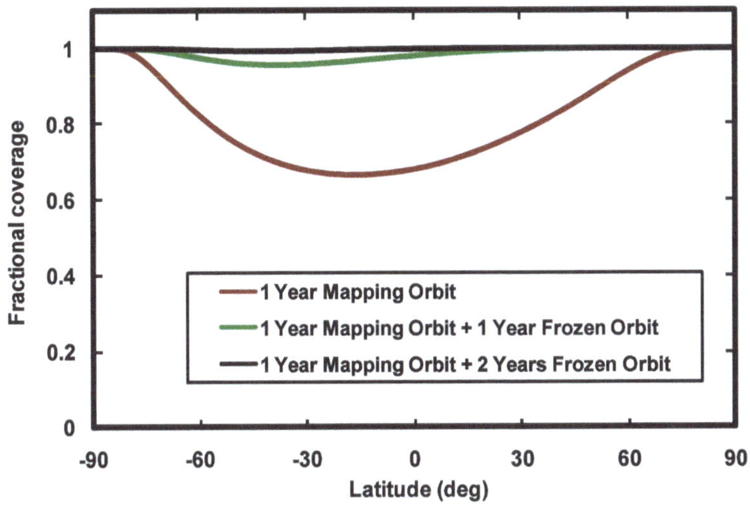

Fig. 14 Statistical model for the accumulation of Diviner nighttime coverage (7 pm to 5 am local time) during the LRO mapping and extended missions

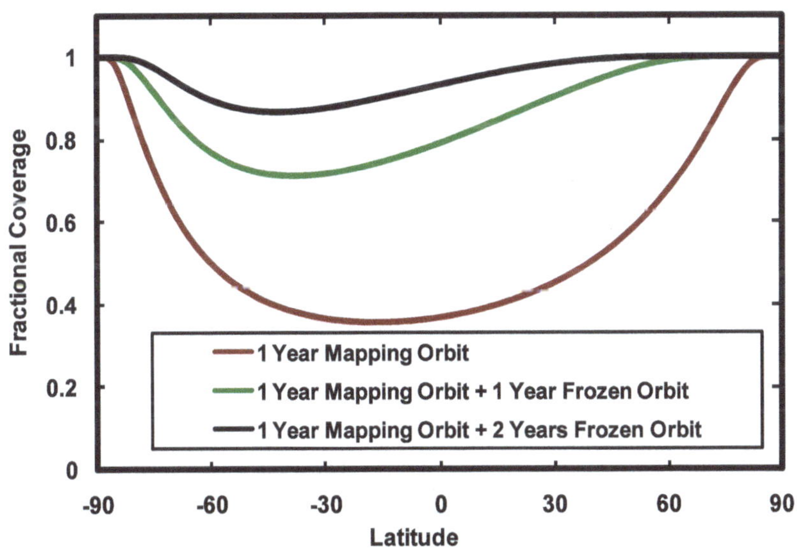

Fig. 15 Statistical model for the accumulation of Diviner noon-time coverage (10 am to 2 pm local time) during the LRO mapping and extended missions

swath width increases in proportion to the orbital altitude, higher altitude extended mission orbits will result in an increase in Diviner coverage in the north polar regions at the expense of spatial resolution.

During the one year LRO mapping mission, Diviner will acquire sufficient coverage to map the polar regions in detail, and provide extensive coverage of mid-latitude and equatorial regions. The nighttime coverage curves provide estimates of the percentage of the

Moon's surface area over which Diviner will acquire observations suitable for estimating bulk thermal properties and rock abundances. The noon coverage curves provide estimates of the percentage of the Moon's surface area over which Diviner will acquire observations suitable for estimating surface emissivity and regions of permanent shadow. Averaged over a year, latitudes with good noontime coverage can also be interpreted as having good coverage at all local times. Diviner's extended mission coverage will enable the creation of more complete maps of the polar regions. Due to the higher eccentricity of LRO's extended mission orbit, Diviner's extended mission coverage of the north polar region will be at lower spatial resolution than during the primary mission, but it will still enable improved characterization of seasonal and diurnal variability. At equatorial and mid-latitudes, Diviner's enhanced extended mission coverage will fill most of the remaining coverage gores at the end of the primary mission, producing a nearly complete global dataset.

4 Instrument Testing, Characterization and Calibration

A full set of ground calibration and characterization testing was performed on the Diviner instrument, including measurements of radiometric response, spectral response, and the instantaneous field of view of every pixel. These measurements confirm that the flight instrument performs at levels that meet or exceed original design goals.

4.1 Radiometric Response

Diviner's laboratory radiometric calibration used two external blackbodies, which were scale copies of those used to calibrate the Improved Stratospheric and Mesospheric Sounder (ISAMS) instrument (Nightingale and Crawford 1991). The blackbody temperatures were monitored to an accuracy better than 0.1 K. These blackbodies were positioned with their apertures within a few cm of the Diviner instrument such that movement of the elevation actuator could point the instrument into either external blackbody (Fig. 16). Each of the external blackbody apertures filled both of the instrument apertures. One of the external blackbodies (labeled Fixed Temperature Blackbody) was cooled to 90 K with liquid nitrogen. The second blackbody (labeled Variable Temperature Blackbody) was varied in temperature from 20 to 410 K using liquid helium, liquid nitrogen, and electrical heating. At each temperature setpoint of the variable blackbody, the instrument was scanned between the two

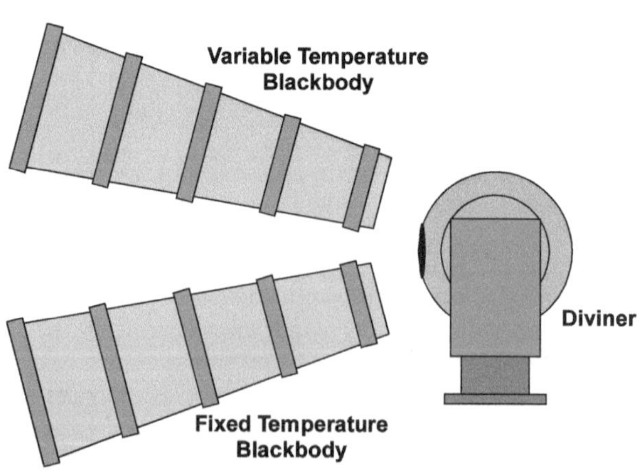

Fig. 16 Configuration used for laboratory radiometric calibration. The fixed temperature blackbody was maintained near 90 K using liquid nitrogen. The variable temperature blackbody varied in temperature from 20 to 415 K

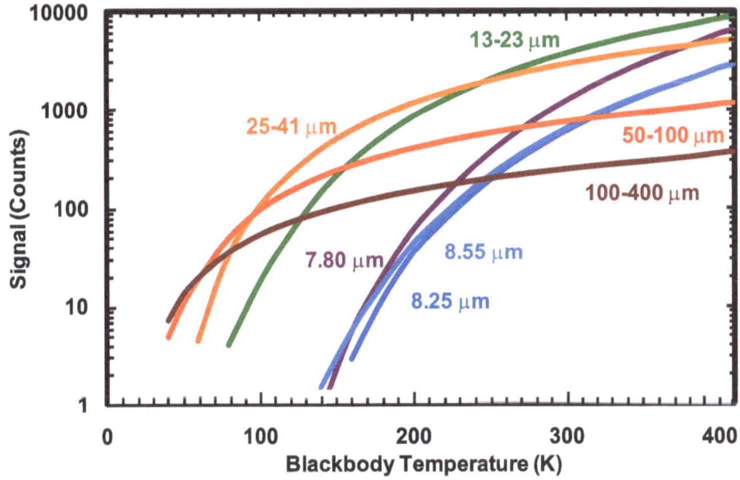

Fig. 17 Signal-to-noise ratio as a function of blackbody temperature for the thermal and 8 μm channels

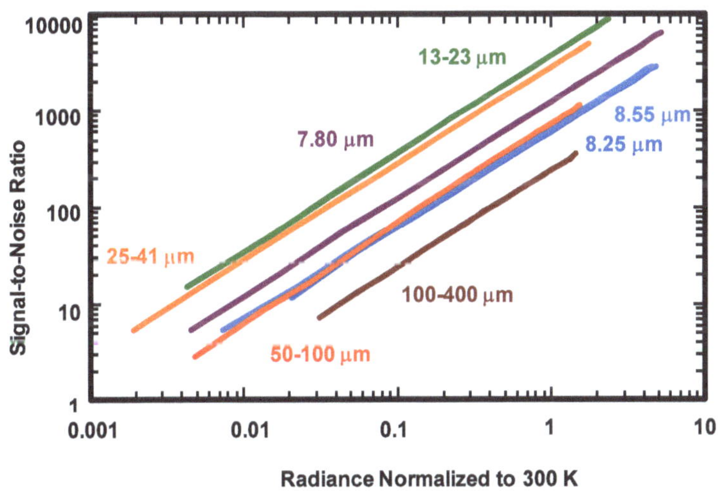

Fig. 18 Signal-to-noise ratio as a function of in-band radiance normalized to 300 K for the thermal and 8 μm channels

external blackbodies and the internal blackbody multiple times. Using the fixed blackbody signal as a transfer standard, the signal at each variable blackbody temperature was referenced to the signal for a variable blackbody temperature of 20 K, for which the signal in all channels is effectively zero.

The output signal-to-noise ratio from the instrument as a function of blackbody temperature for the thermal and 8 μm channels is shown in Fig. 17. The same data set is plotted in Fig. 18 with radiance normalized to 300 K on the horizontal axis. Each trace in these figures represents the average of all 21 detectors in each spectral channel.

The radiometric calibration was performed over a range of instrument temperatures. The thermopile detectors have little temperature dependence. Changing the instrument tempera-

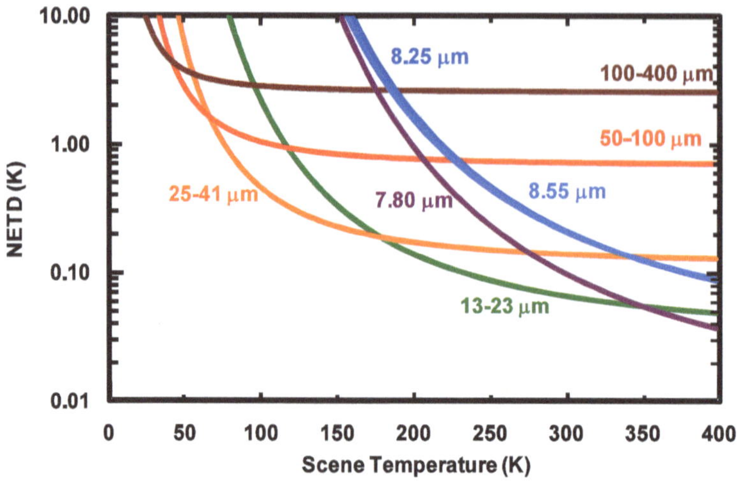

Fig. 19 Noise equivalent temperature for a single detector in each of the 8 μm and thermal channels

ture changes the background radiation exchange between the detectors, the baffles, and the scene. In flight, the rate of this offset change is slow due to the instrument thermal design, and is removed by views of space every 10 minutes. Changes in detector and electronics gain are calibrated and removed by viewing the internal blackbody target in conjunction with the space views.

4.2 Noise and Noise Equivalent Temperature Difference

The detector noise level at the output of the signal chain is 2.5 counts with a standard deviation between pixels of 0.1 counts. The noise-equivalent temperature difference (NETD) for a single pixel in each of the thermal and 8 μm channels is shown in Fig. 19.

4.3 Radiometric Accuracy

The uncertainty in a single radiometric measurement has components due to both offset and gain. Offset uncertainties in flight will be due to a combination of detector/electronics noise plus uncertainties in the interpolation of signals from space and blackbody looks at 10 minute intervals. A simulation of orbital conditions during thermal vacuum testing suggests that the uncertainty due to interpolation between space looks is roughly 2.5 counts. Assuming this uncertainty to be random, it can be combined with the uncertainty in detector/electronics noise (2.5 counts) by taking the square root of the sum of the squares, to result in a total offset uncertainty of 3.5 counts. This offset uncertainty is the dominant uncertainty source when looking at cold scenes.

There are numerous sources of gain uncertainty. During the ground radiometric calibration, uncertainty contributions include blackbody temperature measurements, non-unity blackbody emissivity, detector/electronics noise, and interpolation between points at different blackbody temperatures. In flight, additional uncertainties will occur from temperature measurement and non-unity emissivity in the internal calibration blackbody, detector/electronics noise and nonlinearity, and interpolation between gain measurements (space and blackbody looks) at 10 minute intervals. A complete and conservative analysis of these

Table 3 Estimated in-flight gain uncertainties in the 8 μm and thermal channels

Channel number	Wavelength (μm)	Gain uncertainty (%)
3	7.80	1.7
4	8.25	1.6
5	8.55	1.6
6	13–23	0.9
7	25–41	0.6
8	50–100	0.6
9	100–400	1.1

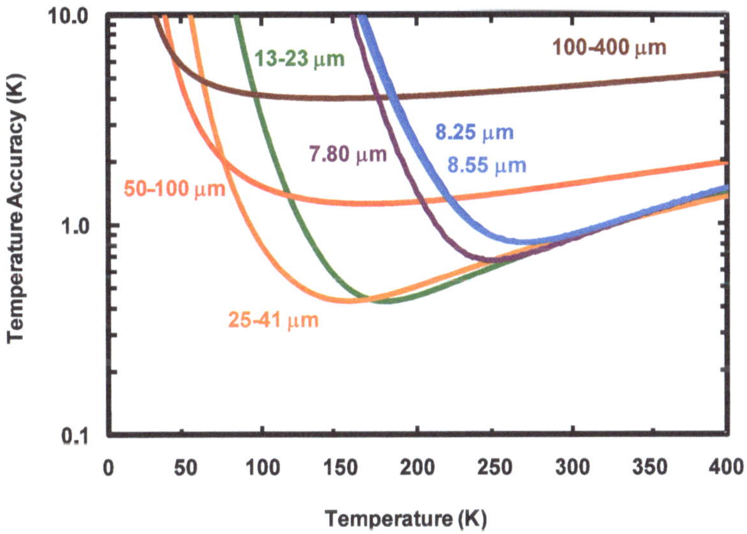

Fig. 20 Estimated temperature accuracy in 8 μm and thermal channel measurements

various sources of uncertainty for the 8 μm and thermal channels results in the estimated gain uncertainties listed in Table 3. Gain uncertainty dominates when looking at hot scenes.

The gain uncertainties in the 8 μm channels will track each other because they are mostly due to ground and in-flight calibration uncertainty. Thus the uncertainty in the ratios of signals between these three channels will be much lower than the values shown in Table 3. As determining the location of the Christiansen feature relies on these ratios rather than the absolute signals, the values used in this determination will be much more accurate than indicated in Table 3.

Figure 20 shows estimated temperature accuracy in the 8 μm and thermal channels. The decrease in accuracy at low temperatures is due to offset uncertainty, while the decrease in accuracy at high temperatures is due to gain uncertainty. The results shown in Fig. 20 suggest that Diviner will acquire low noise, high accuracy radiometric measurements in multiple spectral channels over the full range of expected lunar surface temperatures. Higher signal-to-noise ratios may be obtained by averaging signals spatially or temporally, though accuracy is not always improved by this process due to systematic errors.

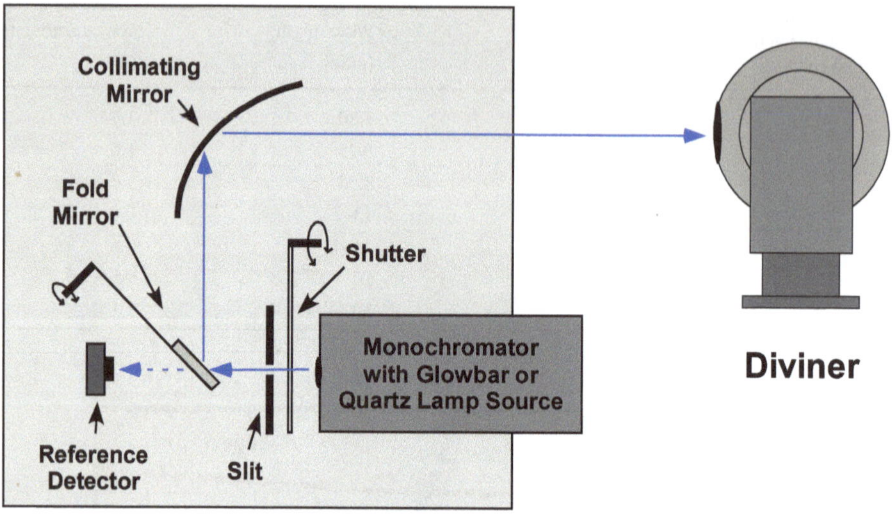

Fig. 21 Schematic diagram of spectral response measurement setup. The signal from the hot glowbar or quartz lamp goes through a slit, reflects off a fold mirror, and is collimated. A shutter behind the slit is used to block the source for background correction. The fold mirror can move to allow the signal to fall on a spectrally flat reference detector

4.4 Spectral Response

The spectral response in the solar channels, the 8 μm channels, the 13–23 μm channel, and the 25–41 μm channel was measured in the laboratory at the instrument level using a grating monochromator source (Fig. 21). The signal from the monochromator output slit was collimated, and then focused with the instrument optics onto a single detector in the instrument. A shutter behind the monochromator exit slit allowed subtraction of background signals. A fold mirror directed the monochromator signal either to the Diviner instrument or to a spectrally flat reference detector. A quartz lamp radiation source allowed measurements for 0.2–4.2 μm. A hot glowbar source was used for 2.6–56 μm.

For measurements of the solar channels, eight scans were needed to cover the spectral range 0.2–4.2 μm using the quartz lamp source. Between scans the gratings and/or order-sorting filters were changed. An additional scan from 3.6–6.8 μm was performed with the glowbar source. Within a scan, the relative spectral response shape is accurate. However, due to changing the gratings and order-sorting filters, adjacent scans are not expected to match exactly in the overlap region. To correct these mismatches, the data from each scan were manually shifted by as much as 50% in amplitude to form a complete seamless data set. The resulting errors of a few percent at the seams will introduce some error in the overall passband shape. Three detectors in the solar channels were measured over the full spectral range. These measurements were combined to form a single spectral response curve, shown in Fig. 22.

Figure 22 shows a significant spectral leak in the solar channels above 3 μm. Thus for warm scenes a fraction of the signal in the solar channels will be from thermal emission. Figure 23 shows an estimate of this thermal leak as a function of temperature. For this calculation, the lunar surface was assumed to be in radiative equilibrium with albedo 0.1. At 400 K roughly 3% of the signal in the solar channels will be due to thermal emission. To achieve accuracy in measurements of reflected solar radiation, a partial correction for this

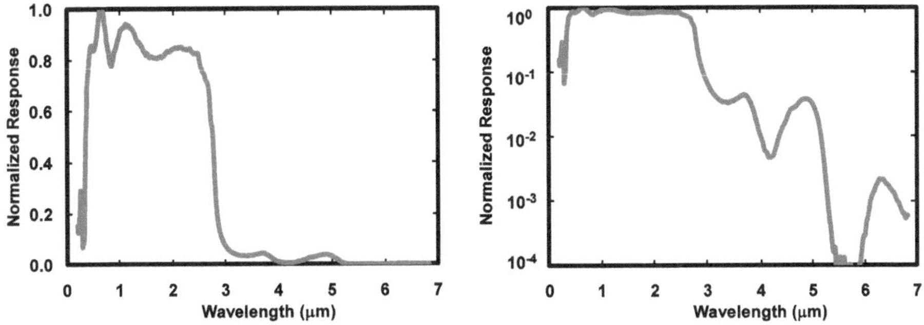

Fig. 22 Normalized spectral response of Diviner solar channels, *linear vertical scale* on *left* and *logarithmic vertical scale* on *right*

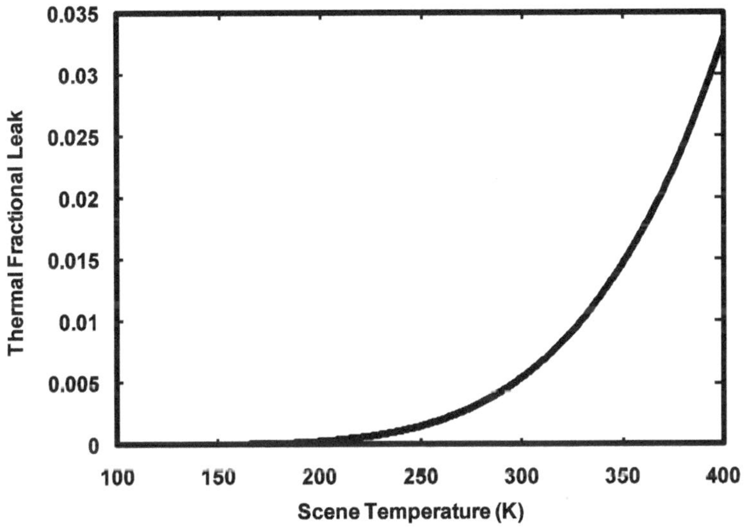

Fig. 23 Estimated contribution of thermal emission to signal in solar channels. Lunar surface is assumed to be an isotropic emitter and reflector in radiative equilibrium, with a solar spectrum averaged albedo 0.1

effect can be made using knowledge of the surface temperature from the thermal and 8 μm channel signals.

The spectral response of the 8 μm channels was measured from 6.5–9.8 μm using a single monochromator grating and order-sorting filter, eliminating the need to stitch together data from multiple scans. Five equally-spaced detectors across each of the three arrays were measured. The passband variation across the detectors in a single channel was 25 nm or smaller. Because accurate spectral knowledge in these bands is critical for the determination of Christiansen feature location, a calibration check was performed on the monochromator in this wavelength region. A thin sheet of polypropylene was used as a transfer standard. The polypropylene was first measured with a Fourier transform spectrometer using an internal helium-neon laser reference. Then the output of the monochromator was measured using its reference detector, both with and without the polypropylene sheet in the beam path. The spectral features in the ratio of these two monochromator data sets matched the spectrometer

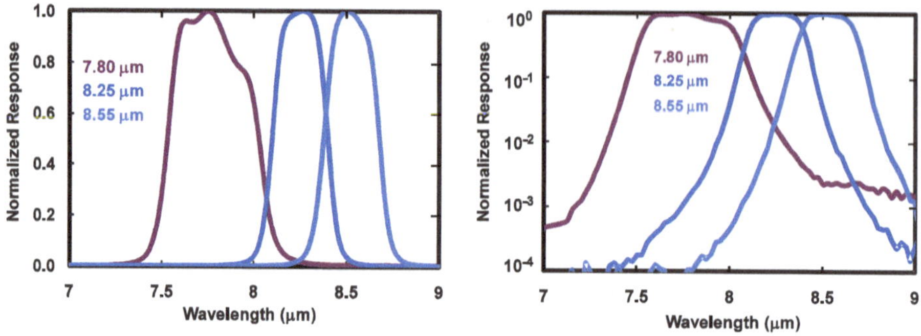

Fig. 24 Normalized spectral response of Diviner 8 μm channels, *linear vertical scale* on *left* and *logarithmic vertical scale* on *right*

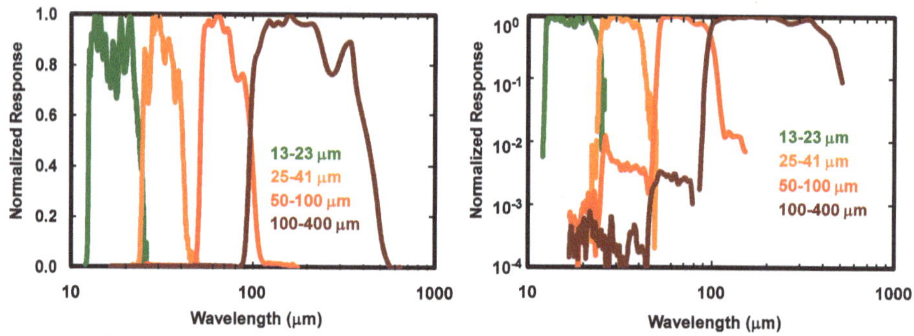

Fig. 25 Normalized spectral response of Diviner thermal channels, *linear vertical scale* on *left*, *logarithmic vertical scale* on *right*

polypropylene data to within a few nanometers, demonstrating sufficient monochromator spectral accuracy for this purpose. Figure 24 shows the spectral response of the three 8 μm channels.

Spectral response measurements for the thermal channels were challenging due to limitations in monochromator signal at long wavelength. For this reason, only the 13–23 and 25–41 μm channels were measured at the instrument level. For these channels, data from multiple spectral ranges and multiple detectors were stitched together as in the solar channel measurements. For the 50–100 μm and 100–400 μm channels, the spectral transmission of the assembled filter stacks was measured with an f-number beam similar to the instrument. The resulting spectral response curves are shown in Fig. 25. The detector response is uncertain beyond 200 μm and may very well roll off in this region. Thus the actual spectral response of the 100–400 μm channel is likely to roll off at long wavelength more quickly than shown in Fig. 25. Because the Planck distribution also rolls off in this range, the uncertainty in instrument response at long wavelength does not significantly affect its response to scene temperature.

The shapes of Diviner's radiometric response curves are in good agreement with model calculations of Diviner's radiometric response based on its measured spectral response. This provides independent evidence that there are no significant unanticipated spectral leaks that might complicate the interpretation of Diviner's thermal emission measurements.

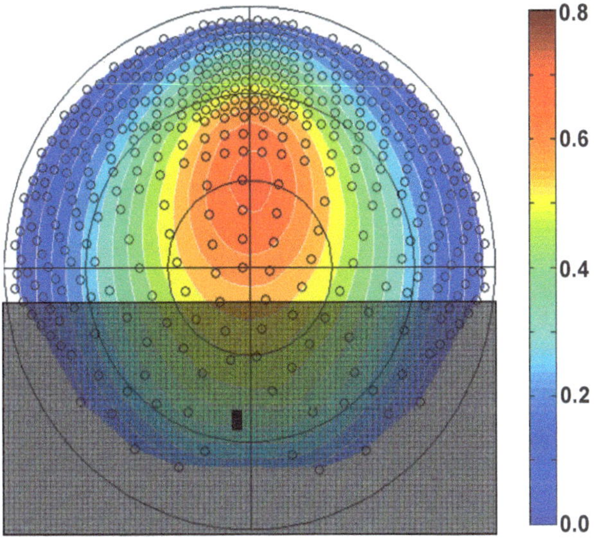

Fig. 26 Polar plot showing the solar-spectrum averaged reflectance of the Diviner solar calibration target relative to a normally illuminated perfectly reflecting diffuse reflector as a function of solar incidence angle (shown from 0 to 90°), and the solar azimuth angle. The *open circles* show the angles sampled by the goniometer measurements. The *gray shaded region* approximates the incidence and azimuth angles that are blocked by the Diviner instrument. The *solid black rectangle* shows the fixed viewing geometry of Diviner's solar channel detectors relative to the solar calibration target in emission and azimuth angle space. The solar calibration target was measured using six different fixed viewing geometries during pre-flight calibration

4.5 Solar Target Calibration

The hemispheric spectral reflectance of the Diviner solar calibration target was measured using a spectrophotometer coupled to an integrating sphere. The bi-directional reflectance of the solar calibration target was measured with a goniometer in 10 spectral channels in the wavelength region of 0.402–1.061 μm at the same solar illumination angles and viewing geometry anticipated during flight. Diviner views the solar calibration target with constant viewing geometry so emission angle is effectively fixed, while solar incidence and azimuth angles can vary widely over the course of an orbit, and over the course of the mission. Figure 26 shows the solar spectrum-integrated bi-directional reflectance of the Diviner solar calibration target derived from the spectrophotometer and goniometer observations.

4.6 Instantaneous Field of View

The instantaneous field of view of each detector was measured in the laboratory using a target projector (Fig. 27). The target projector signal is generated by a hot target with a black painted surface. A slit wheel in front of the target allows one to change between narrow and wide vertical and horizontal slits without breaking vacuum. The image of the slit is collimated by a set of two mirrors. The target and slit are mounted on a sled, which has high-precision 3-axis actuators.

Two types of scans were performed with the Diviner instrument. For low-resolution scans the target projector sled stayed in a fixed position. These scans were performed by stepping the instrument actuators along the axis being measured. Each actuator step corresponds to 0.101° or 1.76 mrad. High-resolution scans were performed over a narrower range of angles by fixing the Diviner actuator position at the center of the desired range and stepping the sled along the desired direction. Step sizes for high-resolution scans corresponded to 0.39 mrad for in-track scans and 0.19 mrad for cross-track scans.

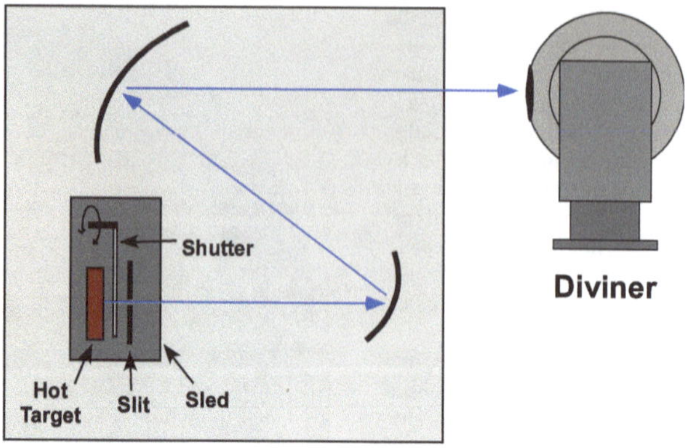

Fig. 27 Target projector setup for instantaneous field-of-view measurements

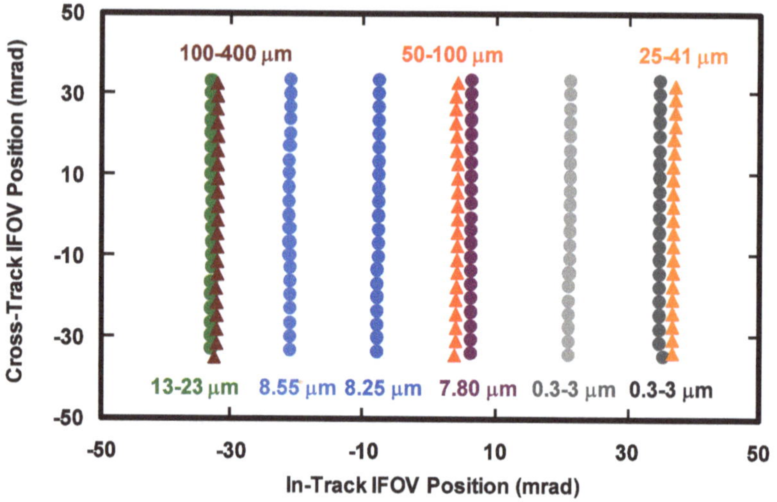

Fig. 28 Relative positions of field-of-view peaks for all Diviner detectors

Figure 28 shows a pixel position map in angular space, constructed from the peak positions determined from the low-resolution scans.

Figures 29 through 34 show the results of the high-resolution field-of-view scans. Each of these curves represents an average of all 21 detectors in a channel. For interpretation of lunar data, however, the unique instantaneous field of view for each detector will be used except in the 100–400 μm channel, where the signal-to-noise ratio in the field-of-view characterization is insufficient. For each group of channels (solar, 8 μm, and thermal), the fields of view are plotted both cross-track and in-track with both linear and logarithmic vertical scales.

In nearly all of the cross-track scans (see Figs. 29, 31 and 33), shoulders can be seen with a magnitude roughly 5% of the central peak. These shoulders are due to thermal crosstalk between adjacent detectors. Multiple shoulders of decreasing magnitude are present in some of the cross-track plots. The in-track fields of view of the reduced-sensitivity solar channel (Fig. 30) and the 13–23 μm channel (Fig. 34) are narrowed due to the reduced apertures over these spectral filters.

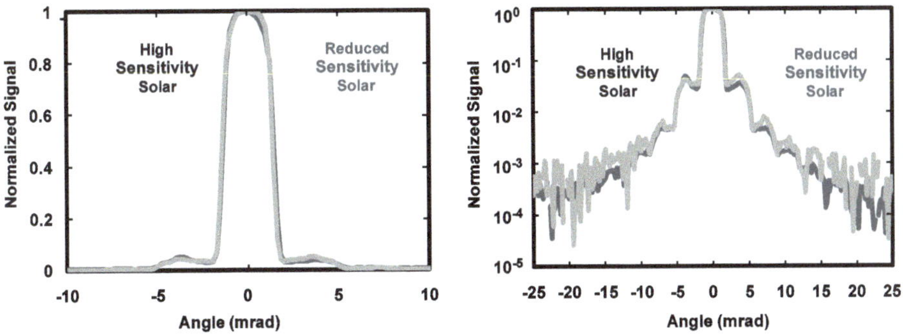

Fig. 29 Cross-track instantaneous field of view of Diviner solar channels, plotted with both linear and logarithmic scales

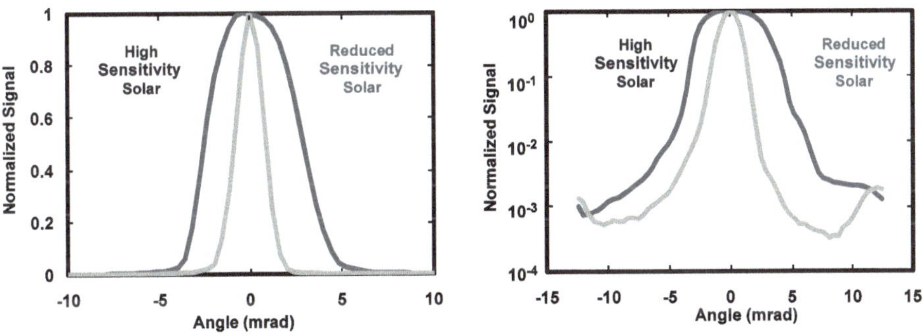

Fig. 30 In-track instantaneous field of view of Diviner solar channels, plotted with both linear and logarithmic scales

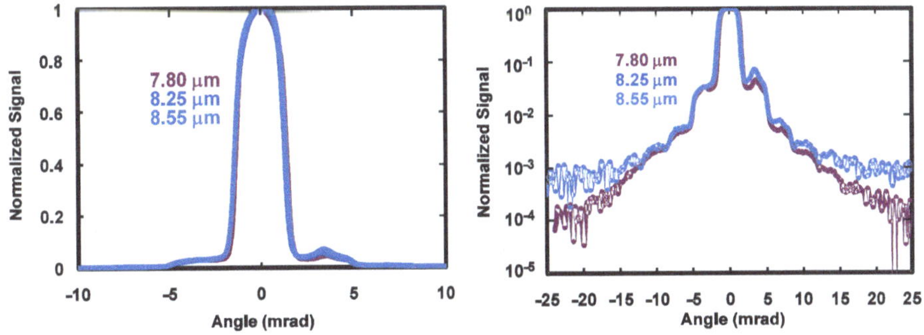

Fig. 31 Cross-track instantaneous field of view of Diviner 8 μm channels, plotted with both linear and logarithmic scales

The three longest wavelength channels have the broadest fields of view (Figs. 33 and 34). Two effects in the mesh filters used in these channels contribute to this broadening—scattering and filter heating. Radiation transmitted through these mesh filters is scattered somewhat due to the filtering mechanism. In addition, the poor thermal conductivity of the polypropylene mesh support structure allows heating and cooling of the filters due to

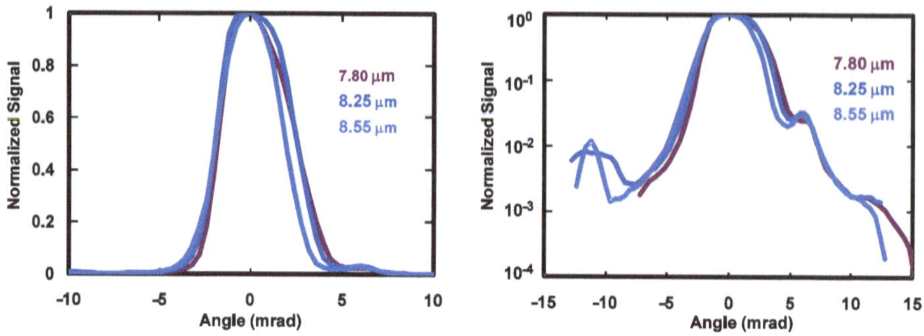

Fig. 32 In-track instantaneous field of view of Diviner 8 μm channels, plotted with both linear and logarithmic scales

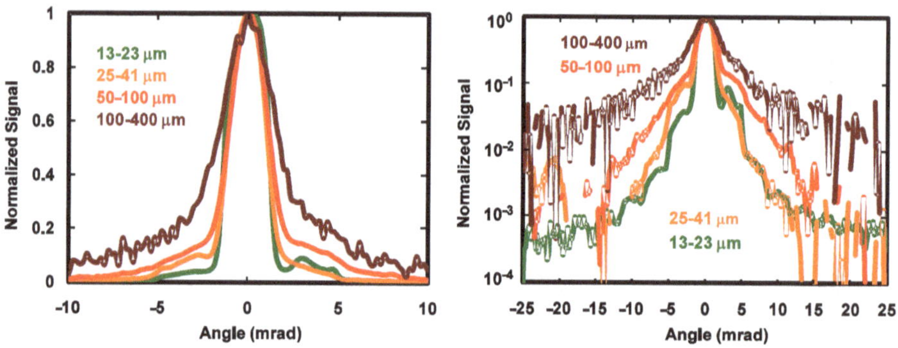

Fig. 33 Cross-track instantaneous field of view of Diviner thermal channels, plotted with both linear and logarithmic scales

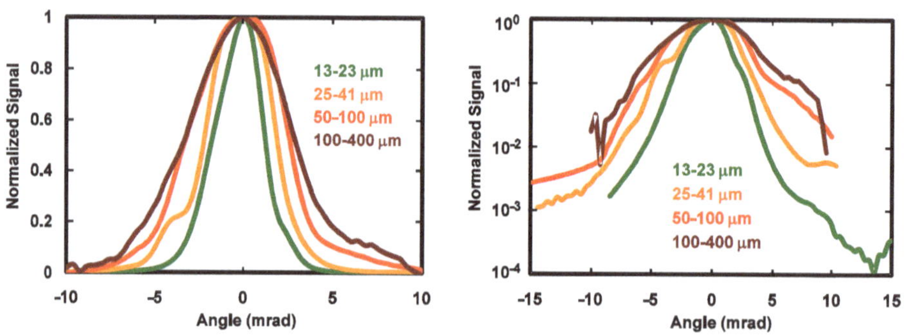

Fig. 34 In-track instantaneous field of view of Diviner thermal channels, plotted with both linear and logarithmic scales

changes in incident radiation. These temperature changes are conducted along the mesh structure, producing re-emission into neighboring pixels. Short-wave blocking elements in these filter stacks decrease these two effects significantly but do not completely eliminate them. Note that the increased broadening at the longest wavelengths is significantly more than would be expected due to diffraction.

Additional field-of-view information will be obtained in orbit by scanning across the lunar limb.

4.7 Effective In-Track Field of View

The effective field of view in the in-track direction is broadened by two effects: spacecraft movement during the 128 ms sample integration time, and the approximately 110 ms exponential detector thermal response time. The effective in-track field of view is the convolution of the instantaneous field of view (IFOV, shown in Figs. 30, 32 and 34) with a triangular function representing the integration time and with an exponential function representing the detector response time. The ground speed is approximately 1.66 km/s, varying only slightly with orbital altitude. Thus the smear caused by the integration time and detector response time corresponds to roughly 200 m on the ground. The footprint of the IFOV also varies linearly with orbital altitude. Therefore broadening of the in-track field of view is relatively larger at low altitude than it is at high altitude.

Note that in the cross track direction no field-of-view broadening occurs. The two-dimensional effective field of view is oval, elongated in track due to the combination of a rectangular pixel, integration time, and detector response time.

Figure 35 shows the in-track instantaneous field of view for the 8.25 μm channel as well as the two functions representing broadening due to integration time and detector response time for a 50 km altitude. All three functions are normalized to have a maximum value of one. The direction of the response time broadening depends on the spacecraft velocity direction. When the spacecraft performs a 180° yaw maneuver every six months to keep the Sun on the same side, the velocity reverses with respect to the IFOV. The normalized

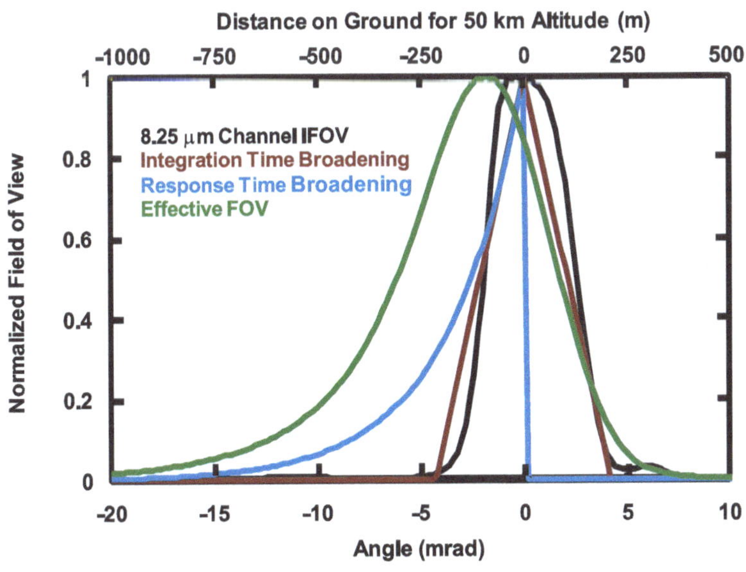

Fig. 35 8.25 μm channel in-track IFOV plotted with broadening functions due to integration time and detector response time for a 50 km altitude. The *green curve* is the effective field of view resulting from the convolution of the other three functions

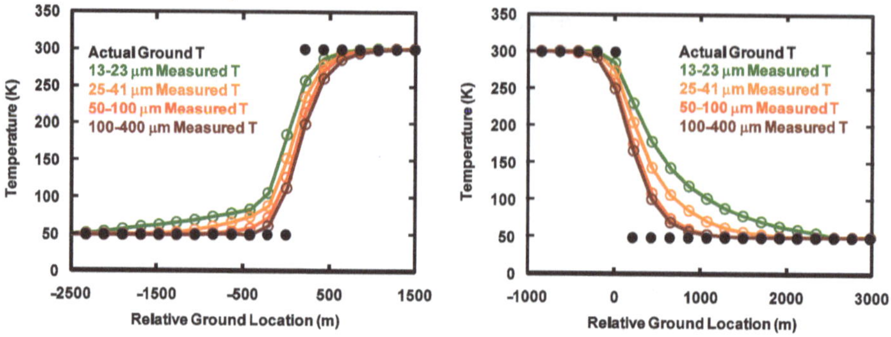

Fig. 36 Simulated Diviner thermal-channel response to a temperature step from 50 to 300 K (*left*) and from 300 to 50 K (*right*). Direction of travel is from *left* to *right*

convolution of these three functions (effective FOV) is also shown in Fig. 35. Note that the effective field of view for a given pixel is not fixed with time, but rather depends on the spacecraft altitude and velocity vector.

4.8 Response to a Step Function in Scene Temperature

With knowledge of the effective field of view of the pixels in each channel, one can calculate the instrument response to a step function in temperature. A large, sudden step in scene temperature could occur at a crater rim for example, in which one side is illuminated by the Sun and the other side is in shadow. While looking at one side of this step, part of each pixel's effective field of view sees the much different temperature on the other side of the step. The contamination caused by the nearby scene depends both on the magnitude of the field-of-view tails and on the way the response of that spectral channel depends on wavelength.

Figures 36 shows simulated Diviner data of in-track temperature transitions from 50 to 300 K (left) and from 300 to 50 K (right). The two results are not mirror images of each other because of the asymmetry in effective field of view due to detector response time (Fig. 35). The shortest wavelength channel (13–23 μm) has the most leakage of the warm signal into the cold region. This channel is the most nonlinear as a function of temperature, with response strongly weighted towards the higher temperatures (see Fig. 17). The longest wavelength channel, in contrast, has nearly linear response as a function of temperature. This channel has roughly equal leakage in both directions.

The Diviner's spatial response has been sufficiently well characterized to enable confident localization of surface thermal features at spatial scales that approach the field of view of the instrument. Enhanced ground processing may make it possible to deconvolve the angular and temporal smear that is inherent in the raw observations to provide even more spatially resolved and more accurate temperature maps of regions of high scientific interest.

5 Data Analysis and Interpretation

5.1 Standard Data Products

The Diviner team will create a series of standard archived data products for NASA's Planetary Data System as shown in Table 4. Level 0 and Level 1 data products will be archived

Table 4 Diviner archived data products

Archived data product	Description
Pre-flight Calibration Data	Pre-flight calibration data (Spectral response, blackbody response, solar target reflectance, fields of view, alignment)
Experimenter's Notebook	Chronologic text description of instrument operation and performance
Level 0	Time-sequenced raw science and housekeeping data
Level 1	Calibrated radiances and housekeeping data merged with project-supplied geometry and timing data
Level 2	Gridded global brightness temperature (latitude, longitude, local time, season)
Level 3	Grided derived global fields: Lambert albedo, surface temperature, fine-component thermal inertia, rock abundance, spectral emissivity
Level 4	Special polar products: Maps of permanent shadow, illumination levels, surface and subsurface temperatures

within 6 months of acquisition. Level 2, 3 and 4 data products will be archived after the end of the one-year LRO mapping mission. Diviner's archived data products will be in the form of ASCII text files.

5.2 Calibrated Radiances

Diviner's Level 1 calibrated radiance product will be the starting point for the creation of all higher-level data products. Diviner's calibration approach is similar to that employed by MCS.

Diviner radiance measurements in 42 visible and 147 infrared detector channels will be acquired in the form of raw signal counts once every 128 msec. Raw counts are converted to calibrated radiances using a two-point calibration procedure, based on views of space, instrument blackbody targets, and solar calibration targets. Space views provide good zero signal, or offset, calibration for all channels, whereas blackbody views provide an accurate radiance standard for the infrared channels, allowing gain to be calculated. For the visible channels, the reflectivity along the line of sight can be related directly to solar target reflectivity, which has been characterized as a function of solar elevation and azimuth angle during laboratory calibration.

For the infrared detector channels, space and blackbody views are obtained at approximately ten equally-spaced times per orbit, and are used to calculate signal offset and gain. Interpolation between these discrete calibration points will be done using a simple polynomial fit. Signal counts for each channel are then converted linearly to a preliminary radiance by interpolating offsets and gains between calibration times. Finally, a simple polynomial expression, derived from laboratory linearity measurements on the flight instrument, is used to convert preliminary to calibrated radiance. A similar approach is used for the visible channels. Again space views are available ten times per orbit, but the solar calibration target is observed only once or twice per orbit when the position of the Sun is favorable. Linearity measurements are not available for this channel, but its radiometric accuracy and precision requirements are met by assuming a linear response. Both gain and offset are influenced by instrument temperature variations around the LRO orbit, and the calibration approach is designed to correct for this, in addition to longer term drifts.

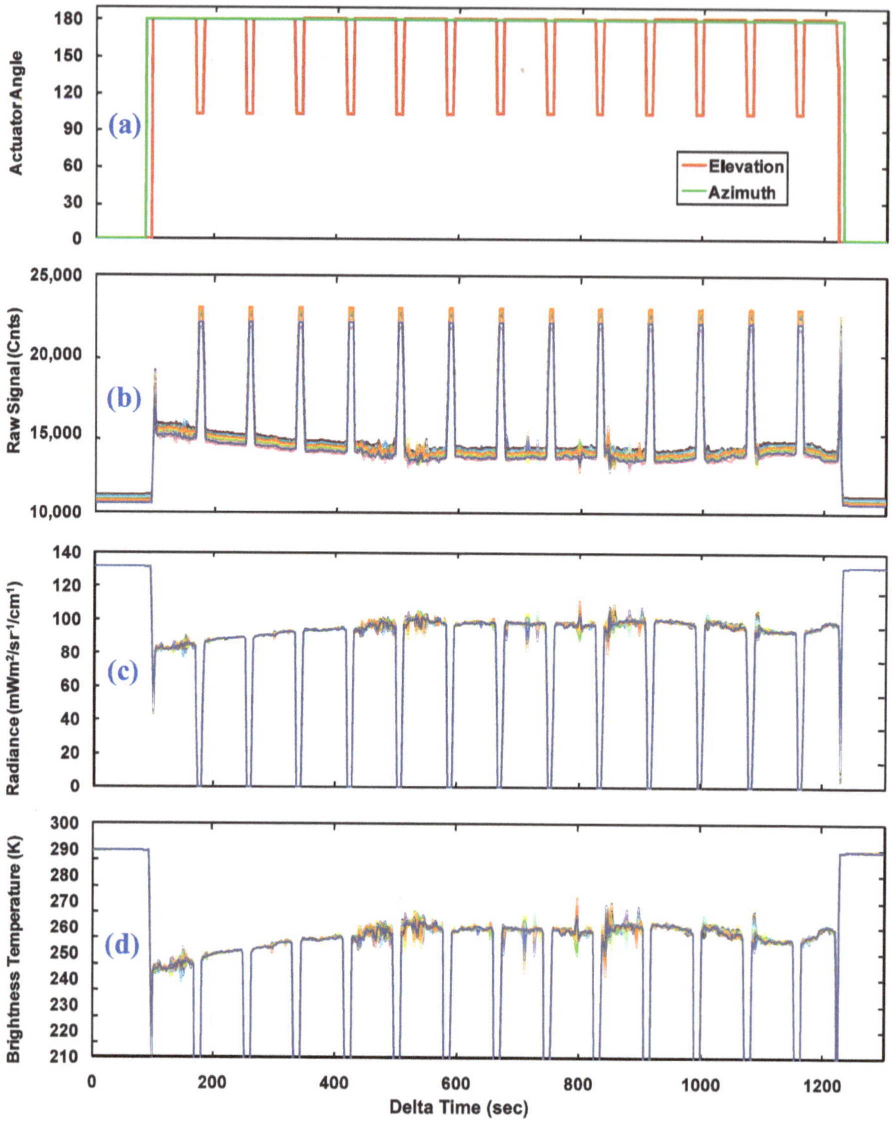

Fig. 37 Raw and calibrated MRO MCS 21.6 μm channel observations of the surface of Mars acquired in nadir pushbroom mapping mode. The data shown here were acquired during a 20 minute period in November, 2006 along a single MRO orbit track along ~150° east longitude extending from 8.3° S to 50.9° N latitude at approximately 15.25 hours Mars local time. Panel (**a**) displays instrument actuator azimuth (*green*) and elevation (*red*) angles. Elevation angles of 0°, 103° and 180° correspond respectively to blackbody, space and nadir views. Panel (**b**) shows uncalibrated signals in counts for the 21 detectors of the MCS 21.6 μm micron channel. The contrasting blackbody, space, and on-planet signals are evident, and the shifting of the curves relative to each other results from the different offsets and gains associated with each detector before calibration. Panel (**c**) shows calibrated radiances for the same time period and detector channels. The curves are now superimposed except for surface temperature structure seen in some of the nadir views. Panel (**d**) shows brightness temperatures corresponding to the radiances of panel (**c**). Martian surface temperatures vary over the range 240–270 K in this time period. Contrast across the detector arrays is caused primarily by solar illumination angle. Flat, featureless terrain produces very little contrast

The calibration procedure that Diviner will employ for the Moon is similar to that illustrated in Fig. 37, which shows actuator angles, raw counts, radiances and brightness temperatures acquired by MCS for the surface of Mars in its 21.7 μm channel.

5.3 Higher-Level Data Products

Diviner's Level 2 data product is a spatially and temporally averaged version of the global calibrated dataset for mapping and correlation with mapped data products provided by other instruments. The Level 3 and 4 data products are created by fitting the Diviner data with the results of models for the thermal behavior of the lunar surface and subsurface, the spectral emission and solar reflectance of the lunar surface, and maps of the Moon's topography and roughness. The approaches that will be employed to derive the Level 3 and 4 products will be described in detail in future publications.

6 Summary

The Diviner lunar radiometer experiment will provide comprehensive and rich global datasets that will support the exploration and science goals of the LRO mission. The instrument will be the first to systematically map the thermal emission of the lunar surface over the complete range of latitude, longitude, local time and season. The results of the investigation will be applicable to better defining and understanding the lunar thermal environment, the geologic history and composition of the lunar surface, and the distribution and history of lunar volatiles.

Acknowledgements We thank the following individuals for their contributions to the Diviner experiment: Henry Awaya and Nick Emis—Thermal Design, John Bousman—Mechanical Integration, Scott Nolte—Electronics and Integration support, Bradley Drake—Electronics assembly, Robert Hughes—Thermal Analysis, Peter Barry, Kim Plourde—Mission Assurance, Nick Taylor—Reliability, James Aragon—Quality Assurance, Robert Stephenson—Structures, Mike O'Connell—Dynamics, Todd Newell—Environmental Requirements, Valsamual Galaske—Mechanical Design, Paul Willis—Materials, Yuri Beregovski—Optics, Orland Harrison, Lee Wigglesworth—Electrical Ground Test Equipment, Albert Chen—Electronics, Cami Vongsouthy—Safety, Kelly Stanford—Electronic Parts, Tom Pierce—Electrical Compatibility, John Diehl—Ground Data Systems, Glenn Aveni—Contamination Control, Joyce Grunwald—Contract Manager, Laurie Guay—budgeting, Tarek Baayon and Amy Liu—Scheduling, Paul Cate—Procurement Support, Jon Temple—Design and Manufacture, Jason Perry—Filter Frame Development, Gary Hawkins and Richard Sherwood—Multilayer Filters, Peter Ade and Carol Tucker—Mesh Filters, Stewart Biggar—Goniometry, Mike Jakobson—Spectrophotometry, Paul Hayne, Michaela Shopland, Matt Siegler, Tim Tran and Joann Zhang—Thermal Vacuum Testing Support, Young Park and Thomas Luchik—Management Support, Arlin Bartels—Payload Management. The research described in this paper carried out at the Jet Propulsion Laboratory, California Institute of Technology, was done so under a contract with the National Aeronautics and Space Administration.

References

C.J. Cremers, R.C. Birkebak, in *Proc. Lunar Sci. Conf. 2nd* (1971), pp. 2311–2315

M.B. Duke, L.R. Gaddis, G.J. Taylor, H.H. Schmitt, Rev. Mineral. Geochem. **60**, 597–655 (2006)

W.C. Feldman, S. Maurice, D.J. Lawrence, R.C. Little, S.L. Lawson, O. Gasnault, R.C. Wiens, B.L. Barraclough, R.C. Elphic, T.H. Prettyman, J.T. Steinberg, A.B. Binder, J. Geophys. Res. **106**, 23,231–23,251 (2001)

B.T. Greenhagen, D.A. Paige, Lunar Planet. Sci. **37** (2006), abstract 2406
B.T. Greenhagen, D.A. Paige, Lunar Planet. Sci. **40** (2009), abstract 2255
R.R. Hodges Jr., in *Proc. Lunar Planet. Sci. Conf. 11th* (1980), pp. 2463–2477
J.C. Jaeger, A.F.A. Harper, Nature **166**, 1026 (1950)
S.J. Keihm, M.G. Langseth Jr., in *Proc. Lunar Sci. Conf. 4th* (1973), pp. 2503–2513
S.L. Lawson, B.M. Jakosky, H.S. Park, Lunar Planet. Sci. **28** (1997), abstract 1375
S.L. Lawson, B.M. Jakosky, H.S. Park, M.T. Mellon, J. Geophys. Res. **105**, 4273–4290 (2000)
L.M. Logan, G.R. Hunt, J.W. Salisbury, S.R. Balsamo, J. Geophys. Res. **78**, 4983–5003 (1973)
D.J. McCleese, J.T. Schofield, R. Zurek, J.V. Martonchik, R.D. Haskins, D.A. Paige, R.A. West, D.J. Diner, J.R. Locke, M.F. Chrisp, W. Willis, F.W. Taylor, Appl. Opt. **25**(23), 4232–4245 (1986)
D.J. McCleese R.D. Haskins, J.T. Schofield, R.W. Zurek, C.B. Leovy, D.A. Paige, F.W. Taylor, J. Geophys. Res. **97**, 7735–7757 (1992)
D.J. McCleese, J.T. Schofield, F.W. Taylor, S.B. Calcutt, M.C. Foote, D.M. Kass, C.B. Leovy, D.A. Paige, P.L. Read, R.W. Zurek, J. Geophys. Res. **112**, E05S06 (2007). doi:10.1029/2006JE002790
W.W. Mendell, Ph.D. Thesis. Rice University, Houston, TX, 1976
W.W. Mendell, F.J. Low, Moon **9**, 97–103 (1974)
I. Mitrofanov, Space Sci. Rev. (2009, this issue)
F.H. Murcray, D.G. Murcray, W.J. Williams, J. Geophys. Res. **75**, 2662–2669 (1970)
B.C. Murray, M.J. Wildey, Astrophys. J. **139**, 734–750 (1964)
T.J. Nightingale, J. Crawford, Metrologia **28**, 233–237 (1991)
A.M. Ono, Ph.D. Thesis. University of California, Los Angeles, CA (1999)
E. Pettit, S.B. Nicholson, Astrophys. J. **71**, 102–135 (1930)
J.W. Salisbury, R.K. Vincent, L.M. Logan, G.R. Hunt, J. Geophys. Res. **75**, 2671–2682 (1970)
J.W. Salisbury, G.R. Hunt, L.M. Logan, in *Lunar Planet. Sci. Conf. 4th* (1973), pp. 3191–3196
L.D. Stimpson, J.W. Lucas, J. Spacecr. Rockets **7**, 1317–1322 (1970)
A.R. Vasavada, D.A. Paige, S.E. Wood, Icarus **141**, 179–193 (1999)
K. Watson, B.C. Murray, H. Brown, J. Geophys. Res. **66**, 3033–3045 (1961)

Space Sci Rev (2010) 150: 161–181
DOI 10.1007/s11214-009-9578-6

LAMP: The Lyman Alpha Mapping Project on NASA's Lunar Reconnaissance Orbiter Mission

G. Randall Gladstone · S. Alan Stern · Kurt D. Retherford · Ronald K. Black · David C. Slater · Michael W. Davis · Maarten H. Versteeg · Kristian B. Persson · Joel W. Parker · David E. Kaufmann · Anthony F. Egan · Thomas K. Greathouse · Paul D. Feldman · Dana Hurley · Wayne R. Pryor · Amanda R. Hendrix

Received: 10 October 2008 / Accepted: 6 October 2009 / Published online: 12 December 2009

Abstract The *Lyman Alpha Mapping Project* (LAMP) is a far-ultraviolet (FUV) imaging spectrograph on NASA's *Lunar Reconnaissance Orbiter* (LRO) mission. Its main objectives are to (i) identify and localize exposed water frost in permanently shadowed regions (PSRs), (ii) characterize landforms and albedos in PSRs, (iii) demonstrate the feasibility of using natural starlight and sky-glow illumination for future lunar surface mission applications, and (iv) characterize the lunar atmosphere and its variability. As a byproduct, LAMP will map a large fraction of the Moon at FUV wavelengths, allowing new studies of the microphysical and reflectance properties of the regolith. The LAMP FUV spectrograph will accomplish these objectives by measuring the signal reflected from the night-side lunar surface and in PSRs using both the interplanetary HI Lyman-α sky-glow and FUV starlight as light sources. Both these light sources provide fairly uniform, but faint, illumination. With the expected LAMP sensitivity, by the end of the primary 1-year LRO mission, the SNR for a Lyman-α albedo map should be >100 in polar regions >1 km^2, providing useful FUV constraints to help characterize subtle compositional and structural features. The LAMP instrument is

G.R. Gladstone (✉) · K.D. Retherford · R.K. Black · D.C. Slater · M.W. Davis · M.H. Versteeg · K.B. Persson · T.K. Greathouse
Southwest Research Institute, 6220 Culebra Road, San Antonio, TX 78238, USA
e-mail: rgladstone@swri.edu

S.A. Stern · J.W. Parker · D.E. Kaufmann · A.F. Egan
Southwest Research Institute, 1050 Walnut Street, Suite 400, Boulder, CO 80302, USA

P.D. Feldman
The Johns Hopkins University, 3400 North Charles Street, Baltimore, MD 21218, USA

D. Hurley
The Johns Hopkins University Applied Physics Laboratory, 111000 Johns Hopkins Road, Laurel, MD 20723, USA

W.R. Pryor
Central Arizona College, 8470 North Overfield Road, Coolidge, AZ 85228, USA

A.R. Hendrix
Jet Propulsion Laboratory, 4800 Oak Grove Drive, Pasadena, CA 91109, USA

based on the flight-proven *Alice* series of spectrographs flying on the *Rosetta* comet mission and the *New Horizons* Pluto mission. A general description of the LAMP instrument and its initial ground calibration results are presented here.

Keywords Lunar · Ultraviolet · LRO · Lyman-α

1 Introduction

The *Lunar Reconnaissance Orbiter* (LRO) is the first of a series of missions planned by NASA to explore and characterize the Moon in anticipation of future human missions. The LRO mission is tasked with satisfying precursor exploration objectives that will support future robotic and human landed missions. The specific goals of LRO are to: (1) establish the lunar radiation environment, (2) obtain a 3-D topographic grid with high spatial resolution, (3) assess the resources available near the lunar poles, and (4) obtain a global inventory of elemental and mineralogical composition at high spatial resolution (Chin et al. 2007). With a nominal reconnaissance mission of 1 year, LRO will have a 3-axis stabilized, nadir-pointing attitude and will maintain a circular polar orbit at an altitude of 30–70 km above the lunar surface. The orbital period of LRO will be about 113 minutes, and the east longitude of the ascending node will decrease by $\sim 1^\circ$/orbit (~ 12–13°/day). A second year of science observations with LRO is planned by NASA's Science Mission Directorate (SMD); however, the spacecraft orbit during this mission phase is not yet finalized. Figure 1 illustrates LAMP's nominal nadir view of the lunar surface from the LRO spacecraft.

The objectives of the LAMP investigation are to (1) identify and localize exposed water frost in permanently shadowed regions (PSRs), (2) characterize landforms and albedos in PSRs, (3) demonstrate the feasibility of using natural starlight and sky-glow illumination for future lunar surface mission applications, and (4) characterize the lunar atmosphere and its variability. The LAMP instrument is a "close cousin" of the sensitive ($\sim$1 Rayleigh level), lightweight ($\sim$4 kg), low-power ($\sim$4 W) *Alice* imaging ultraviolet spectrograph (UVS) currently in flight on the ESA/NASA *Rosetta* comet orbiter (Stern et al. 1998; Slater et al. 2001) and NASA *New Horizons* missions (Stern et al. 2005, 2008; Slater et al. 2005). LAMP will be the first UVS to operate in lunar orbit since that on the Apollo 17 mission in 1972.

2 Scientific Objectives

It is commonly believed that the coldest sub-units of PSRs, where exposed volatiles are most likely to exist, are completely dark, since they are not even exposed to earthshine. However, this standard assumption that the coldest PSRs must be devoid of light is incorrect. Starlight provides a faint but detectable light source at UV, visible and infrared wavelengths. Moreover, at far-ultraviolet (FUV) wavelengths the entire sky is aglow, faintly illuminating the PSRs. The FUV light source illuminating all PSRs comes from resonance fluorescence of solar Lyman-α emission by neutral H atoms in the interplanetary heliosphere. This fluorescence generates a diffuse glow at the 121.6 nm H Lyman-α spectral line. The Lyman-α sky-glow has been studied extensively by spacecraft dating back over 35 years (e.g., Ajello et al. 1987; Pryor et al. 2008), and its deleterious (sputtering and photolyzing) effect on exposed lunar H_2O-ice has been appreciated for a long time as well (Morgan and Shemansky 1991). The brightness of the Lyman-α sky-glow as seen from Earth orbit varies

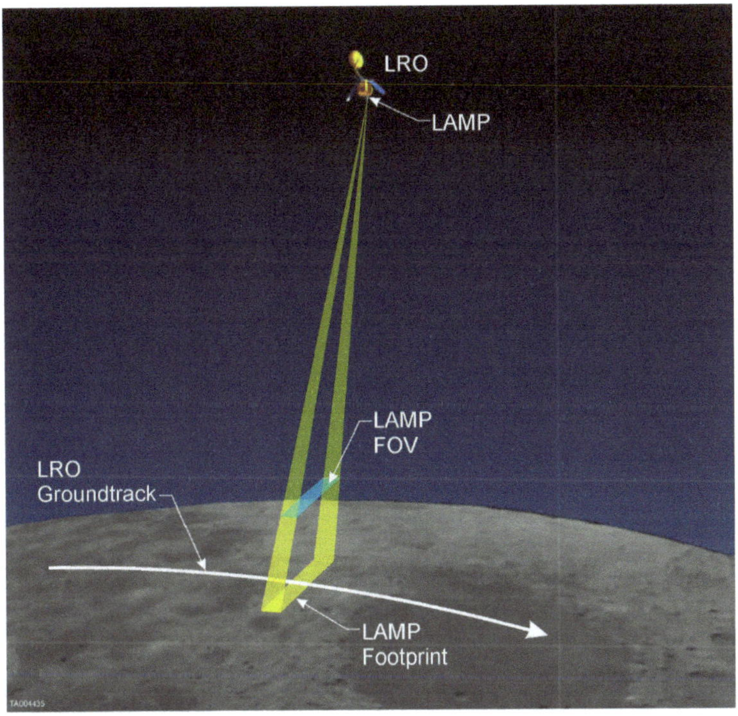

Fig. 1 LAMP's nadir-viewing geometry on LRO. The LAMP field of view of 0.3° × 6° results in a footprint of 260 m × 5.2 km from an altitude of 50 km

seasonally and with the Sun's activity, but is typically a few hundred Rayleighs, which corresponds to a characteristic flux of $\sim 10^8$ Lyman-α photons falling on each square centimeter of the lunar surface every second. Occasionally (e.g., when the Sun is active), the Lyman-α sky brightness can be 2–3 times brighter. Even at a level well below the typical 10^8 photons/cm^2/sec, the Lyman-α sky line produces sufficient signal for even the small LAMP instrument to see terrains in the Moon's permanently shadowed regions. In addition to the Lyman-α sky-glow, the cumulative light from UV-bright (typically B spectral class) stars also provides weak but detectable night-side illumination (e.g., Mathis et al. 1983; Henry 2002). These two sources of illumination, along with the ability to observe the sunlit lunar atmosphere above a dark surface, allow us to address LAMP's four main objectives.

2.1 Searching for Frost

The presence of an excess H signature in the lunar regolith, as detected by neutron spectroscopy (Feldman et al. 2001), cannot guarantee the hydrogen is in the form of water (Crider and Vondrak 2003; Hodges 2002), especially as there is no supporting radar signature (Simpson and Tyler 1999; Campbell et al. 2006). LAMP provides a way to spectrally fingerprint exposed H_2O frost in the lunar polar regions for the first time. This will be accomplished using observations illuminated purely by Lyman-α sky-glow and broadband FUV starlight. The lifetime of exposed water frost in polar cold traps is predicted to be short, owing to regolith gardening and sputtering processes (Morgan and Shemansky 1991; Crider and Vondrak 2003). However, even if this is true, thin surface films of recently deposited frost must exist in equilibrium set by the balance between source and loss rates to the pole, and their discovery would be a major find for LRO and an attractive prize for landed polar missions. Water ice has a diagnostic spectral reflectance signature at FUV wavelengths, as evidenced by laboratory studies (Wagner et al. 1987) and observed from icy satellites (Hendrix 1996; Hendrix and Hansen 2008). Depending on grain size and other factors, water frost shows a strong absorption edge at ~160–180 nm. At wavelengths less than the edge the reflectivity is ~2%, while at longer wavelengths the reflectivity is ~35%. Owing to this signature, FUV spectroscopy provides a way to directly fingerprint H_2O-frost exposed on the lunar surface, even at the relatively small (~4%) mixing ratios predicted by volatile transport models (Crider and Vondrak 2003) and the 1.5% level implied by *Lunar Prospector* neutron spectroscopy (Feldman et al. 2001). To make these detections LAMP will obtain spectral maps across the lunar surface, identify regions of suspected frost presence through their Lyman-α albedo (which will be lower than in drier surrounding regions), and retrieve H_2O frost abundances (or strong upper limits) by rationing PSR spectra above and below the 160–180 nm absorption edge to those of waterless ("dry") regions nearby the PSRs. Figure 2a shows the estimated flux from the lunar surface at LRO's nominal altitude, broken out by its major illumination sources. Figure 2b shows the calculated nightside observing time LAMP achieves after a 1-year LRO mission, as a function of resolution element size and latitude. Figure 2c shows a conservative estimate of the count spectra LAMP will achieve, as a function of wavelength, on surface units with and without H_2O frost. Note that neither the aggregate FUV starlight, nor the Lyman-α sky-glow, normally vary rapidly in time (solar flare times will be excluded from the data). The opposition effect (e.g., Hapke et al. 1998) may occasionally be important if UV-bright stars pass overhead in the final LRO orbit. Sky illumination brightness variations can be accounted for using data available from instruments now in space (e.g., SWAN on SOHO; Bertaux et al. 1997), supported by models of sky-glow Lyman-α (e.g., Pryor et al. 2008) and stellar FUV (e.g., Henry 2002) emissions.

2.2 PSR Mapping and Night Vision

It is well known that the brightness of starlight in the visible is sufficient to allow military and other dark-night applications using night vision technology. This fact can be exploited for lunar exploration, and LAMP will serve as a first demonstration of this capability, using UV starlight and Lyman-α sky-glow as its light source. Because the combined flux of UV starlight and Lyman-α sky-glow in LAMP's 57–196 nm passband is only about 17% of the total due to visible starlight, LAMP will prove the technique's feasibility for both UV *and* visible/NIR lunar polar/night vision systems with considerable margin. Such a low-level natural light vision system could provide vision capability for future rover and human exploration at night or in PSRs, and might do so with fewer of the power demands associated with operating artificial lighting at night or in polar regions.

The same data that LAMP collects to achieve the polar/night vision objective also satisfies its objective of obtaining landform maps of the PSRs. Using only FUV starlight and Lyman-α sky-glow, LAMP is capable of mapping even the darkest PSRs—where no reflected sunshine or earthshine is available whatsoever. Figure 3 shows signal-to-noise ratio (SNR) calculations that demonstrate the feasibility of achieving these LAMP objectives. These estimated SNRs were computed for typical lunar surface reflectivities (~4%) measured by the Astro-2 HUT and Apollo 17 UV instruments (Henry et al. 1995). Figure 3 shows that for illumination by Lyman-α sky-glow alone (i.e., neglecting FUV starlight, and in the complete absence of reflected sunshine or earthshine), LAMP will achieve SNRs of >30 per 200 × 200 m^2 resolution element near the poles (using the predicted effective area curve of Fig. 7). The SNR achieved by LAMP in regions illuminated by scattered sunlight

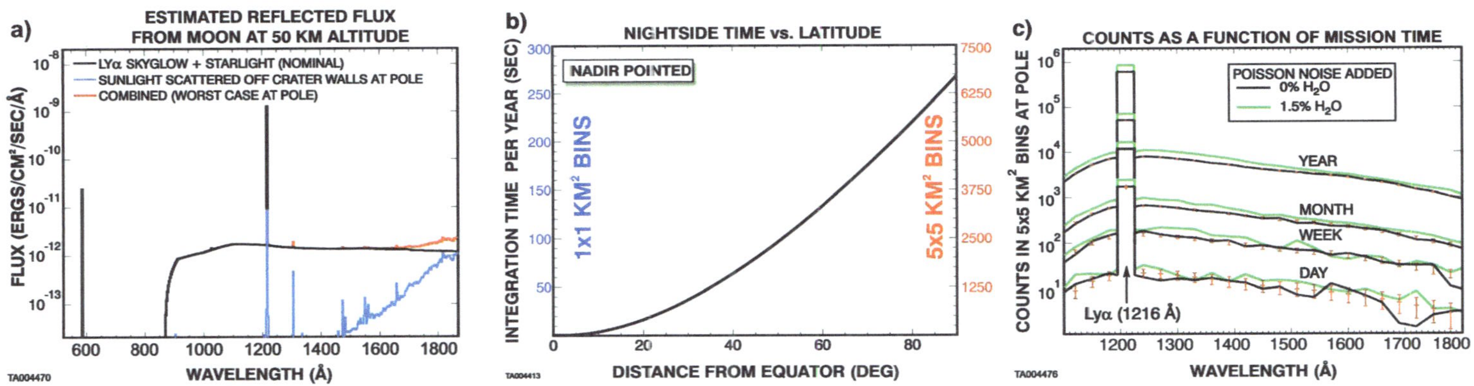

Fig. 2 (**a**) Estimated reflected FUV flux spectrum from the lunar night-side at LRO's nominal 50 km altitude. Sources of night-side flux assumed in this calculation are a conservatively low 300 R of Lyman-α sky-glow (Ajello et al. 1987), and the composite sky spectrum of UV starlight (Mathis et al. 1983). Regions near the terminator will be slightly contaminated with reflected UV sunlight (*blue curve*), producing a total estimated flux spectrum shown in *red*. (**b**) Calculated night viewing integration time for two mapping product resolutions, as a function of latitude, after 1-year of accumulated exposure time in an orbit at 50-km altitude. (**c**) Expected counts obtained in 5×5 km^2 bins, as a function of wavelength, after various elapsed mission times. *The black curves* are for dry soil; *the green curves* include 1.5% water frost abundance on the surface, using FUV frost albedos from Wagner et al. (1987). These count estimates assume the predicted effective area as shown in Fig. 7. Error bars are shown, but become difficult to see for mission times >1 month. After 1 month, the data quality is sufficient to allow regions with 1.5% areal mixing of water frost to be statistically distinguished from nearby dry lunar regolith. The signal at Lyman-α is plotted as a histogram for clarity

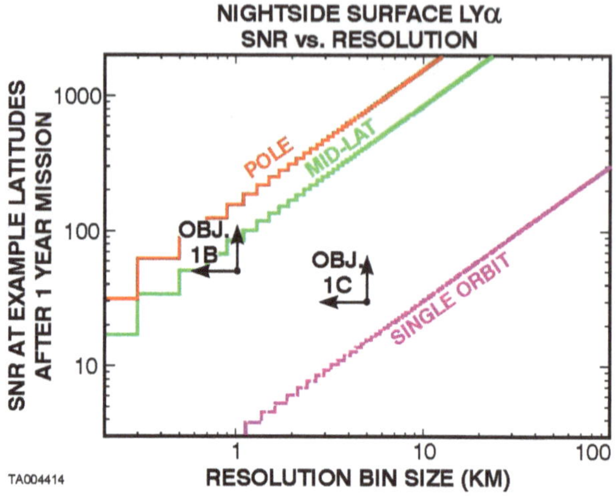

Fig. 3 Expected SNR at Lyman α, obtained after 1 year, as a function of resolution mapping bin size. These calculations assume the LAMP effective area curve shown in Fig. 7 below, the mission integrated observing time per latitude from Fig. 2b, and include the expected detector dark count rate at Lyman-α and Poisson noise. A single orbit's integration time curve is shown for comparison. Also shown are LAMP measurement requirements for SNR and spatial resolution for the landform mapping (1B) and night vision demonstration (1C) objectives

will be far higher, but other instruments on LRO can map these regions (albeit with possible complexities owing to the difficulties of scattered light control in mixed bright-dark scenes). LAMP data will also be used to compile FUV albedo maps of the entire Moon for the first time.

2.3 Lunar Atmosphere

The lunar atmosphere, with a total mass of order 40 tons, is one of many Surface Boundary Exosphere (SBE) atmospheres in the solar system (Stern 1999; Mendillo 2001; Killen 2002). Other SBEs of note include the atmospheres of Mercury, Europa, and Ganymede (plus regions of Io and Enceladus).

Species detected in (or from) the lunar atmosphere include Ar, Na, K, Al, Si, and O. Of these, Ar, detected by the Apollo 17 LACE mass spectrometer, appears to be an internal release product generated by K decay; Na and K are sputter- or meteoritic-vaporization products detected as atomic neutrals; O, Si, and Al are probably similarly derived but have to date only been detected as ions by instruments on the AMPTE and ISEE missions downwind of the Moon. LAMP's wavelength range allows it to follow already-detected atmospheric Ar (104.8, 106.6 nm) over repeated lunations to better characterize the sources and sinks. Argon is particularly interesting, since it mostly comes from the lunar interior and may be tied to specific release sites that could be of interest to future landing site selections (Schultz et al. 2006).

Because the atmospheric H distribution itself is a tracer of water vapor transport to the lunar poles (Crider and Vondrak 2003), maps of the time-averaged H distribution as a function of latitude may reveal an H gas deficit ("polar holes") at high latitude which could constrain the rate of water transport to the polar regions.

Although LAMP's sensitivity per unit time is somewhat less than the Apollo 17 UVS, LAMP is a staring spectrograph, whereas Apollo 17's UVS had a single-pixel photometer detector that had to time-share as the grating rotated across the spectral passband. Assuming that only 50% of all possible twilight atmospheric observing opportunities are realized, LAMP's total lunar atmospheric observing time in twilight will be ~250 hours/year, more than an order of magnitude increase over what Apollo 17 achieved at any given wavelength. LRO's year-long nominal mission allows LAMP to make a far more extensive survey for neutral species than achieved previously. LRO's orbit near 50 km altitude is well suited to observing many species in the lunar atmosphere, which have characteristic scale heights of 50–100 km (meaning most of the atmospheric column of these species will be below the spacecraft).

Most of LAMP's atmospheric observations will be made looking to the nadir near the terminator, as illustrated in Fig. 4. At these geometries, which occur ~6% of the time for LRO, the lunar surface is dark but the atmosphere between the surface and LRO's altitude is sunlit, allowing atoms there to resonantly fluoresce. During these times, normal LAMP data taking will include emissions from the atmosphere along LAMP's line of sight to the surface. Surface background (e.g., from reflected Lyman-α sky-glow and the interstellar radiation field) can be accurately removed by comparing observations of the same locations made when the surface and atmosphere there are both in shadow. In addition, about once per month observations will be made with LAMP staring at the atmosphere above the lunar limb. With the largest possible line of sight atmospheric column density, these limb observations will provide LAMP's most sensitive look at the lunar atmosphere. Such limb viewing observations are expected to be made more often during the extended mission and along the dawn terminator, where the largest column densities are expected (Hodges 1975). Table 1 lists the expected LAMP effective area, the *g*-factor (i.e., the scattering rate for a ground state atom or ion illuminated by the Sun at 1 AU), the existing measured or upper limit on the lunar atmospheric density, and the expected LAMP density limit (for a single limb pointing of 30 minutes duration) for representative resonance lines (within the LAMP bandpass) of several important species. Note that in only one orbit LAMP should provide improved upper limits for certain species.

Figure 5 depicts model results showing LAMP's ability to detect (or tightly constrain) emission levels from the lunar atmosphere compared to existing knowledge, assuming only nadir-pointed observations. For targeted observations of possible outgassing regions, similar limits apply as in Table 1. Assuming argon is vented at a rate of $Q_{\mathrm{Ar}} \sim 10^{21}$ atoms/s (as estimated by Hodges et al. 1974) with a velocity of $v_{\mathrm{Ar}} \sim 0.3$ km/s (appropriate for argon atoms at the subsurface temperature of 250 K; Vasavada et al. 1999), the localized Ar density would exceed the global maximum observed argon density of $n_{\mathrm{Ar}} \sim 10^5$ cm^{-3} within a region of $r_{\mathrm{Ar}} \sim \sqrt{Q_{\mathrm{Ar}}/2\pi n_{\mathrm{Ar}} v_{\mathrm{Ar}}} \sim 2$ km from the vent, which is a large enough region for

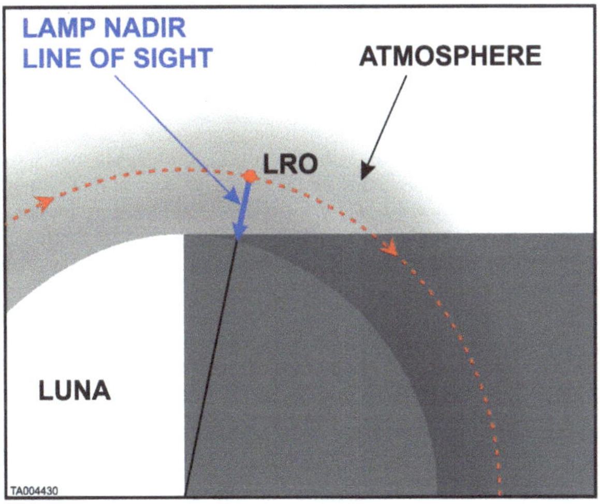

Fig. 4 Nadir observing geometry used in atmospheric exploration at twilight. LRO's orbit is illustrated with a *dashed red curve*. The illuminated atmospheric column near the terminator will emit photons detectable by LAMP. Because the surface is dark, this technique mitigates reflected sunlight background, achieving a more sensitive result than possible with dayside viewing. Monthly limb observations of the atmosphere are also planned

Table 1 LAMP atmosphere species detection

Feature (nm)	LAMP Aeff (cm^2)	g (photons s^{-1})	n_{obs} (cm^{-3}) [day, night]	n_{LAMP}[a] (cm^{-3})
He 58.4	0.011	1.1E−5	2E3, 4E4	8.7E2
Ne 63.0	0.018	?	?	?
O^+ 83.4	0.042	1.5E−6	?	1.7E3
C^+ 90.4	0.040	?	?	?
Ar 104.8	0.047	8.9E−8	1E5,4E4	2.5E4
N 113.4	0.429	?	<6E2	?
H 121.6	0.645	1.2E−3	<17	1.4E−1
O 130.4	0.481	4.6E−6	<5E2	4.8E1
H_2 146.2	0.244	4.0E−8	<9E3	1.1E4
S 147.4	0.233	6.0E−6	<150	7.6E1
CO 151.0	0.204	1.7E−7	<14E3	3.1E3
C 165.7	0.119	1.7E−5	<2E2	5.2E1

[a]Assumes 30 min, 3σ observation & path length of 300 km

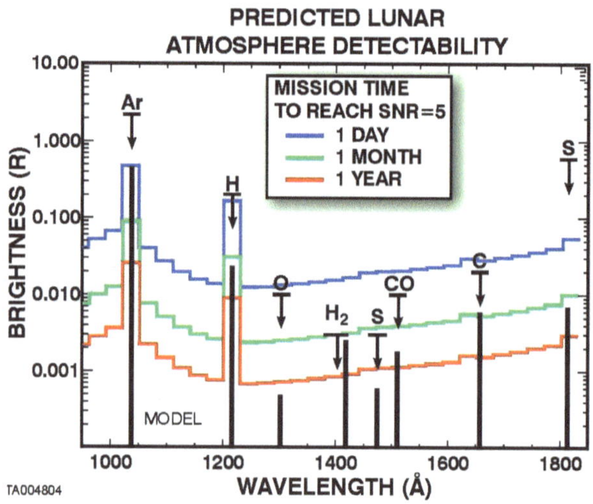

Fig. 5 Upper limits and models of lunar atmospheric FUV feature brightnesses due to atmospheric species of interest. Optically thin emissions are assumed, with nadir viewing at twilight (as illustrated in Fig. 4). The most constraining upper limits for the species of interest (Feldman and Morrison 1991; Parker et al. 1998) are shown as *downward-facing black arrows*. Expected brightnesses are indicated with *vertical black bars*. LAMP detectable brightnesses with SNR = 5 are shown after different mission elapsed times (*various lines*), assuming that only nadir-pointed observations are made. These count estimates assume the predicted effective area curve shown in Fig. 7

LAMP to detect it. LAMP's characterization of the lunar atmosphere will also be particularly timely with LRO, since this mission will occur before major increases in mission traffic, and therefore rocket motor firings, disturb the natural environment. As shown decades ago (Vondrak et al. 1974; Vondrak 1988; NRC 2007), even modest exploration of the Moon will significantly impact the lunar atmosphere.

3 Instrument Description

3.1 Overview

The photon-counting LAMP imaging spectrograph is closely based on the *Pluto Alice* (*P-Alice*) instrument design (Stern et al. 2005, 2008; Slater et al. 2005) which is, in turn, based largely on the *Rosetta Alice* (*R-Alice*) instrument design (Stern et al. 1998; Slater et al. 2001), with only relatively minor changes in baffling, software, and the addition of a lunar terminator sensor (see below). In this section we provide a description of the instrument, concentrating on the differences with the *P-Alice* design.

The LAMP spectrograph comprises a telescope and Rowland-circle spectrograph. An opto-mechanical layout of the instrument is shown in Fig. 6. LAMP has a single $40 \times 40\ \text{mm}^2$ entrance aperture that feeds light to the telescope section of the instrument. Entering light is collected and focused by an $f/3$ off-axis paraboloidal (OAP) primary mirror at the back end of the telescope section onto the instrument's $0.3° \times 6.0°$ entrance slit.

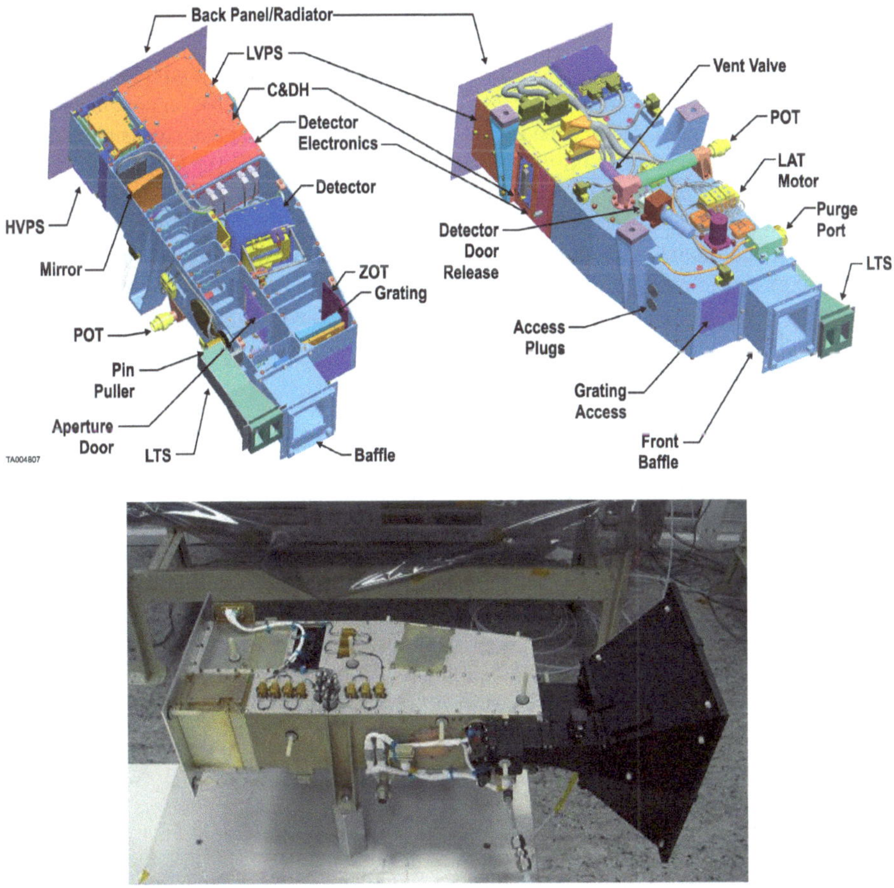

Fig. 6 LAMP design (with the top cover and entrance baffle removed) as seen from above (*upper left*) and below (*upper right*). Photograph of the flight instrument just prior to shipment to LRO (*lower*). The overall dimensions of LAMP are 55 cm × 16 cm × 12 cm. Acronyms: *LAT* = limited angle torque; *POT* = pump out tube; *ZOT* = zero-order trap

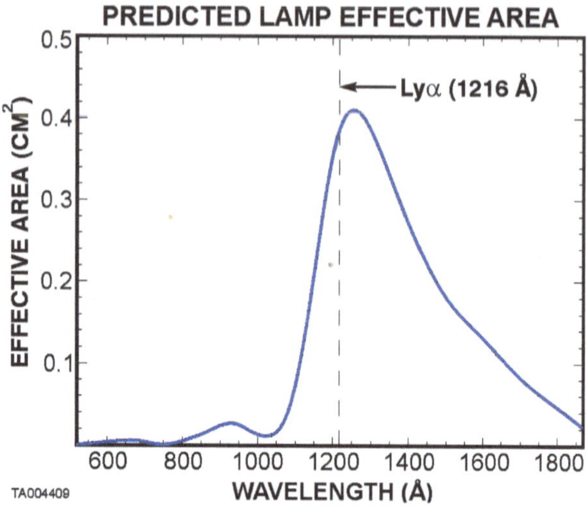

Fig. 7 Expected effective area as a function of wavelength, based on P-Alice results, but with Al/MgF_2-coated optics and a CsI-coated photocathode

After passing through the entrance slit, the light falls onto a toroidal holographic diffraction grating, which disperses the light onto a double-delay line (DDL) microchannel plate (MCP) detector (Siegmund et al. 1992). The 2-D (1024 × 32)-pixel format detector is coated by a CsI solar-blind photocathode and has a cylindrically-curved MCP-stack that matches the Rowland-circle. The predicted effective area as a function of wavelength is shown in Fig. 7. LAMP is controlled by an Intel 8052 compatible microcontroller, and utilizes lightweight, compact, surface-mount electronics to support the science detector, as well as the instrument support and interface electronics. The resulting design is highly systems engineered to minimize mass and complexity, and enjoys strong parts-level heritage from previous UV spectrographs.

3.2 Opto-mechanical Design Overview

The OAP mirror and diffraction grating are constructed from monolithic pieces of aluminum, coated with electroless nickel and polished using low-scatter polishing techniques. The aluminum optics, in conjunction with the aluminum housing, form an athermal optical design. Both the OAP mirror and the grating are overcoated with sputtered Al/MgF_2 for optimum reflectivity within LAMP's spectral passband. Besides using low-scatter optics, additional control of internal stray light is achieved using internal baffle vanes within both the telescope and spectrograph sections of the housing, a holographic diffraction grating with low scatter and near-zero line ghost problems, and an internal housing with alodined aluminum surfaces (Jelinsky and Jelinsky 1987; Moldosanov et al. 1998). In addition, a zero order light trap is treated with a nickel-phosphorus (Ni–P) black coating with very low surface reflectance at EUV/FUV wavelengths (Brown et al. 2002).

3.3 Detector and Detector Electronics Overview

The 2-D imaging photon-counting detector located in the spectrograph section of the instrument utilizes an MCP Z-stack that feeds the DDL readout array (Siegmund et al. 1992). The input surface of the Z-stack is coated with an opaque photocathode of CsI (Siegmund 2000). The detector tube body is a custom design made of a lightweight brazed alumina-Kovar structure that is welded to a housing that supports the DDL anode array.

To capture the entire 57–196 nm passband and 6° spatial FOV, the size of the detector's active area is 35 mm (in the dispersion direction) by 20 mm (in the spatial dimension), with a pixel format of (1024 × 32)-pixels. The 6° slit-height is imaged onto the central 22 of the detector's 32 spatial channels; the remaining spatial channels are used for dark count monitoring. Our pixel format allows Nyquist sampling with a spectral resolution of 0.36 nm, and an angular resolution of ∼0.6°.

The MCP Z-stack is composed of three 80:1 length-to-diameter (L/D) MCPs which are cylindrically curved with a radius-of-curvature of 75 mm to match the Rowland-circle for optimum focus. The total Z-Stack resistance is ∼300 MΩ. The MCPs are rectangular in format (46 × 30 mm^2), with 12-μm diameter pores. Above the MCP Z-Stack is a repeller grid that is biased ∼900 V more negative than the top of the MCP Z-Stack. This repeller grid reflects electrons liberated in the interstitial regions of the MCP back down to the MCP input surface to enhance the detector quantum efficiency. The MCP Z-stack requires a high voltage bias of ∼−3 kV; an additional −600 V is required between the MCP Z-stack output and the anode array (the anode array is referenced to ground). The overall detector gain is $\sim 2E7(+/-25\%)$. At an expected average count rate of 1000 count/s, the amount of charge as a function of time is ∼0.1 Coulomb/year.

3.4 Instrument Electrical Design Overview

The LAMP instrument support electronics are largely single-string, but include redundant features for certain important functions. The LAMP electronics include redundant low-voltage power supplies (LVPS) and actuator electronics, the single-string command and data handling (C&DH) electronics, the redundant optics decontamination heater system, and the redundant detector high-voltage power supplies (HVPSs). All of these sub-systems are controlled by a radiation-hardened version of the Intel 8052 microprocessor with 32 kB of fuse programmable PROM, 128 kB of EEPROM, 32 kB of SRAM, and 128 kB of acquisition memory. The C&DH electronics are contained on four circuit boards located just behind the detector electronics (see Fig. 6). A detailed block diagram of these electronics is shown in Fig. 8.

3.5 Data Collection Modes Overview

The predicted sensitivity of LAMP is such that we expect a data rate of ∼1000 counts/s, both on the nightside with the aperture door open and on the dayside through the pinhole. As with *P-Alice*, LAMP has two detector data collection modes: (i) *pixel list mode*; and (ii) *histogram mode*. However, because of its nature, nearly all the LAMP data will be obtained in pixel list mode (exceptions will be made for flat-field and MCP pulse height distribution measurements). The science data from the detector is collected in a dual port acquisition memory that consists of two separate 32 k × 16-bit memory channels. In *pixel list mode* each memory channel can hold up to 32 k detector and/or time-hack events, where each detector event consists of a 16-bit word—an x-position word 10 bits in length, a y-position word 5 bits in length, and a single status bit that distinguishes between a time hack word and a detector event word. A time hack word includes a single status bit plus 15-bits that encodes the instrument time. When 32 k detector address and time-hack events have accumulated in one of the two acquisition memories, that acquisition memory stops accumulating event data and begins to read the data out to a parallel-to-serial converter and a low-voltage differential signaling (LVDS) interface to the spacecraft on-board memory. At the same time that this LVDS readout is taking place, the other side of the dual acquisition memory continues to

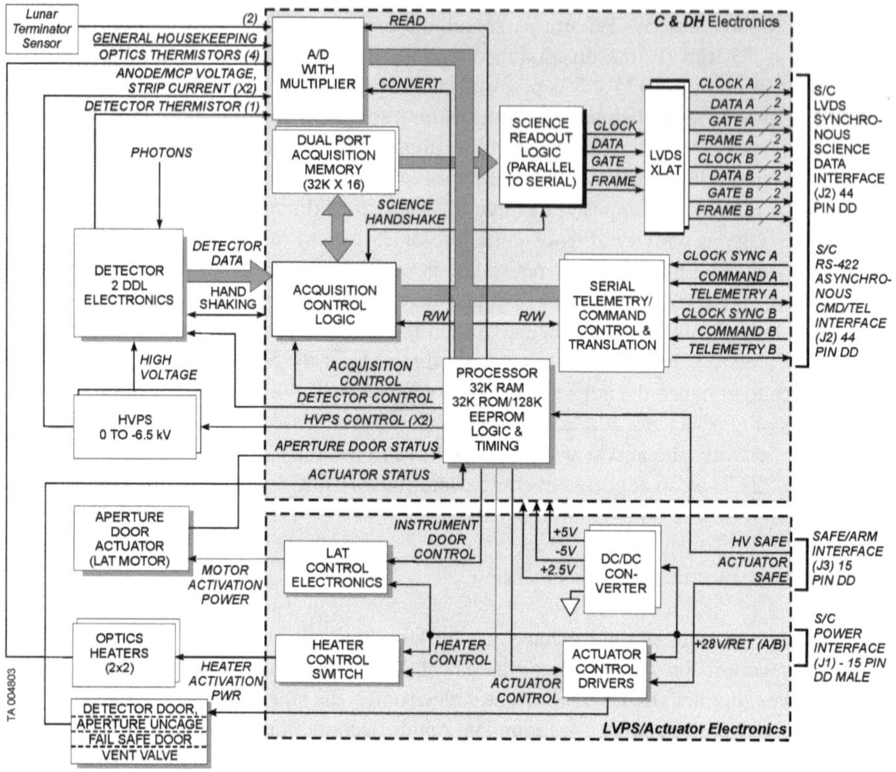

Fig. 8 Detailed block diagram of LAMP

collect detector and time-hack data until it fills up, whereupon it reads out to the LVDS interface to the spacecraft memory, while the first acquisition memory takes over again in the collecting of detector and time-hack data. This back-and-forth data collection flow between both acquisition memories is called "ping-pong" acquisition—it allows contiguous readout of detector event data as long as the data event rate does not exceed the rate at which the data can be read out of memory to the LVDS interface. The "ping-pong" acquisition process is controlled using logic encoded in one of two FPGAs within the C&DH electronics.

LAMP also includes a flight software mechanism to filter out events from selected areas of the detector. This may be used to suppress 'hot pixels' that could develop in the detector, especially in the pixel list mode, which might otherwise consume a large fraction of the available acquisition memory. This filtering is performed before events are processed in either histogram or pixel list mode. The system defines up to 8 rectangular regions of 32 by 4 pixels that suppress any events from the selected regions from being processed. Configuration parameters allow for the placement of these filters areas at any desired positions within the detector area.

3.6 Changes from previous *Alice* builds

Table 2 lists the main differences between LAMP and the earlier *Alice* flight instruments.

Apart from the optics & detector coatings and the slit shape, the only important differences between LAMP and *P-Alice* are the removal of the solar occultation channel (SOCC)

Table 2 *R-Alice*, *P-Alice*, and LAMP comparison

Attribute	*Rosetta-Alice*	*Pluto-Alice*	LAMP
Spectral Passband	70–205 nm	52–187 nm	57–196 nm
Input Channels	Airglow Only	Airglow & Solar Occultation Channel	Airglow Only
Entrance Slit FOV	$0.1° \times 2.0° + 0.05° \times 2.0° + 0.1° \times 1.5°$	$0.1° \times 4.0° + 2.0° \times 2.0°$	$0.3° \times 6.0°$
OAP Mirror & Grating Specs	Identical Specifications (except for optical coating)		
Optical Coating	SiC	SiC	Al/MgF_2
DDL Detector	Identical Specifications (except for photocathode and Lyman-α gap)		
Detector Photocathodes	KBr (70–120 nm)	KBr (52–118.1 nm)	
	CsI (123–205 nm)	CsI (125.1–187 nm)	CsI (52–196 nm)
Lyα Gap (i.e., bare MCP)	3 nm	7 nm	No Gap
Detector Door Actuator	Dimple Motor	Wax Pellet	Wax Pellet
Detector MCP Pulse Height Distribution Bins	16	64	64
HVPS	1 (non-redundant)	2 (redundant)	2 (redundant)
Lunar Terminator Sensor	None	None	Included

and the addition of the Lunar Terminator Sensor (LTS). The SOCC was included in the *New Horizons P-Alice* for solar occultation observations, which are not needed on LRO. The LTS has been added to protect the highly-sensitive LAMP instrument from observing the relatively bright daylight regions of the Moon with its aperture door open, either by lowering the detector high voltage or closing the aperture door. Note that the LRO spacecraft enforces a solar avoidance zone of 60° (primarily for the LRO cameras, which have no shutters). In any case, LAMP will safe itself within 1 second if the count rate exceeds a certain amount (typically set at ~30,000 counts/s). The design of the LTS uses redundant Si PN photodiode detectors and amplifier electronics which are fed by an externally mounted mechanical collimator that over-samples the LAMP field of view. The LTS field of view is 3° × 10° for both a "forward" photodiode (parallel to the LAMP slit but canted at 1.5° ahead of the track of the LAMP boresight) and an "aft" photodiode (parallel to the LAMP slit but canted at 1.5° behind of the track of the LAMP boresight). The signal from these two diodes are each differenced at a 10-Hz rate with "dark" photodiodes (to correct for temperature effects) and combined (in a programmable way) to provide LAMP's response to the crossing of the terminator (from day to night or vice versa). Depending on the orientation of the LRO spacecraft, the forward and aft LTS photodiodes will switch roles several times during the mission. The LAMP aperture door has a small (2-mm diameter) hole in it which allows dayside observations at about the same count rate with the door closed as are observed on the night side when the aperture door is open. Currently, the LRO mission plan calls for using the LTS to lower and raise the detector high voltage during the first month of operations, then using the LTS to close and open the aperture to allow continuous LAMP day and night side observations for the remainder of the mission.

4 Ground Calibration Radiometric Performance

After it was assembled and through with environmental testing, just prior to delivery to GSFC for integration on the LRO spacecraft, LAMP underwent a series of tests to characterize its radiometric performance. Specifically, the dark count, spatial and spectral resolution, wavelength calibration, scattered light sensitivity, and effective area were measured. While useful for estimating the expected on-orbit instrument performance, each of these measurements will be superceded by further tests during commissioning of LAMP in lunar orbit. In addition, some important calibration data (e.g., flat field measurements) are only obtained after launch.

Table 3 lists the LAMP requirements and radiometric calibration results for several key performance metrics.

4.1 Dark Count Rate

The dark count rate measured for the LAMP detector, as averaged over the three most recent 1000-s dark exposures, is 5.7 counts/s. The on-orbit count rate will be considerably larger (probably ~20 counts/s or so) due to the energetic particle environment in space. For comparison, the *R-Alice* count rate measured during ground calibration was only 3.2 counts/s, while after launch it became ~20 counts/s.

4.2 Spatial/Spectral Resolution

The point-spread function (PSF) is undersampled in the spatial direction, as the LAMP pixels (0.3°) are substantially larger than the PSF expected from raytracing. Since the slit width

Table 3 LAMP minimum required and measured performance

Attribute	Minimum performance requirement[a]	Measured performance
Effective area	>0.2 cm^2 at 125 nm	1.3 ± 0.4 cm^2 at 121.6 nm
Field of view	$(0.3° \pm 0.02°) \times (6.0° \pm 0.1°)$	$(0.31° \pm 0.01°) \times (6.00° \pm 0.01°)$
Spectral passband	120–180 nm (first order)	57.5–196.5 nm (first order)
Spectral plate scale	4.1667±0.0140 nm/mm	4.1667±0.0006 nm/mm
Spatial resolution element	$<0.3°$ per spatial pixel	$0.29° \pm 0.03°$ per spatial pixel
Spectral resolution element	0.18±0.01 nm	0.182±0.001 nm
Total spectral point spread function	<2 nm FWHM	0.39–0.57 nm FWHM
Total spatial point spread function	$<1°$ Nyquist sampled	$0.30° \pm 0.18°$ Nyquist sampled
Filled slit spectral resolution	<4.0 nm FWHM	3.64±0.18 nm FWHM
Detector flat field uniformity	<25% (1σ) at 253.7 nm	17.2% (1σ) at 253.7 nm
Stray light rejection	$<10^{-5}$ at 7° off boresight	$<8\times10^{-6}$ at 7° off boresight
Detector dead-time metrics	>5 kHz at 10% DQE loss	6.172 kHz at 10% DQE loss
Detector global background rate	Total array: <50 counts/s	Total array: 5.7 counts/s

[a]LAMP's minimum performance requirements were derived to satisfy the Group 1 Instrument and Dataset Requirement

is also 0.3°, the nominal LAMP spatial pixels are 0.3° × 0.3°, or 260 m × 260 m from an altitude of 50 km. In the spectral direction, the measured PSF varies from 0.39 nm to 0.57 nm FWHM over the LAMP bandpass. However, since LAMP's field-of-view is filled by the Moon during normal nadir-viewing operations, the actual spectral resolution is substantially worse than the spectral PSF. The measured filled-slit spectral resolution of LAMP is 3.64 nm, which is sufficient to separate the key resonance line emissions of expected atmospheric species (see Fig. 5).

4.3 Wavelength Calibration

From ground radiometric measurements using Ar, Ne, and H/He lamps, the LAMP wavelength dispersion is 0.1827 ± 0.0005 nm/pixel, with good linearity ($R^2 = 0.9999$). The final wavelength calibration will be determined on orbit using standard UV stars.

4.4 Scattered Light Characteristics

The scattered light performance of LAMP was determined using Lyman-α emissions to be $<8 \times 10^{-6}$ at 7° off axis. An optical layout shown in Fig. 9 illustrates light paths through the instrument. These measurements were made without the entrance baffle attached, so the actual on orbit performance should be even better.

4.5 Effective Area

As with the wavelength calibration, the final effective area will be determined on orbit using standard UV stars. The effective area measured during radiometric calibration, shown in Fig. 10, exceeds the LAMP requirement by a factor of 5 or more. Since LAMP's optics will face many possible sources of contamination before and during launch, the actual on orbit calibration could well end up to be lower than it is currently.

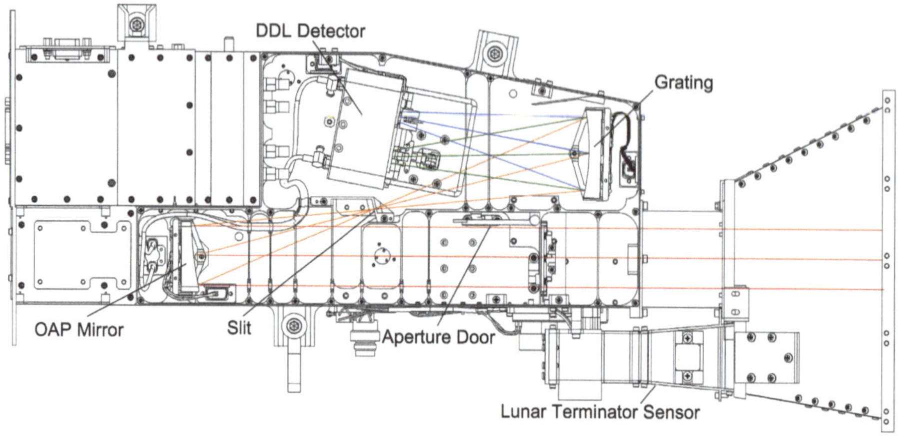

Fig. 9 LAMP optical layout showing example ray traces through the instrument

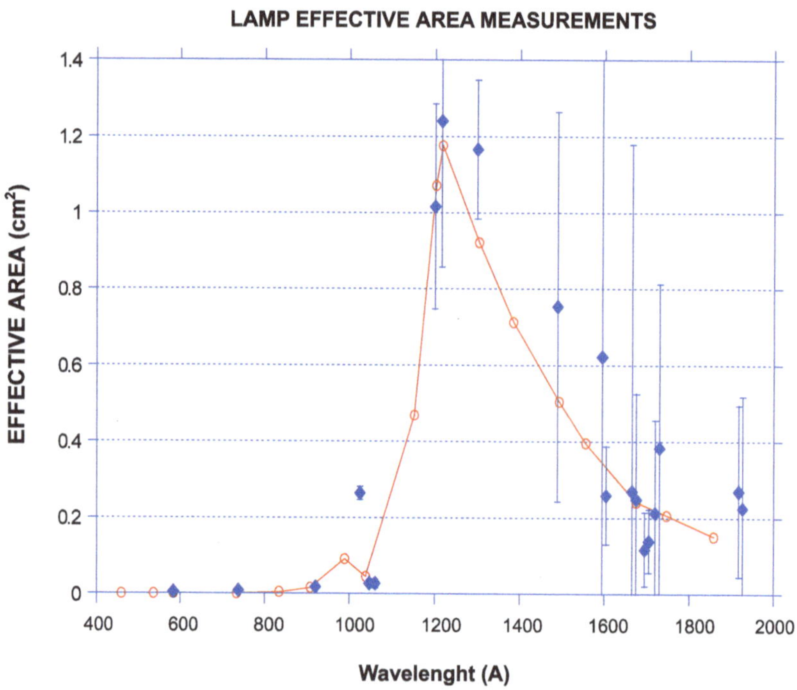

Fig. 10 LAMP effective area as a function of wavelength (*blue diamonds with error bars*) measured during radiometric calibration before installation on LRO, with component-based expectation for comparison (*red*)

4.6 On-Orbit Commissioning and Calibration

Once in orbit around the Moon, LAMP will be calibrated in a number of ways, during an initial 60-day commissioning period and throughout the mission. During commissioning and monthly calibrations, standard UV-bright stars will be observed to monitor effective area.

Both pointed (histogram) and scanned (pixel list) observations will be made, and a few very bright stars will be viewed through the aperture door pinhole. Besides pointed observations, the stellar sources are also rastered across and along the LAMP slit to provide accurate slit dimensions, vignetting information, and row-to-row variations in detector sensitivity. To characterize scattered light performance, we plan to slew the closed door across the bright limb, and follow up these observations with similar slews in which the aperture door is opened as soon as possible after the slit has moved off the bright limb. The wavelength scale will be determined through dayside observations of reflected sunlight, where there will be numerous well-known emission lines spaced throughout the LAMP bandpass. Filled-slit line shapes are acquired by looking at the interplanetary Lyman alpha signal, flat field information is obtained by summing dayside data (with the LRO spacecraft temporarily rotated at a 90° yaw so that each row of the LAMP detector covers exactly the same surface regions). Also during commissioning the optimal high voltage and discriminator settings are found, by inspection of a matrix of test observations and pulse height distributions (PHDs). The PHD is only obtained during histogram observations (which will be acquired during monthly calibrations for monitoring purposes).

Another important part of our commissioning activities will be to determine the analog and digital deadtimes which can decrease the measured signal at large count rates (the digital deadtime is likely to be larger because of the extra processing required by the electronics). Ground tests by the detector provider indicate a digital deadtime of 18 μs, which corresponds to a 10% lower measured count rate at an input count rate of 6173 counts/s. From in-flight observations on New Horizons, which uses an identical detector and detector electronics, we expect that paralyzable deadtime effects are seen until the input count rate is >50,000 counts/s. All lunar observations should be well within the range where only non-paralyzable deadtime is important. We plan to confirm the deadtime by observing several well-known UV stars with a variety of brightnesses.

The monthly calibrations are handled separately from the normal data pipeline (i.e., they will be planned, reduced and analyzed by hand).

5 Data Products

LAMP will deliver (at intervals of 3–6 months after the data is acquired) both low-level (Committee on Data Management, Archiving, and Computing—CODMAC—level 2 and 3) and high-level (CODMAC level 5) data products to the PDS Imaging Node for permanent storage and public access. In normal science operations, the raw LAMP low-speed housekeeping (HK) files will consist of one 128-byte packet per second for each ~113 minute LRO orbit (i.e., ~6780 packets), plus a 64 byte file header. The raw LAMP science data files will consist of a number of instrument-generated 65536-byte frames, each preceded by a 6-byte acquisition time field, plus a 64-byte file header. The exact number of science data frames will vary from file to file. When operated in the nominal pixel list mode with a 4-ms hack rate, LAMP is expected to generate approximately 285 data frames per orbit, when taking data over the entire orbit. Low-level data products generated from a typical orbit are expected to average ~450 MB (cf., Table 4). The actual data volume will depend on the solar, stellar, and Lyman-α sky-glow illumination, the lunar FUV reflectivity, and LAMP's in-orbit sensitivity.

As mentioned above, the total daily volume of LAMP pipeline low-level data products produced will be approximately 5.5 GB (i.e., ~2 TB over the nominal 1-year LRO mission). Each file of these low-level data products contain the following nine types of data as FITS

Table 4 LAMP data products

Name	Description	CODMAC level	Inputs	Format	Data volume (GB)
Engineering level data	Instrument value science data (raw photon events, pixel list & histogram) + instrument housekeeping data	2	Raw + engineering calibration data	FITS (science), ASCII (HK), binary (memory dumps)	0.275 per day
Science level data	Calibrated value science data (tabulated pixel list and histogram) + ancillary data	3	Engineering level + science calibration data	FITS	5.5 per day
Far-UV brightness map	2D lunar surface map of UV 2D brightness + statistical errors, 2D data quality	5	Science level data	FITS	36 total
Landform albedo map	2D lunar surface map of UV albedos + statistical and systematic errors and data quality	5	Science level data	FITS	36 total
H_2O Absorption feature depth map	2D lunar surface map of H_2O absorption feature depths, first-order H_2O surface mixing ratios + statistical and systematic errors and data quality	5	Science level data	FITS	0.5 total
Atmospheric nadir spectra	Near-terminator spectra of atmospheric constituents + error vectors	5	Science level data	FITS	<0.1 total
Atmospheric limb spectra	Atmosphere limb brightness profiles + error vectors	5	Science level data	FITS	<0.1 total

extensions:

1. Calibrated Histogram (Aperture Door Open): A reconstructed histogram generated from raw pixel list data but with instrumental corrections and calibrations applied (e.g., dead time and dark count corrections, flat-fielding, effective area, and wavelength calibrations). Only the pixel list data acquired when the aperture door was open are included in this dataset.
2. Calibrated Histogram (Aperture Door Closed): Similar to the first dataset, but the reconstructed histogram here is generated from data acquired when the aperture door was closed.
3. Acquisition List: A list of the generated acquisitions (i.e., the frames of data of which the file is comprised) as determined from the housekeeping data file.
4. Calibrated Pixel List Mode Data: A calibrated version of the complete pixel list science dataset from the raw data product, plus propagated estimated errors introduced by the separate calibration steps and ancillary spatial location information.
5. Calibrated Histogram Mode Data: A calibrated version of the complete histogram science dataset from the raw data product, plus propagated estimated errors introduced by the separate calibration steps. Since LAMP will be operated nearly entirely in pixel list mode, this extension will generally be empty.
6. Calibrated Calculated Count Rate: A high-resolution sequence of UV photon count rates computed from the calibrated pixel list data for each acquisition (nominally the whole orbit). The resolution of this count rate sequence depends on the commanded hack rate. If no consecutive pixel list science data are available, this dataset will contain lower resolution (1 Hz) count rate data from the housekeeping data file.
7. High-rate LTS Data: LAMP housekeeping data contain up to 10 values per second of gain and offset corrected measurement values from each of the Lunar Terminator Sensors. This dataset provides the full 10 Hz LTS dataset.
8. Housekeeping Data: This dataset contains the complete housekeeping dataset, both in raw and (where applicable) calibrated engineering units.
9. Wavelength Lookup Table: This dataset contains a 1024×32 image whose floating-point pixel values are the wavelengths corresponding to the pixel locations on the detector.

LAMP high-level data products include brightness maps (global and of selected features) over specific wavelength ranges, similarly constructed albedo maps (i.e., brightness maps normalized by the illumination), band depth maps (i.e., maps of color ratio comparing albedo maps for wavelength ranges on and off important absorption bands), and atmospheric spectra of global and specific features (e.g., the dawn terminator, the LCROSS impact site, regions of possible recent outgassing). The high-level data products will use the ensemble low-level pixel lists as input, from which individual photon events will be selected for inclusion based on criteria such as timing, wavelength, emission and incidence angles, and spacecraft location and attitude. For the mapping products an important intermediate product is the exposure map, which provides the time that LAMP actually observed a given surface element under the conditions of the selected criteria. For the albedo and band depth maps, another important intermediate product is the sky illumination (from the Lyman-α sky-glow, UV starlight, and, for near-terminator regions, scattered sunlight) as a function of wavelength and time for each surface element of interest. The high-level data products will be produced regularly during the nominal mission, but will be refined after the end of the nominal mission to include important inputs from other LRO instruments (e.g., surface roughness and slope estimates from LOLA).

6 Summary

LRO is planned for launch in the spring of 2009, with a 1-year nominal mission in lunar orbit, followed by an additional year of science operations. Monthly in-flight calibration observations of hot UV stars are planned for LAMP during the mission to characterize its in-flight performance. The LAMP design is based on the simple and successful *Alice* spectrographs, and will be used to search for exposed frosts, to map landforms in permanently shadowed regions of the Moon, and to survey the tenuous lunar atmosphere; it will also demonstrate Lyman-α assisted vision techniques for possible application in future missions to the lunar poles or in night side operations. During the additional year of science, depending on the orbit, LAMP will make further atmospheric observations (of Earth as well as the Moon) and will perform specific observations related to the study of space weathering and mineralogy of certain lunar craters.

Acknowledgements We wish to thank the entire LAMP proposal team at SwRI, as well as Ossy Siegmund, John Vallerga and Rick Raffanti at Sensor Sciences, and Craig Tooley, Arlin Bartels, Mary Reden, Jo-Anne Baker, Gordon Chin, John Keller, and Rich Vondrak of the LRO Project Office at NASA GSFC. This work was supported by NASA contract NNG05EC87C.

References

J.M. Ajello, A.I. Stewart, G.E. Thomas, A. Graps, Solar cycle study of interplanetary Lyman-alpha variations—Pioneer Venus Orbiter sky background results. Astrophys. J. **317**, 964–986 (1987)

J.-L. Bertaux, E. Quémerais, R. Lallement, E. Kyrölä, W. Schmidt, T. Summanen, J.P. Goutail, M. Berthé, J. Costa, T. Holzer, First results from SWAN Lyman alpha solar wind mapper on SOHO. Sol. Phys. **175**, 737–770 (1997)

R.J.C. Brown, P.J. Brewer, M.J.T. Milton, The physical and chemical properties of electroless nickel-phosphorus black surfaces. J. Mater. Chem. **12**, 2749–2754 (2002)

D.B. Campbell, B.A. Campbell, L.M. Carter, J.-L. Margot, N.J.S. Stacy, No evidence for thick deposits of ice at the lunar south pole. Nature **443**, 835–837 (2006)

G. Chin, S. Brylow, M. Foote, J. Garvin, J. Kasper, J. Keller, M. Litvak, I. Mitrofanov, D. Paige, K. Raney, M. Robinson, A. Sanin, D. Smith, H. Spence, P. Spudis, S.A. Stern, M. Zuber, Lunar Reconnaissance Orbiter Overview: The instrument suite and mission. Space Sci. Rev. **129**, 391–419 (2007)

D.H. Crider, R.R. Vondrak, Space weathering effects on lunar cold trap deposits. J. Geophys. Res. **108**(E7), 5079 (2003). doi:10.1029/2002JE002030

P.D. Feldman, D. Morrison, The Apollo 17 Ultraviolet Spectrometer: Lunar atmosphere measurements revisited. Geophys. Res. Lett. **18**(11), 2105–2108 (1991)

W.C. Feldman, S. Maurice, D.J. Lawrence, R.C. Little, S.L. Lawson, O. Gasnault, R.C. Wiens, B.L. Barraclough, R.C. Elphis, T.H. Prettyman, J.T. Steinberg, A.B. Binder, Evidence for water ice near the lunar poles. J. Geophys. Res. **106**, 23231–23251 (2001)

B. Hapke, R. Nelson, W. Smythe, The opposition effect of the Moon: Coherent backscatter and shadow hiding. Icarus **133**, 89–97 (1998)

A.R. Hendrix, The Galileo Ultraviolet Spectrometer: in-flight calibration and the ultraviolet albedos of the Moon, Gaspra, Ida and Europa, Ph.D. thesis, University of Colorado, Boulder, 1996

A.R. Hendrix, C.J. Hansen, Ultraviolet observations of Phoebe from the Cassini UVIS. Icarus **193**, 323–333 (2008)

R.C. Henry, P.D. Feldman, J.W. Kruk, A.F. Davidsen, S.T. Durrance, Ultraviolet albedo of the Moon with the Hopkins Ultraviolet Telescope. Astrophys. J. **454**, L69–L72 (1995)

R.C. Henry, The local interstellar ultraviolet radiation field. Astrophys. J. **570**, 697–707 (2002)

R.R. Hodges, Formation of the lunar atmosphere. Moon **14**, 139–157 (1975)

R.R. Hodges, Reanalysis of Lunar Prospector neutron spectrometer observations over the lunar poles. J. Geophys. Res. **107**, 5125 (2002). doi:10.1029/2000JE001483

R.R. Hodges, J.H. Hoffman, F.S. Johnson, The lunar atmosphere. Icarus **21**, 415–426 (1974)

P. Jelinsky, S. Jelinsky, Low reflectance EUV materials: a comparative study. Appl. Opt. **26**(4), 613–615 (1987)

R.M. Killen, Source and maintenance of the argon atmospheres of Mercury and the Moon. Meteorit Planet. Sci. **37**(9), 1223–1231 (2002)

J. Mathis, P. Mezger, N. Panagia, Interstellar radiation field and dust temperatures in the diffuse interstellar matter and in giant molecular clouds. Astron. Astrophys. **128**, 212–229 (1983)

M. Mendillo, The atmosphere of the Moon. Earth Moon Planets **85/86**, 271–277 (2001)

K.A. Moldosanov, M.A. Samsonov, L.S. Kim, R. Henneck, O.H.W. Siegmund, J. Warren, S. Cully, D. Marsh, Highly absorptive coating for the vacuum ultraviolet range. Appl. Opt. **37**(1), 93–97 (1998)

T.H. Morgan, D.E. Shemansky, Limits to the lunar atmosphere. J. Geophys. Res. **96**, 1351–1367 (1991)

National Research Council, *The Scientific Context for Exploration of the Moon* (The National Academies Press, Washington, 2007), 120 pp.

J. Parker, S.A. Stern, G.R. Gladstone, J. Shull, The spectroscopic detectability of argon in the lunar atmosphere. Astrophys. J. **509**, L61–L64 (1998)

W.R. Pryor, P. Gangopadhyay, B. Sandel, T. Forrester, E. Quemerais, E. Moebius, L. Esposito, A.I.F. Stewart, B. McClintock, A. Jouchoux, J. Colwell, V. Izmodenov, Y. Malena, W.K. Tobiska, D.E. Shemansky, J.M. Ajello, C. Hansen, M. Bzowski, P. Frisch, Radiation transport of heliospheric Lyman-alpha from combined Cassini and Voyager data sets. Astron. Astrophys. **491**, 21–28 (2008)

P.H. Schultz, M.I. Staid, C.M. Pieters, Lunar activity from recent gas release. Nature **444**, 184–186 (2006)

O.H.W. Siegmund, Microchannel plate imaging detector technologies for UV instruments, in *Conference Proceedings From X-rays to X-band—Space Astrophysics Detectors and Detector Technologies* (Space Telescope Science Institute, 2000)

O.H.W. Siegmund, J. Stock, R. Raffanti, D. Marsh, M. Lampton, Planar delay line readouts for high resolution astronomical EUV/UV spectroscopy, in *UV and X-Ray Spectroscopy of Astrophysical and Laboratory Plasmas, Proceedings from the 10th International Colloquium*, Berkeley, CA, 3–5 February 1992 (1992), pp. 383–386

R.A. Simpson, G.L. Tyler, Reanalysis of Clementine bistatic radar data from the lunar south pole. J. Geophys. Res. **104**, 3845–3862 (1999)

D.C. Slater, S.A. Stern, T. Booker, J. Scherrer, M.F. A'Hearn, J.L. Bertaux, P.D. Feldman, M.C. Festou, O.H.W. Siegmund, Radiometric and calibration performance results for the Rosetta UV imaging spectrometer *Alice*, in *UV/EUV and Visible Space Instrumentation for Astronomy and Solar Physics*, ed. by O.H.W. Siegmund, S. Fineschi, M.A. Gummin. Proceedings of SPIE, vol. 4498 (SPIE, Bellingham, 2001), pp. 239–247

D.C. Slater, M.W. Davis, C.B. Olkin, S.A. Stern, J. Scherrer, Radiometric performance results of the *New Horizons' Alice* UV imaging spectrograph, in *X-Ray, UV, Visible, and IR Instrumentation for Planetary Missions*, ed. by O.H.W. Siegmund, G.R. Gladstone. Proceedings of SPIE, vol. 5906B (SPIE, Bellingham, 2005)

S.A. Stern, The lunar atmosphere: history, status, current problems, and context. Rev. Geophys. **37**, 453–491 (1999)

S.A. Stern, D.C. Slater, W. Gibson, J. Scherrer, M. A'Hearn, J.-L. Bertaux, P.D. Feldman, M.C. Festou, Alice: an ultraviolet imaging spectrometer for the Rosetta Orbiter. Adv. Space Res. **21**, 1517–1525 (1998)

S.A. Stern, J.R. Scherrer, D.C. Slater, G.R. Gladstone, L.A. Young, G.J. Dirks, J.M. Stone, M.W. Davis, *Alice*: The ultraviolet imaging spectrograph aboard the *New Horizons* spacecraft, in *X-Ray, UV, Visible, and IR Instrumentation for Planetary Missions*, ed. by O.H.W. Siegmund, G.R. Gladstone. Proceedings of SPIE, vol. 5906B (SPIE, Bellingham, 2005)

S.A. Stern, D.C. Slater, J.R. Scherrer, J.M. Stone, G.J. Dirks, M.H. Versteeg, M.W. Davis, G.R. Gladstone, J.Wm. Parker, L.A. Young, O.H.W. Siegmund, Alice: the ultraviolet imaging spectrograph aboard the New Horizons Pluto-Kuiper Belt mission. Space Sci. Rev. **140**, 155–187 (2008)

A.R. Vasavada, D.A. Paige, S.E. Wood, Near-surface temperatures on Mercury and the Moon and the stability of polar ice deposits. Icarus **141**, 179–193 (1999)

R.R. Vondrak, Lunar base activities and the lunar environment. LPI Contribution **652**, 246 (1988)

R.R. Vondrak, J.W. Freeman, R.A. Lindeman, Measurements of lunar atmospheric loss rate, in *Proc. 5th Lunar Sci. Conf.* (1974), p. 2945

J.K. Wagner, B.W. Hapke, E.N. Wells, Atlas of reflectance spectra of terrestrial, lunar, and meteoritic powders and frosts from 92 to 1800 nm. Icarus **69**, 14–28 (1987)

Space Sci Rev (2010) 150: 183–207
DOI 10.1007/s11214-009-9608-4

Lunar Exploration Neutron Detector for the NASA Lunar Reconnaissance Orbiter

I.G. Mitrofanov · A. Bartels · Y.I. Bobrovnitsky · W. Boynton · G. Chin · H. Enos · L. Evans · S. Floyd · J. Garvin · D.V. Golovin · A.S. Grebennikov · K. Harshman · L.L. Kazakov · J. Keller · A.A. Konovalov · A.S. Kozyrev · A.R. Krylov · M.L. Litvak · A.V. Malakhov · T. McClanahan · G.M. Milikh · M.I. Mokrousov · S. Ponomareva · R.Z. Sagdeev · A.B. Sanin · V.V. Shevchenko · V.N. Shvetsov · R. Starr · G.N. Timoshenko · T.M. Tomilina · V.I. Tretyakov · J. Trombka · V.S. Troshin · V.N. Uvarov · A.B. Varennikov · A.A. Vostrukhin

Received: 11 November 2008 / Accepted: 10 November 2009 / Published online: 19 December 2009

I.G. Mitrofanov (✉) · D.V. Golovin · A.A. Konovalov · A.S. Kozyrev · M.L. Litvak · A.V. Malakhov · M.I. Mokrousov · A.B. Sanin · V.I. Tretyakov · V.S. Troshin · V.N. Uvarov · A.B. Varennikov · A.A. Vostrukhin
Institute for Space Research, Moscow 117997, Russia
e-mail: imitrofa@space.ru

V.V. Shevchenko
Sternberg Astronomical Institute of Moscow State University, Moscow 119992, Russia

A.R. Krylov · V.N. Shvetsov · G.N. Timoshenko
Joint Institute of Nuclear Energy, Dubna, Moscow 141980, Russia

Y.I. Bobrovnitsky · A.S. Grebennikov · S. Ponomareva · T.M. Tomilina
A.A. Blagonravov Institute of Mechanical Engineering, Moscow 117334, Russia

L.L. Kazakov
All-Russian Research Institute of Atomic Reactors, Dimitrovgrad-10 433510, Russia

G.M. Milikh · R.Z. Sagdeev
University of Maryland, College Park, MD 20742, USA

A. Bartels · G. Chin · S. Floyd · J. Garvin · J. Keller · T. McClanahan · J. Trombka
NASA Goddard Space Flight Center, Greenbelt, MD 20771, USA

W. Boynton · H. Enos · K. Harshman
University of Arizona, Tucson, AZ 85721, USA

R. Starr
Catholic University, Washington, DC 20064, USA

L. Evans
Computer Sciences Corporation, Glenn Dale, MD 20706, USA

Abstract The design of the Lunar Exploration Neutron Detector (LEND) experiment is presented, which was optimized to address several of the primary measurement requirements of NASA's Lunar Reconnaissance Orbiter (LRO): high spatial resolution hydrogen mapping of the Moon's upper-most surface, identification of putative deposits of appreciable near-surface water ice in the Moon's polar cold traps, and characterization of the human-relevant space radiation environment in lunar orbit. A comprehensive program of LEND instrument physical calibrations is discussed and the baseline scenario of LEND observations from the primary LRO lunar orbit is presented. LEND data products will be useful for determining the next stages of the emerging global lunar exploration program, and they will facilitate the study of the physics of hydrogen implantation and diffusion in the regolith, test the presence of water ice deposits in lunar cold polar traps, and investigate the role of neutrons within the radiation environment of the shallow lunar surface.

Keywords Moon · Neutrons emission · Surface composition · Radiationbackground

Abbreviations

AC	Anti-coincidence
ACS	Anti-coincidence system
CSEN 1–4	Collimated sensor of epithermal neutrons 1–4
DB	Data base
DSA	Deep space antenna
eV	Electron-volt
FPGA	Field programmable gate array
FOV	Field of view
FRE	Front-end/read-out electronics
FRE&S	Front-end/read-out electronics and selection
FU	Flight unit
FWHM	Full width half maximum
GSFC	Goddard space flight center
HV	High voltage
HVP	High voltage provision
IKI	Russian acronym for institute for space research
LCROSS	Lunar crater observation and sensing satellite
LEND	Lunar exploration neutron detector
LND	Name of manufactory of neutron sensors
LOLA	Lunar orbiter laser altimeter
LRO	Lunar reconnaissance orbiter
LVP	Low voltage provision
MeV	Mega electron-volt
MC	Module of collimation
MHz	Mega-hertz
MLI	Multi-layer isolation
MOC	Mission operation center
MSE	Module of sensors and electronics
OSR	Optical solar reflector
PDS	Planetary data system
PMT	Photo-multiplier tube
ppm	Particles per million
PPS	Pulse per second

RAM	Random access memory
RTAX	Radiation tolerant model of Actel FPGA
S/C	Spacecraft
SHEN	Sensor of high energy neutrons
SETN	Sensor of epithermal neutrons
SOC	Science operation center
STD	Standard
STN 1–3	Sensor of thermal neutrons 1–3
SPE	Solar particle events
TCS	Thermal control system
UMD	University of Maryland
UofA	University of Arizona
QU	Qualification unit

1 Introduction: Objectives of the LEND Investigation

The Lunar Exploration Neutron Detector (LEND) instrument on board the Lunar Reconnaissance Orbiter (LRO) spacecraft will measure both the neutron emission from the lunar surface and the local neutron background in orbit. The primary measurement objectives of the LEND science investigation are to map the spatial distribution of hydrogen over the entire surface of the Moon, as well as to determine the possible presence and abundance of water (ice) at the Moon's poles. The additional measurement of the lunar neutron flux will also help to determine the surface neutron radiation environment, which is of great importance in the planning and operation of future human lunar surface missions.

The LRO mission is a prelude to future human exploration of the lunar surface and to the eventual establishment of a human-tended, sustainable outpost on the Moon. One of the goals of the LRO mission is to locate in situ resources that would be of value for supporting a future human lunar outpost. Perhaps the most valued of such resources is water (ice). There is evidence from previous missions that water ice might exist in permanently shadowed impact craters near the lunar poles; such water ice is theorized to have been deposited by ice-rich cometary or asteroidal materials over aeons of time (Hsieh and Jewitt 2006). Another LRO mission measurement objective is to define the lunar neutron radiation environment, which is also important for optimizing the safety and effectiveness of future human surface operations on the Moon.

The LEND instrument on the LRO spacecraft will measure the neutron emission flux from the entire lunar surface, including many of the known permanently shadowed region (PSR) cold traps, with surface spatial resolution comparable to the diameter of the smaller of these cold traps. Since the characteristic of the neutron signal is strongly affected by the presence of hydrogen (H), analysis of the data to be acquired by the LEND experiment will make a significant contribution to the search for possible water-bearing regions. The measured decrease (by LEND) in the epithermal neutron flux can be associated with the presence of hydrogen and thus possible water ice deposits within the upper 1 meter of the lunar surface.

The Bistatic S-band Radar Experiment (i.e., spacecraft S-band radio pulses transmitted through the high gain antenna received by the 70 meter Deep Space Network antennae on Earth) implemented by means of the lunar-orbiting *Clementine* spacecraft in 1994, suggested the possibility of water ice in one of the permanently shadowed areas of the south pole on the basis of a coherent backscatter opposition effect (Nozette et al. 1996). Four years later the NASA *Lunar Prospector* provided the first orbital map of neutron emission from the entire surface of the Moon (Feldman et al. 1998, 2000; Maurice et al. 2004; Lawrence et al. 2006). These data showed noticeable decreases in the flux of epithermal neutrons within high latitude regions around both poles. However, the *Lunar Prospector* Neutron Spectrometer (LPNS) detectors could not resolve specific impact craters because of their broad spatial resolution, which was typically on the order of 56–80 km (Maurice et al. 2004). The LPNS neutron signals detected represented more of a composite signal over a much larger area then most of the individual polar region impact craters of interest. Both signatures in GHz frequency microwave scattering (i.e., Clementine S-band Bistatic) and LPNS neutron data were interpreted as possible evidence for water ice deposits within the "cold traps" in permanently shadowed region (PSR) craters. The surface physical temperature within these isolated cold traps was computed to be below 90 K; at such temperatures there is very little sublimation of water ice, or of most other volatiles, even in the extreme vacuum of space. Therefore, deposits of water ice and other volatiles may have accumulated in these "cold traps" from ephemeral lunar atmospheres produced by cometary or asteroidal impacts. These putative ice-rich deposits would be valued resources for human exploration if extensive water ice were concentrated at "ore-body" levels. However, it has been suggested that there are other physically-plausible explanations for the radar backscatter and neutron signatures in the lunar polar regions (e.g., see Crider and Vondrak 2000). For example, the solar wind implants nuclei of hydrogen, helium and other ions into the lunar surface rocks and volumetrically-dominant regolith. Analysis of lunar rocks from the Apollo missions provides ground truth data for the elemental concentration. The implanted ion surface concentration is known to increase with increased exposure to the solar wind, and samples with higher maturity have a higher concentration of implanted nuclei. Analysis and calculations (Crider and Vondrak 2003) concluded that the retention efficiency of the cold polar regolith is adequate for preserving volatile deposits over long periods of time. Therefore, the colder regions (PSR's) at the lunar poles may have a higher enrichment of implanted hydrogen simply due to inefficient space weathering processes.

The LEND investigation will address which of these two alternative models for the observed effect of the decrease of the epithermal neutron flux from the polar regions by means of high-sensitivity measurements of neutron emission from the Moon. Spatial resolution is the key to resolving the origin of the signal that is observed. Unlike the Lunar Prospector LPNS sensor, the LEND instrument is a collimated neutron detector system with a $\sim$10 km diameter (FWHM) footprint at the nominal 50 km orbital altitude of the LRO spacecraft. This footprint is smaller than many of the polar impact craters of interest, including, for example, the 18 km diameter Shackleton crater adjacent to the south rotational pole of the Moon. Therefore, the LEND instrument will allow comparison of neutron emissions both inside and outside of numerous craters in the lunar polar regions, as well as comparison of different locations within many permanently shadowed craters, at both poles.

2 Physics of Neutron Remote Sensing of the Moon, Instrument Functional Requirements and Main Measurements by LEND

Remote sensing is defined as the measurement of signals from an orbiting spacecraft. The LEND instrument will measure lunar neutrons at thermal energies (<0.4 eV), at epithermal energies (>0.4 eV), and at the higher-energies (>300 keV). The major source of these neutrons is attributed to galactic cosmic ray spallation interactions within the lunar regolith, producing high-energy neutrons (e.g., see Vinogradov et al. 1966; Metzger et al. 1973; Feldman et al. 1993). These neutrons can be moderated or slowed down in energy by scattering

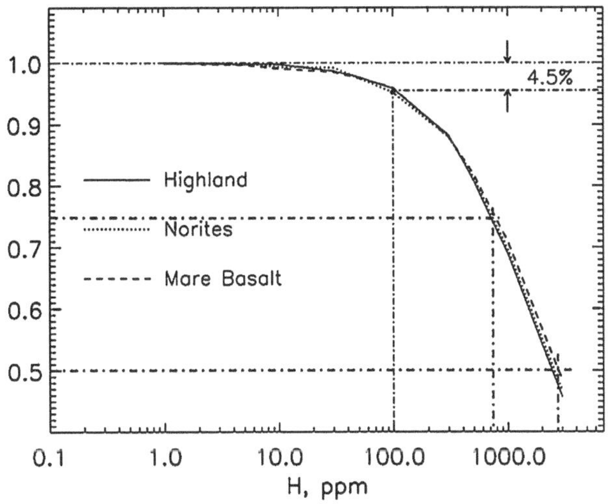

Fig. 1 Decrease of epithermal neutron count rate in a ^{3}He sensor (diameter 5 cm, pressure 20 atm) at 50 km altitude above the Moon as function of hydrogen content (ppm) in the regolith. Three models of lunar regolith composition are shown for comparison. Decreases of 4.5%, 25% and 50% are shown, as corresponding to detection level of 100 ppm, 700 ppm and 3000 ppm of Hydrogen, respectively (these values correspond to about 0.1 weight%, 0.6 weight% and 3 weight% of water ice)

with the various nuclei within the lunar surface, producing epithermal and thermal neutrons. The relative amounts of fast, epithermal versus thermal neutrons strongly reflects the nature of the surface composition. Many of these neutrons diffuse outward and escape the lunar surface. Those that escape constitute the neutron leakage flux and are measured by the LEND detectors. The energy spectrum of the neutron flux is very sensitive to the subsurface hydrogen concentration. Hydrogen nuclei have the same mass as a neutron, and absorb about half of a neutron's energy at each collision. Therefore, hydrogen will strongly moderate the high-energy neutrons and change the epithermal to fast flux ratio.

A numerical simulation program has been used to predict emission of neutrons from the lunar surface at different angles relative to zenith and at different energies. These simulations took into account different regolith compositions and different concentrations of hydrogen. It has been shown that the flux of epithermal neutrons provides the most sensitive physical signature for the concentration of hydrogen in the shallow subsurface. The decrease of epithermal neutrons with increasing hydrogen content is illustrated in Fig. 1. It demonstrates that the relative difference δ between the fluxes of epithermal neutrons from a spot with no hydrogen and one with 100 ppm of hydrogen (100 particles per million) corresponds to about a decrease of 4.5%. Such a decrease of neutron emission from a polar region with a diameter of 10 km should be detectable by LEND after approximately one year of orbital mapping in order to fulfill the LRO mission Level 1 measurement requirement (see instrument requirement (1), Table 1).

To satisfy this LRO mission measurement requirement, the LEND instrument has been designed as a collimated instrument with four high pressure ^{3}He proportional counters each with a Cd foil cover (Figs. 2–4). These individual collimated sensors for measuring epithermal neutrons are identified as CSEN 1 to CSEN 4. They are placed inside four holes within the Collimation Module (Figs. 2 and 3). The fifth collimated sensor utilizes *stilbene* (organic crystal scintillator) and measures high energy neutrons (SHEN). It is placed within the central hole of the Collimation Module of the LEND instrument (Figs. 2 and 3). An active anti-coincidence shield allows the SHEN electronics to reject cosmic rays and other background signals from charged particles. The collimator is constructed from polyethylene and boron-10 (^{10}B). Off-angle neutrons are moderated by the polyethylene and then captured by the ^{10}B. The physical design is such that the collimator defines a surface footprint of ~10 km diameter (FWHM) for an LRO spacecraft mapping orbit altitude of 50 km. Design information is provide in more detail in Sect. 3.

Table 1 Correspondence between LRO Mission Requirements, LEND objectives, and instrument functional requirements

LRO mission requirement	Objectives of LEND investigation	LEND instrument functional requirements
(1) High spatial resolution hydrogen mapping of the Moon's surface	(1) Determination of hydrogen content of the subsurface with spatial resolution of 10 km (FWHM) for an orbital altitude of 50 km and with sensitivity to concentration variations of 100 parts per million (ppm) at the lunar poles	(1) Measure epithermal neutrons by collimated sensors with narrow Field of View with opening angle of about 0.1 rad and capability to detect 5% difference (3 sigma confidence) of epithermal neutron flux for polar footprints with a radius of 5 km; sensors CSETN 1–4 (Fig. 2)
(2) Identification of putative deposits of appreciable near-surface water ice within the Moon's polar cold traps	(2) Characterization of the surface distribution and column density of possible near-surface water ice deposits in the Moon's polar region cold traps with spatial resolution of 10 km (FWHM)	see (1) above (2) Measure high energy neutrons using a sensor with a narrow Field of View at 16 energy channels from 300 keV up to 15 MeV; sensor SHEN (Fig. 2)
(3) Characterization of space radiation environment from a low altitude lunar orbit	(3) Global characterization of lunar neutron emission at altitude of 50 km above Moon's surface with a spatial resolution about 70–100 km at a broad energy range from thermal energies up to 15 MeV	see (1) above see (2) above (3) Measure thermal and epithermal neutrons by omni-directional sensors with capability to distinguish lunar emission from local background of the spacecraft; sensors STN 1–3 and SETN (Fig. 2)

The concept of neutron collimation is summarized as follows. The number of detected neutrons N_{signal} within the field-of-view FOV from any surface element with a 10 km size foot print is described as:

$$N_{\text{signal}} = \pi \theta_{\text{FOV}}^2 F \varepsilon S_{\text{det}} t_{\text{obs}} \tag{1}$$

where θ_{FOV} is the opening angle of the instrument FOV, F is the flux of epithermal neutrons from the spot (in neutrons/cm^2 ster sec), ε is the efficiency of the detector with geometrical area S_{det}, and t_{obs} is the time of observation. The collimation cannot be perfect, so a number of neutrons, N_{bgd}, outside of the FOV will also be detected which gives the total number of all detected counts, $N_{\text{tot}} = N_{\text{signal}} + N_{\text{bgd}}$. The optimal collimator design goal was to have $k = N_{\text{bgd}}/N_{\text{signal}} \sim 1$. Indeed, $k = 0$ would correspond to no neutrons outside of FOV, i.e. "perfect collimator" with infinite mass of absorbing material. Decreasing the parameter k from $\gg 1$ (no collimation case) to 1 corresponds to a decrease of noise that follows $k^{1/2}$, but further decreases of k from $\sim$1 to 0 (perfect collimation case) corresponds to a decrease in noise that behaves as $(1+k)^{1/2}$.

To distinguish between one surface footprint with 100 ppm of hydrogen and a surrounding area without hydrogen, one has to detect a difference of counts of about $\delta \cdot N_{\text{signal}}$ between the measured value and the prediction based on a no-hydrogen hypothesis. This difference has a high statistical significance (i.e., at the level of 3σ) provided $\delta \cdot N_{\text{signal}} > 3(N_{\text{tot}})^{1/2}$. For $\delta = 0.045$ (see Fig. 1) this condition could be converted into the condition: $N_{\text{signal}}^{1/2} > 67$ $(0.045/\delta)(1+k)^{1/2}$. Therefore, the LEND instrument requirement (1) for detection of

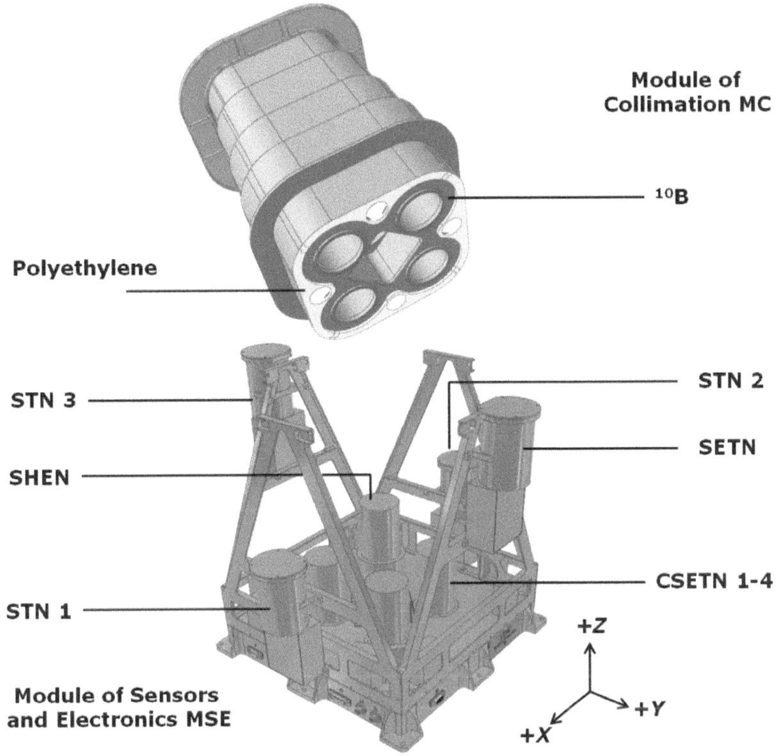

Fig. 2 The LEND instrument is composed of two detachable units; at *the bottom* is the Modules of Sensors and Electronics (MSE) and at *the top* is the Module of Collimation (MC). The definitions of each of the ten detector modules and their functions are described in detail within the body of the text

100 ppm of hydrogen with a surface spatial resolution of ~10 km (see Table 1) corresponds to the following expression, which determines the key parameter $\varepsilon \cdot S_{det}$ associated with the instrument concept:

$$N_{signal} = \pi \theta_{FOV}^2 F \varepsilon S_{det} t_{obs} > 4.5 \times (0.045/\delta)^2 \times 10^3 (1 + k) \qquad (2)$$

The estimated neutron emission flux from the Moon at the nominal LRO orbit (~50 km) is about: $F = 0.5$ neutrons/cm^2/ster/sec according to numerical simulations and observations (e.g., Prettyman 2007). The total observation time t_{obs} of a surface element depends on its latitude on the Moon. The nominal LRO mapping orbit will be in a polar inclination, and circular at 50 km altitude. For a reference surface element at the pole with a size of 10 km, the opening angle of the instrument FOV is 0.1 radians and the total observation time would be about 3×10^4 sec. Using these values, one may derive the necessary effective area $\varepsilon \cdot S_{det}$ from (2) for the simplest version of a neutron telescope, which corresponds to the driving mission requirements (1). This effective area should be about 20 cm^2.

Any neutron detector instrument with this effective area will have the following relationship between the total observation time of particular surface element t_{obs} and the relative decrease δ of the neutron flux at the same surface element in comparison with the surrounding area:

$$t_{obs} > 1.5 \times 10^4 (0.045/\delta)^2 (1 + k) \text{ sec} \qquad (3)$$

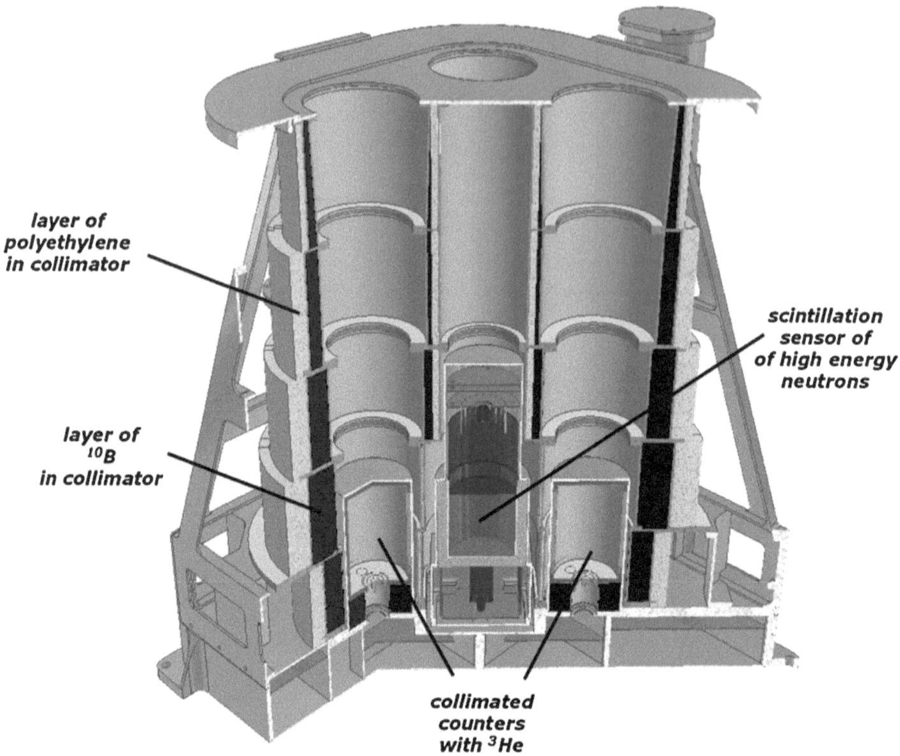

Fig. 3 Cross-sectional view of the LEND Collimation Module that clearly shows the four vertically assembled segments that have four holes for the collimated epithermal detectors CSETN 1–4 and a central hole for the high energy neutron detector, SHEN. These segments contain an inner layer of neutron absorbing material, ^{10}B, surrounding all four holes, as well as an outer layer of neutron-moderating high-density polyethylene. The shape and distribution of absorbing and moderating material were determined from detailed numerical simulations that optimized the collimation efficiency within the highly constrained mass budget of the instrument

The relationship between the relative decrease of the flux of epithermal neutrons and the hydrogen content of the subsurface (see Fig. 1) allows one to relate the detection limit of the instrument for a subsurface water (ice) concentration to the total observation time. This estimate will be discussed for some specific Permanently Shadowed Region (PSR) within the polar region of the Moon (see below).

To obtain the minimum required effective area of about 20 cm^2, four identical sensors are used in the LEND instrument to establish the collimated field of view. Compared with a simple one-sensor configuration (i.e., the LPNS experiment), the four sensors within LEND allow for a decrease in the absorbing mass of the collimator because an absorbing wall between each neighboring sensor works for both of them, and a more compact mass distribution results. The exact shape and structure of the LEND collimator has been optimized by means of numerical simulations that were based on the requirement of achieving the highest sensitivity relative to the instrument measurement objective (1) (Table 1) for the given total instrument mass limit of about 26 kg. For this design, the total mass of absorbing material ^{10}B was about 5 kg.

All of the LEND helium-3 sensors are 7.7 cm long by 5.1 cm in diameter, and consist of 20 atm. proportional counters supplied by the LND company (part number LND253123).

Fig. 4 Photograph of the LEND Engineering Unit that shows the assembled Collimation Module and the Module of Sensors and Electronics

The detection reaction is: $^3He + n \rightarrow ^3H + p + Q = 765$ keV. The lunar viewing surface of the sensors (at nadir) is covered by a Cd foil to absorb thermal neutrons inside the FOV. Emissions from all other surface areas are suppressed by the collimator, which absorbs thermal as well as epithermal neutrons outside of the instrument FOV. The walls of collimators have an external layer of polyethylene, which moderates external neutrons reducing their energy, and an inner layer of ^{10}B, which absorbs the low energy neutrons (Fig. 2).

It is possible to evaluate the vertical distribution of water ice in the lunar regolith, because neutrons of different energies are known to be produced at different depths; for example, larger depths are associated with epithermal neutrons and smaller depths for high energy neutrons (Masarik and Reedy 1996). Therefore, the analysis of neutron data for both epithermal energies and for a range of several MeV energies would make it possible to distinguish between water ice on top of dry regolith and a homogeneous mixture of water ice and soil. On the basis of LEND observations and this type of analysis, it will be possible to formally address measurement objective (2), namely, to characterize the column density of possible near-surface water ice deposits in the Moon's polar cold traps (Table 1). In addition to the detection of epithermal neutrons with the narrow FOV (instrument functional requirement (1), Table 1), it is also necessary to detect the high energy neutrons (instrument functional requirement (2), Table 1). To satisfy this requirement (2), the high energy neutron sensor SHEN has been incorporated into the instrument and placed in the central cylindrical "hole" of the collimation module along its axis of symmetry (Figs. 2 and 3). The position of SHEN inside the collimation module provides a smaller collimation effect for its FOV (total opening angle about 80°), but the structure of collimator has not been specially optimized for this sensor; the driving requirement for LEND called for the most efficient collimation using epithermal neutron sensors CSETN 1–4.

The stilbene sensor ($C_{14}H_{12}$) is an organic molecular single crystal scintillator. It was selected for the SHEN detector because it has optimum efficiency for detecting high energy neutrons by the proton recoil reaction (H + n → p + n′). The energy spectrum of recoiled protons indirectly represents the energy distribution of the original high energy neutrons, because for mono-energetic neutrons with a fixed energy E_o all recoiled protons have energies $<E_o$. This allows spectral analysis of the detected particles by the method of inverse transformation of the measured spectrum of counts in the sensor into the physical spectrum of detected neutrons. External charged particles associated with cosmic rays and the local radiation background are rejected by an active anti-coincidence shield that surrounds the crystal on all sides except in the direction of the lunar surface. Scintillations in stilbene from electrons produced by gamma-rays are rejected by pulse-shape discriminator. This discriminator differentiates long rise-times due to proton interaction with short rise-time due to electron interactions (Gorn and Khazanov 1975).

To characterize the neutron environment at the LRO spacecraft in lunar orbit, the LEND carries four additional ^{3}He detectors external to the collimator. They are identified in Fig. 1 as SETN (sensor of epithermal neutrons), and STN 1–3 (sensors of thermal neutrons). SETN is Cd-shielded to reject all neutrons with energy below 0.4 eV. STN 3 and SETN measure the local background of thermal and epithermal neutrons, which is contributed by the surface emission of the Moon and by secondary neutrons from the spacecraft. They are positioned near the top of collimation module on sides $+Y$ and $-Y$ and have the largest possible views of the lunar surface. The other two sensors of thermal neutrons, STN 1 and STN 2, are positioned on sides $+X$ and $-X$ at the middle of the collimator body (see Figs. 2 and 3). These two sensors constitute a Doppler filter detector (see Feldman and Drake 1986).

The collimator material plays a significant role in the function of the Doppler filter. During the mapping orbit, the LRO flight trajectory may be either along axis $+X$ or $-X$ of the spacecraft (Chin et al. 2007). Therefore, at any time one of the two STN Doppler filter sensors is on the front side of the spacecraft and designated as $\mathrm{STN}_{\mathrm{front}}$, and the other one is on the back side and is designated as $\mathrm{STN}_{\mathrm{back}}$. The collimator obscures one detector from the other (see Figs. 1 and 2). The number of neutron counts from the Moon is different for $\mathrm{STN}_{\mathrm{front}}$ and $\mathrm{STN}_{\mathrm{back}}$. Neutrons on the front side have, in general, a larger relative velocity as measured with $\mathrm{STN}_{\mathrm{front}}$. Neutrons on the back side have, in general, a smaller relative velocity as measured with $\mathrm{STN}_{\mathrm{back}}$. Therefore, the number of counts of thermal neutrons from the Moon is different for $\mathrm{STN}_{\mathrm{front}}$ and $\mathrm{STN}_{\mathrm{back}}$. On the other hand, the number of counts for secondary neutrons from the spacecraft (C_{craft}) is equal for both sensors $\mathrm{STN}_{\mathrm{front}}$ and $\mathrm{STN}_{\mathrm{back}}$, because the angular-velocity distributions of these neutrons are similar at both the front and back sides in the co-moving frame of reference of the spacecraft.

The total numbers of counts in sensors $\mathrm{STN}_{\mathrm{front}}$ and $\mathrm{STN}_{\mathrm{back}}$ are $C_{\mathrm{front}} = C_{\mathrm{craft}} + C_{\mathrm{Moon}}^{(\mathrm{front})}$ and $C_{\mathrm{back}} = C_{\mathrm{craft}} + C_{\mathrm{Moon}}^{(\mathrm{back})}$, respectively. The difference between these two values corresponds to the Doppler filter effect, which is:

$$\Delta C_{\mathrm{Doppler}} = C_{\mathrm{back}} - C_{\mathrm{front}} = C_{\mathrm{Moon}}^{(\mathrm{back})} - C_{\mathrm{Moon}}^{(\mathrm{front})} = \Phi(V_{\mathrm{orb}}, V_{\mathrm{th}}) \times F_{\mathrm{Moon}}^{(\mathrm{th})} \tag{4}$$

The Doppler filter response function, $\Phi(V_{\mathrm{orb}}, V_{\mathrm{th}})$, characterizes the difference in detection efficiency between the front side and the back side sensors with an orbital velocity V_{orb} inside the flux $F_{\mathrm{Moon}}^{(\mathrm{th})}$ of thermal neutrons with thermal velocity V_{th}. This function takes into account the angular-energy distribution of lunar neutrons, the temperature of lunar subsurface, and the parameters of orbital flight. The initial knowledge of this function will be obtained from the pre-flight physical calibration program and numerical simulations. It will be up-dated during the flight segment of the LRO mission during its initial Commissioning stage using in-flight calibration data. Using this function and the measured value of $\Delta C_{\mathrm{Doppler}}$, one may estimate the flux of thermal neutrons from the Moon, $F_{\mathrm{Moon}}^{(\mathrm{th})}$.

The data from all four external omni-directional sensors STN 1–3 and SETN characterize the thermal and epithermal neutrons from the Moon and the local spacecraft background (see objective (3), Table 1). The data from collimated sensors CSETN 1–4 and SHEN may also be used to achieve objective (3) of the LEND investigation. The data for neutrons from the collimated sensors in the nadir direction will correspond with data from omni-directional sensors. Also, the LRO mission will perform several sessions involving planned off-nadir measurements, when the collimated sensors will be directed at 15°, 30°, 45°, 60°and 90° from the nadir. The data from these measurements will provide the validation of the numerical model for the angular distribution of lunar neutron emissions and will facilitate the necessary corrections of this model.

3 Main Aspects of Instrument Design

The LEND instrument is a compact, bolt-on unit as shown in Figs. 2 and 3. It is constructed from two distinct, separable modules; the Module of Sensors and Electronics and the Collimation Module. The primary performance and physical parameters of LEND are presented in the Table 2.

The module of sensors and electronics (MSE) (Fig. 2) has eight helium-3 (^{3}He) proportional counters and one stilbene detector. All eight proportional counters were procured from the LND Corporation and are identical in design. Four of the ^{3}He counters have a cadmium foil cover and are positioned inside the cylindrical holes of the collimator. They are designated as the collimated sensors of epithermal neutrons (CSEN 1–4). The other four ^{3}He proportional counters are external to the collimator. Three of these are designated as the sensors of thermal neutrons (STN 1–3). The fourth counter is covered with a cadmium wrap and referred to as the sensor of epithermal neutrons (SETN).

The analog signal from each proportional counter goes to a 16-channel digitizer. This allows clear identification of the neutron energy capture signals with two well-distinguishable peaks at 191 keV and 764 keV and facilitates the rejection of the lower amplitude charged particle signals, gamma ray signals and the ever-present electronic noise.

Table 2 Primary performance and physical parameters of LEND instrument

Performance and physical parameters of LEND instrument	Values and dimensions
Performance parameters	
Total mass	26.3 kg
Dimensions	461 mm × 460 mm × 440 mm
Total maximal operational power	9 Watts
Total heating power	4 Watts
Maximum data readout per day	32 Mb
Number of commands	31
Physical parameters	
Energy range of epithermal neutron measurements with narrow FOV	0.4–100 eV
Energy range of high energy neutron measurements with moderate FOV	0.3–15.0 MeV
Energy range of epithermal neutron omni-directional measurements	0.4–100 eV
Energy range of thermal neutron omni-directional measurements	<0.4 eV

The stilbene organic crystal scintillator is surrounded by a plastic scintillator that performs the active anti-coincidence shield function. Together they are referred to as the sensor of high-energy neutrons (SHEN). There is one *Hamamatsu* R5611 PMT (photo-multiplier tube) for each scintillator. The SHEN produces three signals which go to three 16 channel digitizers (Fig. 4). The first signal is from the high-energy neutron interactions in the stilbene crystal. It covers the energy range from 300 keV up to 15 MeV. The second signal is from gamma-ray interactions in the stilbene crystal in the energy range of 60 keV up to 3.0 MeV. The gamma ray signal is distinguished from the neutron signal by rise-time discrimination. The third signal comes from galactic cosmic rays or any other charged particles passing through the plastic anti-coincidence shield.

The Collimation Module (Figs. 2 and 3) presented a challenging but critical design effort because it dominates the LEND instrument mass allocation. The collimator's mass comes mostly from the high-density polyethylene and the ^{10}B, which in turn define the ultimate spatial resolution of the instrument. Considerable mechanical and material engineering efforts were made to reduce mass in the structural areas of this module so that it could be used in the collimating materials area. At all times a safety factor of 2.4 was maintained for all structures (the safety factor is the ratio between the damage strength of the material and the value of maximal strength for the structure elements in flight conditions). The LEND instrument mass as delivered was 26.3 kg, well within the design specifications that were formally proposed to NASA in response to the LRO Payload AO in June 2004.

The Collimation Module is assembled from four segments of aluminum enclosures, which have four holes for the CSETN 1–4 detectors and a central hole for the SHEN detector (Fig. 3). The segments contain the layers of neutron absorbing ^{10}B material around the inner walls, which surround all four holes (Fig. 3). This material has a density of about 1.0 g/cm^3 and is made by compressing boron powder with 97% weight fraction of ^{10}B. The thickness of absorbing layers in these segments varies from 0.4 cm at top segment up to 2.4 cm in the bottom segment (see Fig. 3). External to the boron are layers of moderating plastic. This plastic is high-density polyethylene that varies in thickness from 0.4 to 2.1 cm (see Fig. 3). The external surface of the polyethylene is covered by aluminum foil. The shape and distribution of absorbing ^{10}B and moderating polyethylene inside the segments were designed in accordance with the results of numerical simulation for collimation efficiency. Physical calibrations of LEND have demonstrated conclusively that the final design is close to an optimal configuration for achieving the best sensitivity of the instrument for meeting LRO mission measurement requirements (see Sect. 4 below). The Field of view (FOV) for the collimated sensors of epithermal neutrons CSETN 1–4 corresponds to a FWHM = 11.2° and the total FOV for high energy neutrons within SHEN corresponds to 80°.

The electronics is located in the structural base of the Module of Sensors and Electronics as seen in Figs. 2 and 4. It contains the analog and digital signal processing electronics, a low voltage power supply, as well as the ten individually controlled high voltage power supplies for each detector (see Fig. 5). The analog boards each process signals individually from sensors with a dead time of 50 microseconds. All detections of neutrons from each ^{3}He sensor are transformed into their corresponding spectra of counts at 16 energy channels. Counts in the SHEN scintillator from gamma rays and high energy neutrons are selected and recorded separately, but counts with a veto mark from the anti-coincidence shield are not counted and recorded. Each of 10 sensors within LEND has individual HV (high voltage) provision with two commandable levels. The HV levels for the ^{3}He sensors are around 2 kV. The signal processing and instrument control are configured in an Integrated Logic Core centered around an Actel FPGA.

The interface between the LEND instrument and the LRO spacecraft consists of three major components: electrical power, one Pulse-per-Second (PPS) signal from the spacecraft,

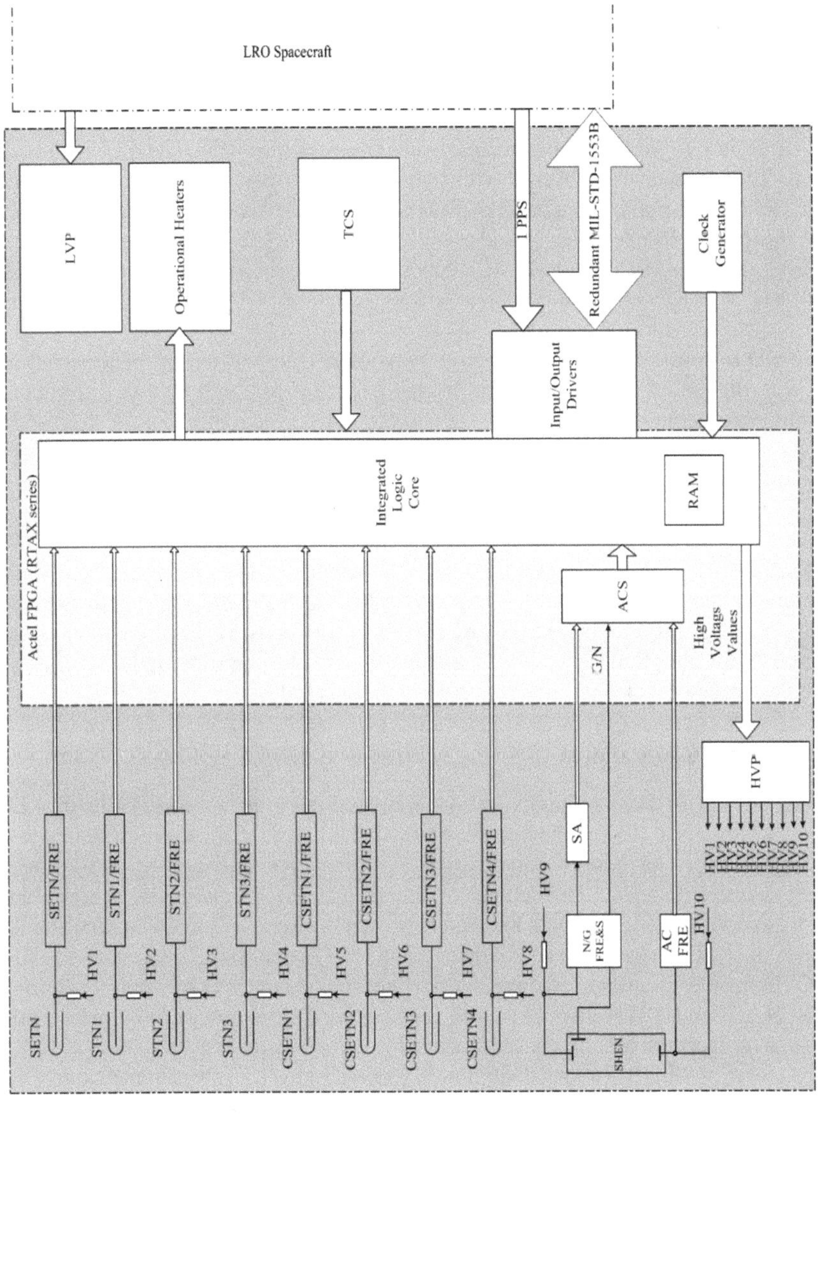

Fig. 5 LEND electronics functional block diagram

and a MIL-STD-1553 data bus. The MIL-STD-1553 is a dual-redundant master-slave data bus that is shared by four instruments and more than 10 spacecraft subsystems. The signaling rate on the bus can be as high as one MHz, but the actual data transfer rate is much less. The LEND instrument responds to 1553 transactions as a Remote Terminal. Control of information transfer on the bus will reside with the LRO spacecraft computer. The LEND 1553 bus interface will always be configured as a Remote Terminal.

Every second, the LRO spacecraft reads science data from the instrument. The LRO Spacecraft will then store the science data, which amounts to approximately 32 megabytes per Earth day. The spacecraft will read the science data nominally at 500 ms after the rising edge of the PPS signal and ensure that all reads occur at least 100 ms after the PPS, and 100 ms before the next PPS.

The thermal design for LEND is shared between the instrument team and the LRO spacecraft team. The instrument is covered with a multilayer insulation (MLI) provided at the time of its delivery from IKI, Russia. The LRO project added a layer of Ge-Black Kapton to complete the thermal blanket. This provides an outer layer that is conductive and has the desired optical properties. Even with the MLI blanket, the four external detectors (SETN and STN 1–3) have supplemental heaters. These heaters are controlled by the LEND electronics box.

The LEND instrument has relatively large mass of 26.3 kg, with most of the mass associated with the collimator. The material in the collimator has high thermal capacitance and therefore temperature changes very slowly. Since the electronics box is structurally adjacent to the collimator, the box has very good temperature stability. The instrument is mechanically and thermally attached to a radiator via eight attachment feet. The radiator is LRO spacecraft hardware provided to the LEND instrument. The radiator is mechanically attached to the spacecraft but thermally detached. The radiator has two heater circuits, one to keep the instrument above operational limits and another to keep the instrument above survival limits. The operational heaters are controlled via the spacecraft's computer and the on and off set points can be changed by software table uploads. This circuit has over-temperature thermostats that protect the instrument in case of a controller failure or thermal runaway. Thermal switches or thermostats control the survival heater circuit.

4 LEND Physical Calibrations

The quality of the LEND investigation results are based upon the accuracy of its physical calibration, which has to provide the full set of physical values to characterize the instrument and its range of responses. These values include the detection efficiencies $\varepsilon_{ij}(E, \mathbf{n})$ of each instrument sensor i; this gives the number of counts detected at a particular energy channel j for a given flux of neutrons with known energy E from known direction $\mathbf{n}$. The angular dependence of $\varepsilon_{ij}(E, \mathbf{n})$ for collimated sensors CSETN 1–4 and SHEN represents the telescope FOV's for these sensors. The angular dependence of $\varepsilon_{ij}(E, \mathbf{n})$ for omni-directional sensors STN 1–3 and SETN also takes into account the distribution of local mass of the instrument and spacecraft around the sensors.

The program of LEND ground calibrations included physical measurements and numerical simulations. Special laboratory radiation sources with known intensity and energy spectra having qualitatively similar, but quantitatively different energy spectrum from lunar neutron emissions were used for physical measurements. A large number of physical measurements were performed with these sources, at many different angles. In parallel, these measurements were also simulated by numerical calculations using the MCNPX code, and good agreement was obtained between the measurements and the corresponding results of numerical simulation.

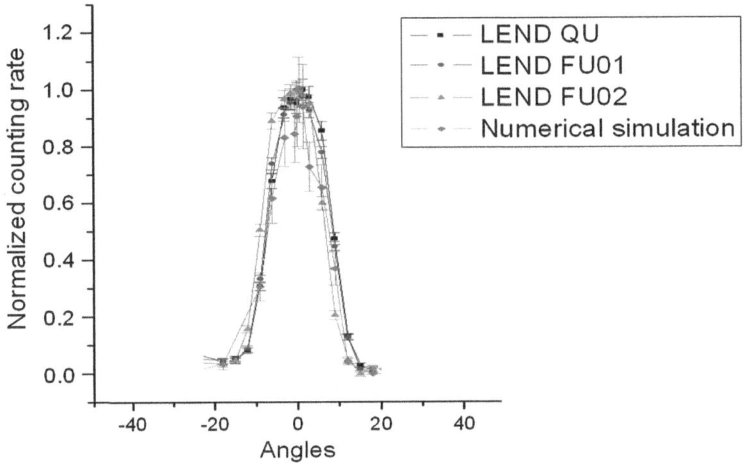

Fig. 6 This figure shows, as an example, the comparison between the angular response of one particular CSETN in flight units FU01, FU02, and that of the qualification unit to a calibration source of ^{252}Cf inside a polyethylene sphere of 3.8 cm radius along with a Monte Carlo numerical simulation of the expected detector response. The precision of the numerical simulations was designed to be comparable to the anticipated instrument measurement errors. All CSETN show similar angular responses, as expected

There are two stages involved with the ground calibration program. Only a limited set of 35 measurements were performed at the first stage for the LEND qualification unit, as well as for the first and the second flight units due to the limited time available for these measurements. The sensors of each unit were irradiated from different angles by one standard emitter of moderated epithermal neutrons from a ^{252}Cf radioactive source inside a polyethylene sphere of 3.8 cm radius. The activity of this isotropic emitter was 3.75×10^6 neutron/sec. The measured angular dependence of counting rates for the collimated sensors within LEND are shown in Fig. 6. The omni-directional sensors of LEND were also characterized by these initial measurements. Such measurements represent the first stage of ground calibration and should be considered as reference calibration data for the performance of LEND flight units FU01 and FU02. Total measurement time of for this reference calibration was about 50 hours. Figure 6 shows the comparison between the measured shape of the FOV function for one of the CSETN collimated sensors and the corresponding results of numerical simulations. It can be seen that the numerical model of the instrument agrees quite well with the experimental data from the calibrations described above. Therefore, one may use the same numerical model for simulation of LEND measurements, provided the source of neutron emission is the surface of the Moon.

One of the two available flight units for the LEND investigation, designated Flight Unit 02 (FU02), is currently operating successfully in lunar orbit onboard LRO. At the second stage of ground calibration, a much more extended program of calibration will be accomplished using the other LEND flight unit, FU01, which has remained on the ground. Three different standard emitters of epithermal neutrons are being used for this program with different ratios of high-energy and low-energy neutrons, and with many more directions for irradiation of the instrument sensors. The total number of measurements with FU01 will be 420 individual irradiations, and the total estimated time of measurements of this program will be about 300 hours. The program will also include additional numerical calculations of instrument response functions for different energies and directions of detected neutrons, which will be validated by the measurements. For each sensor, the final response function for neutrons as a function of energy and direction will be constructed as part of the synergistic analysis between ground calibration measurements and numerical calculations. As the final data product of this comprehensive calibration program, the set of response functions $\varepsilon_{ij}(E, \mathbf{n})$ will be determined for all sensors of the flight unit FU01 on the ground. Reference calibration data for the on orbit flight unit FU02 will be used to create the similar set of response functions for this unit.

LEND has also been calibrated in flight on LRO. There are three different types of measurements for in-flight calibrations. LEND operated continuously during the cruise flight to the Moon and in the Commissioning Phase first elliptical orbits. The cruise data characterized the local background of the LRO spacecraft under the bombardment by galactic cosmic rays from the entire sky. The data for the high-eccentricity elliptical orbit (Commissioning Phase) included the flux from the Moon at variable distance and, correspondingly, local background of the spacecraft under the action of galactic cosmic ray from variable fractions of non-obscured sky. These data from cruise and high-eccentricity orbit have made it possible to estimate the local background of the spacecraft at different distances from the Moon. This constitutes the first data product from the LEND in-flight calibration. During the primary LRO mapping phase (16 September 2009 through 16 September 2010), the estimated counts due to local background are subtracted from the total counting rate for determining the correct count rate for the Moon.

Another type of LEND in-flight calibration is based on measurements at off-nadir pointing angles for the instrument. Several sessions of off-nadir measurements have been conducted as part of the LRO flight program, when the axis of nadir-looking instruments has been directed at $15^\circ, 30^\circ, 45^\circ, 60^\circ$ and 90° from nadir. The data from the LEND collimated sensors during this session are being used to characterize the angular distribution of lunar neutron emissions. The angular distribution function of the lunar emission of neutrons constitutes the second data product from this type of LEND in-flight calibration.

Finally, the third data product for LEND in-flight calibrations will be data from so-called yaw maneuvers, when LRO will change the direction of X axis (the ordinary direction of flight) to $\pm 90^\circ$ from the nominal velocity vector. During these maneuvers the two Doppler filter sensors (STN 1 and STN 2) have symmetrical orientations with respect to the direction of orbital velocity, and the difference of counts due to the Doppler filter effect is absent. The only difference that could be observed during such yaw maneuvers should be attributed to the difference between the local neutron background and that from the LRO spacecraft. Therefore, the data of this in-flight calibration activity will be used to evaluate the response function of the LEND Doppler filter sensors for the standard flight orientation in the primary mapping mission orbit and geometry.

5 Scenario of LEND Observations

The primary data for the LEND measurement investigation in support of lunar exploration will be obtained during the primary LRO orbital mapping phase from a polar inclination orbit at an average altitude of about 50 km. After one year an extended LRO mission will focus upon prioritized lunar science objectives for another two years. LEND will operate continuously during the nominal LRO baseline mission with its FOV pointed primarily at nadir. The orbital velocity of LRO is about 1.7 km/s, so a surface element with a horizontal dimension of 10 km will be traversed in approximately 6 sec. The instrument will perform continuous measurements with a data accumulation time equal to 1 sec; the distance covered during this time frame is 1.7 km which corresponds to the pixel size for neutron mapping by means of data from the LEND collimated sensors.

The Moon rotates with a sidereal period of about 27.3 days with respect to the distant stars, and the orbit of LRO will stay nearly stable with respect to the stellar reference frame. Therefore, the orbital motion of LRO will, in effect, scan the lunar surface along meridians from one pole to another. The cross-track distance at the lunar equator between projections of two successive orbits will be about 31 km. Orbital observations by the collimated sensors will correspond to a coverage strip on the lunar surface with a width of about 10 km. After half of the period of lunar rotation, the observed lunar surface will include the entire lunar globe with limited gaps in a belt between 71° South latitude and 71° North latitude. The total area of the observed surface after half of a lunar day is about 45%. Due to small random components of the ratio between the periods of lunar rotation and LRO orbits, one may expect to cover the entire surface after about 30–40 days of the primary orbital mission.

One may introduce the total time of exposure for each LEND surface element (pixel) on the Moon; this time is equal to the total duration over a point defined on the basis of the LRO projection on the surface inside the boundary of a pixel. It is clear from (2) (see Sect. 2) that LEND sensitivity for hydrogen variations is higher for pixels with a larger accumulation time. In general, accumulation time increases poleward for surface elements with the same area and obtains its maximum at the poles. However, due to small deviation of the actual LRO orbit from a perfect 90° the inclination , there will be local variations of orbital coverage around the poles and along particular latitudes, where some points on the surface will achieve larger or smaller total exposure times. During the one year primary mapping phase of the mission, all data from LEND will be distributed over the pixels that define the global lunar map together with their total exposure time. This LEND experimental measurement data will allow (see Sects. 2 and 6) the development of the specific data products (1) and (3) corresponding to the first and third objectives of LEND investigation (see Tables 1 and 4), which themselves are part of the level one objectives of the LRO mission.

The highest priority measurement activity for LEND data analysis during the primary mapping phase will be testing for the potential existence of water ice deposits by searching for the first available signatures for the presence of water ice within currently known polar cold traps (PSR's). Before launch of LRO, a set of Permanently Shadowed Regions (PSRs) have been be determined for the LEND investigation using all presently available lunar topographic data from the *Clementine* and *Kaguya* missions (see Mitrofanov et al. 2008; McClanahan et al. 2009). Criteria for testing PSR site selection include the condition requiring both permanent shadow and minimal surface physical temperature. On the basis of the first data from LRO instruments in the Commissioning phase, this set of sites have received special scrutiny; after each additional full cycle of coverage of the Moon by means of LRO orbital data, each PSR site within the polar regions are being studied individually to test the presence of statistically significant depressions of the flux of epithermal neutrons in comparison with the neutron flux from the surrounding surfaces. The list of the most probable sites with the highest probabilities of water ice deposits (or with upper limits for them) will be updated by the LEND team routinely during the primary LRO mission. This experimental data will correspond to the LEND data product (3), which is directly traceable to the second objective of LEND investigation (see Tables 1 and 4).

Figure 7 presents the temporal profile of increasing coverage time for a particular PSR, in this case the polar impact crater Cabeus B-1, during the primary mapping phase of the LRO mission. Steeply increasing portions of the coverage curve correspond to time periods when LRO flies directly over the selected PSR region. Horizontal intervals within this curve correspond to time periods when LRO orbit does not cross the diameter of a given PSR.

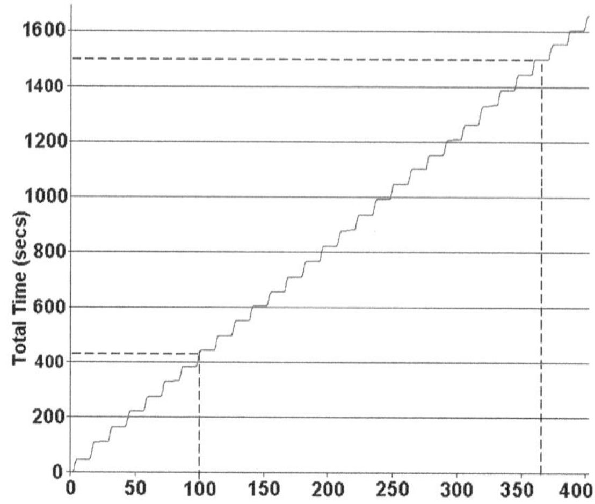

Fig. 7 The graph plots LRO mission duration on the horizontal axis and the increasing time of LEND instrument exposure for the particular case of the Cabeus B-1 PSR near the South Pole (84.9 degrees South latitude). *Intervals of 100 and 360 days* correspond to the duration of the LRO mission at the time of the LCROSS impact (9 October 2009) and the total duration of the primary NASA ESMD mission phase, respectively

One may estimate the sensitivity limits of LEND for this particular PSR using this coverage time profile and (3), which corresponds to the accumulated coverage time with respect to the dimensionless parameter α for a detectable depression.

It is obvious from Fig. 7 and (3) that after 100 days of orbital observations a total coverage time about 400 sec will be sufficient to detect the depression of neutron emissions of a factor of 2 within any given PSR in comparison with the surrounding area. In accordance with the numerical simulations, such a decrease of about 50% corresponds to the presence of 3000 ppm of Hydrogen in the regolith, or about 3 wt% of water ice (equivalent). For a full year of LEND measurements within the primary mapping orbit, the total coverage time for any given PSR such as that of Cabeus B-1 would increase up to 1500 sec, which corresponds to a detectable level of neutron depression of about 25%. Such a depression of neutron emission could be related to the presence of 700 ppm of Hydrogen in the soil (see Fig. 1), or to the presence of about 0.6 wt% of water ice. We estimate that this detection limit is adequate to test for the statistically significant presence of water ice in this particular spot. On the other hand, the PSR associated with Cabeus B-1 lies somewhat distant from the south rotational pole on the basis of its latitude of about -84.9° and the total PSR area for this crater is about 400 km^2. Table 3 lists other PSRs with variable shadowed areas and latitudes at the South and North Pole regions with total exposure times (for LEND observations) larger than 160 seconds. Therefore, we conclude that LEND will provide a comprehensive and sensitive survey for the emission of epithermal neutrons from the Moon that will adequately test for the presence of appreciable water ice within already recognized lunar cold traps (PSRs).

In performing this planned program of flight observations, the LEND team will closely cooperate with the LOLA team (Smith et al. 2009, this issue) who will provide improvements in the precise knowledge of the surface boundaries of PSRs, and with the Diviner team (Paige et al. 2009, this issue) for identification of the surface isotherms associated with cold traps on the lunar surface. During the initial stage of the LRO mission, the LEND observational program for evaluation of the potential for water ice deposits (on the basis of LEND-derived hydrogen mapping) will provide an opportunity for identifying the best available sites for directing the LCROSS impactors (NB: this was achieved successfully on 9 October 2009, and LEND measurements helped to redirect the LCROSS Centaur impact to the Cabeus PSR). On the other hand, during the primary LRO exploration mission, the program of planned LRO observations will determine the most conservative upper limits for the presence of ice deposits at every possible cold trap within the lunar polar regions. This

Table 3 Expected LEND exposure times for various North and South Pole PSRs at varying latitudes

South polar region					North polar region				
	Latitude	Longitude	Area (km^2)	Total exposure time (sec)		Latitude	Longitude	Area (km^2)	Total exposure time (sec)
1	−89.9	111.1	380	47 066	1	89.1	122.5	96	3237
2	−88.5	220.0	400	8093	2	89.0	291.2	87	2640
3	−87.6	38.0	575	7272	3	88.6	32.0	170	3685
4	−87.4	260.2	183	2137	4	88.2	250.6	100	1686
5	−86.8	75.8	257	2438	5	87.9	309.2	92	1330
6	−85.2	48.1	99	627	6	86.7	140.3	92	847
7	−85.1	184.5	58	360	7	85.7	44.8	34	240
8	−84.7	323	100	573	8	85.1	269	150	930
9	−84.7	12.5	116	665	9	84.6	151.2	68	383
10	−84.4	54.8	140	760	10	84.2	60.8	44	231
11	−83.7	84.7	140	676	11	83.8	251.3	82	402
12	−83.5	164.5	70	328	12	81.9	213.6	43	162

measurement program and its results will be necessary for planning more extended observations of these sites during the scientific phase of the extended mission of LRO, which will commence in September 2010.

LRO began its primary mission orbital measurements at the rising stage of the new 24th solar cycle. It is known that powerful solar flares and solar particle events occur quite often during this stage. The LEND measurements during solar events will be excluded from the main instrument data stream for mapping, which should be associated with the period of a quiet Sun only, when neutron emission from the Moon is produced by the stable flux of galactic cosmic rays (GCRs). Indeed, the surface variations of secondary neutron emissions between any two surface elements may only be resolved by LEND investigation provided the total exposure by primary particles remains the same for the data accumulation over each. On the other hand, the LEND data accumulated during the periods of an active Sun (with enhanced emission of solar protons) will also be a very important contribution for lunar exploration; such data will characterize the variable radiation environment from a low lunar orbit during solar flares, which are known to be a major factor in the radiation risk for future human space missions to the Moon. To achieve the data required for the most comprehensive model of the lunar radiation environment during solar flares (SPE's and CME's) of different magnitudes, a special program of observations of the target-of-opportunity type may be implemented during the mission, at which time the LRO spacecraft will perform a set of off-nadir measurements with differing inclination angles during the time of the flare event. Such data from LEND will serve to characterize the angular distribution of secondary neutrons during the solar flare over both the day and night sides of the Moon.

LEND operations and data processing will be supported by different ground segments around the world (Fig. 8). The instrument data from the project Mission Operation Center (MOC) at NASA Goddard Space Flight Center (GSFC) will be distributed between two Science Operation Centers (SOCs); one at the University of Arizona and another at GSFC. The SOC at the University of Arizona will be the primary one and it is responsible for further

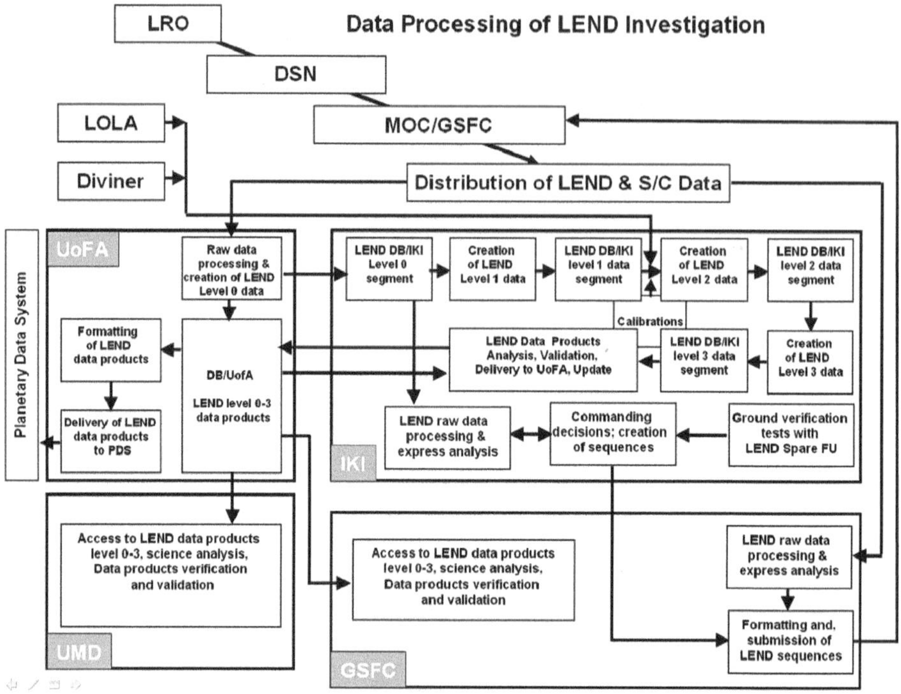

Fig. 8 LEND measurement investigation data processing flow

distribution of the data to all other instrument SOCs including the one located at the Space Research Institute (IKI) in Russia, as well as others at the University of Maryland (College Park), and at NASA's GSFC. Using raw LEND data, the Russian SOC at IKI will perform express data analysis for instrument status evaluation and, when necessary, for development of operational commands. The SOC at GSFC will also perform express data analysis in parallel with one in IKI, and will be responsible for fast feedback communications with the LRO Project to send commands to the instrument. Decisions for commanding will be generated in cooperation between IKI and GSFC taking into account results of from the express or quick-look data evaluation and analysis.

The final responsibility for the development of the required and deliverable LEND data products will be distributed between the SOCs at the University of Arizona and at the Space Research Institute [IKI] (see Fig. 8). The SOC at IKI will be mainly responsible for the generation of the scientific content of data products, levels 1–3, and the SOC at the Univ. of Arizona will be mainly responsible for formatting, final validation and delivery of all LEND data products to the Planetary Data System (PDS). The SOC in the Univ. of Arizona will also be responsible for the primary instrument database with direct access to both raw and reduced data products for all science team members.

6 Definition of LEND Data Products

The Lunar Reconnaissance Orbiter (LRO) is designed to derive geophysical properties of the Moon for use in developing knowledge and technologies needed for future human planetary

Table 4 Correspondence between LEND objectives and final data products

LEND objectives	LEND final data products (level 3 of PDS data)
(1) Determine hydrogen content of the lunar subsurface within the polar regions with spatial resolution of 10 km (FWHM) and with sensitivity to concentration variations of 100 parts per million (ppm, 3 sigma) at the poles	(1) Global maps of hydrogen content of the subsurface with spatial resolution of 10 km with sensitivity to concentration variations of 100 ppm at the poles
(2) Characterization of the surface distribution and column density of possible near-surface water ice deposits in the Moon's polar cold traps	(2) Maps of putative water ice concentration within lunar polar regions with spatial resolution of 10 km, or upper limits on the content of water ice in potential cold traps within polar craters
(3) Global mapping of lunar neutron emissions at an altitude 30–50 km above Moon's surface with a spatial resolution of 5 km (HWHM) at the spectral range of thermal energies up to 15 MeV	(3) Global distribution of neutrons at the Moon's orbit with a spatial resolution of 10 km (FWHM) at different energy ranges for thermal energy up to 15 MeV, separately for periods of quiet Sun and for periods of Solar Particle Events

exploration, and to inform decisions that pertain to cost-effective development of human flight systems. Its objectives are focused on deriving specific data products required for the initial robotic stage of NASA's Constellation Program (C × P). LEND is a contributed instrument to the NASA LRO mission by Russian Federal Space Agency, and the scientific goals of LEND investigations also corresponded to objectives of the Federal space program of Russia. LEND was competitively selected on the basis of the NASA LRO Payload Announcement of Opportunity, which was released in June 2004.

The objectives of the LEND instrument and its measurement investigation are consistent with the LRO mission level one requirements (Table 1), and correspond directly with the definition of required LEND data products (Table 4). There is also a hierarchy of LEND data products, from level 0 up to the level 3, which is in accordance with the degree of knowledge involved, with the highest level 3 datasets corresponding to the most significant scientific knowledge of the lunar environment produced by the LEND investigation (Table 4).

The level 0 data product for the LEND investigation are the raw LEND measurements in the form of counts from each sensor accumulated during specific measurement time intervals, which correspond to the LRO mission event time and to the spatial ephemeris of LRO. These data also include the engineering parameters of the instrument and spacecraft, which themselves characterize measurement conditions, and orbital parameters of the mission. The level 0 data product represents the full archive of basic measurements and engineering configurations of LEND, and will not change after delivery to the PDS.

The level 1 data product contains two different types of data sets. The first type of the level 1 data product is represented by time profiles of counts from each of 8 sensors of epithermal and thermal neutrons (CSETN 1–4, SETN and STN 1–3) with estimates of corresponding background for each sensor, and by time profiles of counts at each of 16 energy channels of the scintillation sensor SHEN with estimates of corresponding background at each channel. Local solar time at the sub-spacecraft point is also presented along time profiles together with spaceraft orbital parameters and Universal Time. Therefore, a segment of 3 months of continuous orbital measurements will be represented by 24 time profiles of LEND counts. Time intervals containing solar flares and solar particle events will be flagged.

The second type of level 1 data products represents 25 global maps: eight maps contain counts from each of 8 sensors of epithermal and thermal neutrons (CSETN 1–4, SETN and

Table 5 The schedule of LEND data products delivery to the Planetary Data System (PDS)

LEND data products	Periodicity of data delivery to PDS	Time delay of data delivery to PDS
Level 0 data products: raw data of measurements with technical parameters of observations	Four segments of LEND level 0 data will be delivered to PDS with a periodicity of 3 months, each segment contains the raw data for a period of 3 months of observations	Each segment of LEND level 0 data will be delivered to PDS two months after the last portion of telemetry data of this segment was obtained by LEND SOC from LRO MOC
Level 1 data products: Sets of averaged data in form of time profiles and maps of average counting rate for time intervals of lunar synodic period	Four segments of LEND level 1 data will be delivered to PDS with periodicity of 3 months, each segment contains time profiles of counts and maps of data in counts for the period of 3 months of observations	Each segment of LEND level 1 data will be delivered to PDS six months after the last portion of telemetry data of this segment was obtained by LEND SOC from LRO MOC
Level 2 data products: Derived data package in form of maps of derived neutron emission	Four editions of maps of derived lunar neutron emission will be delivered to PDS: 1st edition will be based on the data of segment 1; 2nd edition will be based on the of segments 1 and 2; 3rd edition will be based on the data of segments 1–3; 4th edition will be based on the of segments 1–4	Each edition of maps of derived lunar neutron emission will be delivered to PDS three months after he delivery of level 2 data product with data used for this edition
Level 3 final data products: developed data package in form of maps of hydrogen abundance (see (1) in Table 4), maps of putative water ice in polar cold traps (see (2) in Table 4) and maps of neutron component of radiation environment (see (3) in Table 4)	Four editions of level 3 final data products will be delivered to PDS: 1st edition will be based on the data of segment; 2nd edition will be based on the data of segments 1 and 2; 3rd edition will be based on the data of segments 1–3; 4th edition will be based on the data of segments 1–4	Each edition of level 3 data products will be delivered to PDS three months after delivery to PDS of corresponding edition of level 2 data product

STN 1–3) after background subtraction; sixteen maps contain accumulated counts of neutrons at each of 16 energy channels from the scintillation sensor SHEN after subtraction of detector background. The twenty fifth map contains the total observation time for each individual surface-mapped LEND pixel. The ratio of accumulated counts and total observation time for each pixel is used as an estimated average LEND counting rate for a particular time interval of mapping. Pixels have sizes $1^\circ \times 1^\circ$ at latitudes from -80° to $+80^\circ$, and $0.5^\circ \times 0.5^\circ$ at both polar regions above 80° latitudes. One set of maps corresponds to one synodic period of lunar rotation around the Sun of 29.53 earth days.

For each 3 month segment of LRO mapping observations, one delivery package of level 1 data products will be produced for delivery to PDS (see Table 5 and Fig. 9). One delivery package contains 24 time profiles for counts from LEND sensors, and 3 sets of 25 maps, which display the distribution of counts and total observation time over the lunar surface (each map corresponds to an accumulation time of 1 synodic period).

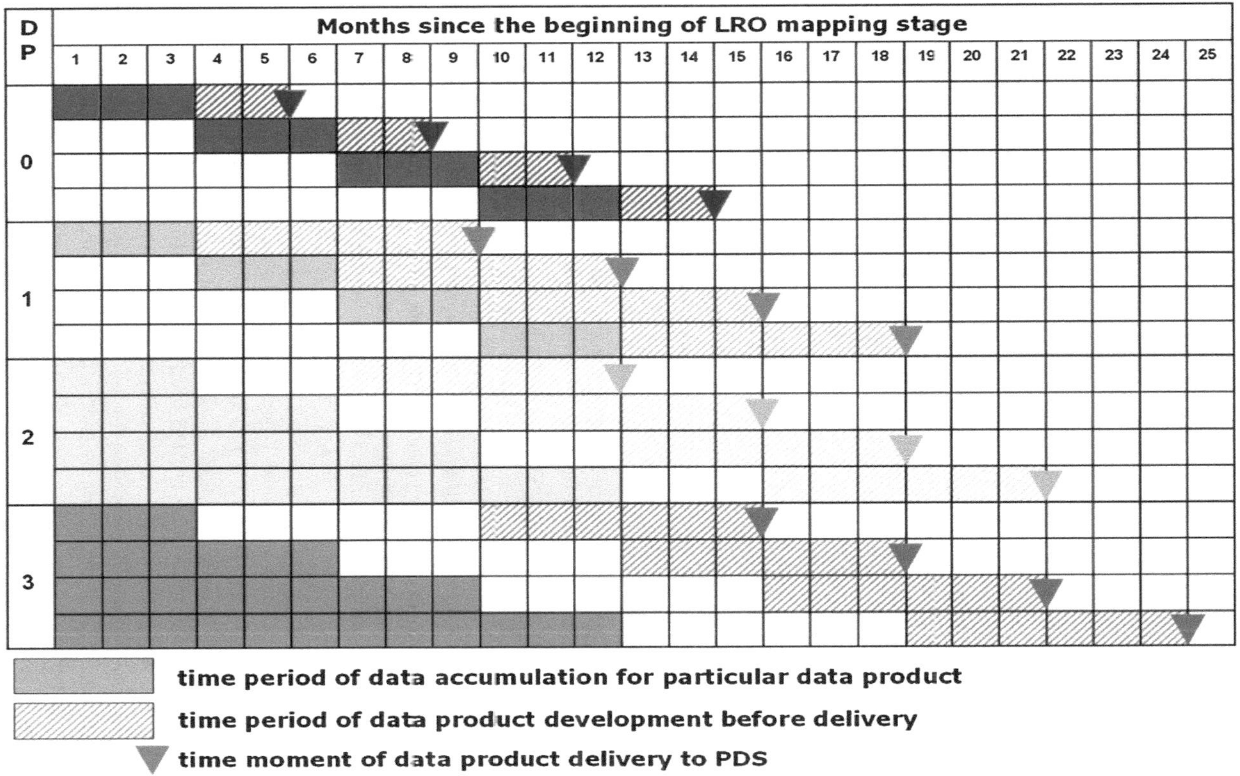

Fig. 9 (Color online) Time schedule for LEND data accumulation and development for PDS data products of level 0 (*red*), level 1 (*green*), level 2 (*bright blue*) and level 3 (*blue*). Four lines for levels 0 and 1 corresponds to four segments of corresponding data products. Four lines for levels 2 and 3 corresponds to four editions of corresponding data products

The level 2 data products contain maps of neutron emission from the Moon, which characterize the flux of neutrons at each pixel of the lunar surface. Maps for fluxes of neutrons will be presented for the energy ranges <0.4 eV, 0.4 eV–100 eV, 0.1 keV–10 keV, 10 keV–1.0 MeV, 1.0 MeV–2.5 MeV, 2.5 MeV–7.5 MeV and 7.5 MeV–15.0 MeV. One set of 7 maps will correspond to one synodic period of lunar observations. The data products of level 2 also contain the results of physical calibrations of each sensor of the instrument. The maps of the level 2 data products may have different sizes of surface elements (pixels) depending on sensors, which have different FOVs. These maps may also be summed up for several synodic periods to achieve the highest accuracy for estimates of spatial variations of neutron emission over the lunar surface.

However, estimated fluxes of neutrons for level 2 data products are *not* model-independent. For counts-to-neutrons deconvolution, the detector specific angular-energy response function is required. Secondly, for orbit-to-surface inverse projection, the model of the angular-energy distribution of neutron emissions from the surface of the Moon is also required. This is based upon numerical simulations as well as the previous iteration of data analysis. Therefore, the level 2 data will have several editions (see Fig. 9 and Table 5). The previous data will be updated, when new improved models of surface composition will be available, or when improved statistics are available to update the counts-to-neutrons data deconvolution.

The level 3 data are ready-to-use final LEND data products, which directly correspond to the instrument objectives (Table 4). All data types of level 3 are developed by means of different analysis from the level 2 data product. The first type (see (1) in Table 4) of level 3 data product is the global map of the Hydrogen distribution over the entire Moon with the spatial resolution of 10 km above $+/-80$ degrees and with lower resolution at mid-latitudes. A complementary map of the statistical errors for Hydrogen content will also be presented. The values for Hydrogen concentrations in the lunar regolith will be obtained by the process of the model-dependent neutrons-to-hydrogen deconvolution, which takes into account the available knowledge of surface elemental composition and layering structure.

The second type (see (2) in Table 4) of level 3 data product is applicable to the second LEND objective (Table 1). These products are maps of lunar polar regions above $+/-80$ degrees of latitude, which will display all known cold traps (PSRs) with estimated column density of any water ice deposit, or with conservative upper limits for the amount of water ice (if any) within each. One may expect that data from several LRO instruments will be studied in an ensemble to provide the most conclusive test of the existence of appreciable water ice deposits, and the LEND level 3 data product, type (2), will be an important part of this analysis.

The third type (see (3) in Table 4) of level 3 data product is broadly similar to the maps of neutron emission associated with the level 2 product described above. The main difference between these data products is the selection of data used for mapping. Maps of neutron emission for the level 2 data product are created from the data accumulated during periods of quiet Sun, when bombardment by galactic cosmic rays is the only source of neutron emission. However, maps of neutron emission for the level 3 data product will characterize all observed conditions of lunar irradiation by energetic particles, including periods of quiet Sun, periods of solar particle events, and periods of crossing the Earth's magnetospheric tail. In addition, the maps associated with level 3 data products will be presented with the same spatial resolution over the lunar globe of 50 km, which will allow the combination of data from collimated and omni-directional sensors of the LEND instrument.

The preliminary time schedule of PDS data delivery for one year of LRO mapping stage is presented in Fig. 9 and Table 5. Data products of lower levels (0 and 1) will be provided segment by segment, as 4 distinct segments. Each segment will contain measurements

for 3 months (or 3 synodic lunar periods) of orbital observations (see red and green solid blocks in Fig. 7). Delivery of each segment of data of levels 0 and 1 will take place after 2 and 6 months, respectively, since the last data of this segment have been recorded on orbit (Fig. 9). Raw data products of level 0 will be delivered to PDS only once, as scheduled. The level 1 data products may require more development after official delivery of the first one or two segments; additional releases of previously delivered data segments may be necessary. New releases of the same data segment may takes place together with the delivery of the next segments of this data product. One may expect that no change of PDS data products of the level 1 will occur after the fourth segment delivery.

The concept for provision of PDS data products of higher levels (2 and 3) is necessarily different; four editions of these data products will be delivered to PDS, which will be on the basis of the data from 3, 6, 9 and 12 months (lunar synodic periods) of lunar observations, respectively (see bright blue and dark blue boxes in Fig. 9). Each new edition of the data products will be based upon a greater number of counting statistics, and could also be based on improved models and/or more sophisticated techniques for data deconvolution. The first delivery of the 1st editions of PDS level 3 data products is scheduled at the end of the entire mapping stage, or after 9 months from the moment when the last data for this edition was obtained (Fig. 9). This edition will represent results from the analysis of the 1st segment of mapping stage only. The final delivery of the 4th edition of PDS level 3 data products is scheduled after one year from the end of mapping stage (Fig. 9). In this edition, the data of all four segments 1–4, or 12 months, will be used, and this delivery will complete the exploration stage of LEND investigations onboard LRO (primary Exploration mapping mission).

References

G. Chin et al., Space Sci. Rev. **129**, 391 (2007)

D.H. Crider, R.R. Vondrak, J. Geophys. Res. **105**, 26773 (2000)

D.H. Crider, R.R. Vondrak, J. Geophys. Res. **108**, 5079 (2003)

L. Gorn, B. Khazanov, *Selective Radiometers* (Atomizdat, Moscow, 1975) (in Russian)

W.C. Feldman, W.V. Boynton, D.M. Drake, in *Remote Geochemical Analysis: Elemental and Mineralogical Composition*, ed. by C. Pieters and P. Englert (1993), p. 213

W. Feldman, D. Drake, Nucl. Instrum. Methods Phys. Res. A **245**, 182 (1986)

W.C. Feldman, S. Maurice, A.B. Binder, B.L. Barraclough, R.C. Elphic, D.J. Lawrence, Science **281**, 1496 (1998)

W.C. Feldman, D.J. Lawrence, R.C. Elphic, B.L. Barraclough, S. Maurice, I. Genatay, A.B. Binder, J. Geophys. Res. **105**, 4175 (2000)

H.H. Hsieh, D. Jewitt, Science **321**, 561 (2006)

D.J. Lawrence, W.C. Feldman, R.C. Elphic, J.J. Hagerty, S. Maurice, G.W. McKinney, T.H. Prettyman, J. Geophys. Res. **111**, E08001 (2006)

J. Masarik, R. Reedy, J. Geophys. Res. **101**, 18 891 (1996)

S. Maurice, D.J. Lawrence, W.C. Feldman, R.C. Elphic, O. Gasnault, J. Geophys. Res. **109**, E07S04 (2004)

T. McClanahan et al., Abstract # 2092 at 40th Lunar and Planetary Science Conference (2009)

A.E. Metzger, J.I. Trombka, L.E. Peterson, R.C. Reedy, J.R. Arnold, Science **179**, 800 (1973)

I.G. Mitrofanov et al., Astrobiology **8**(4), 793 (2008)

T.H. Prettyman, *Remote Chemical Sensing Using Nuclear Spectroscopy*. Encyclopedia of the Solar System, 2e (Academic Press, San Diego, 2007, this issue), 765 pages

D. Paige et al., Space Sci. Rev. (2009). doi:10.1007/s11214-009-9529-2

S. Nozette, C.L. Lichtenberg, P. Spudis, R. Bonner, W. Ort, E. Malaret, M. Robinson, E.M. Shoemaker, Science **274**, 1495 (1996)

D.E. Smith et al., Space Sci. Rev. (2009). doi:10.1007/s11214-009-9512-y

A.P. Vinogradov, Yu.A. Surkov, G.M. Chernov, F.F. Kirnozov, G.B. Nazarkina, Cosmic Res. **4**, 751 (1966)

Space Sci Rev (2010) 150: 209–241
DOI 10.1007/s11214-009-9512-y

The Lunar Orbiter Laser Altimeter Investigation on the Lunar Reconnaissance Orbiter Mission

David E. Smith · Maria T. Zuber · Glenn B. Jackson · John F. Cavanaugh · Gregory A. Neumann · Haris Riris · Xiaoli Sun · Ronald S. Zellar · Craig Coltharp · Joseph Connelly · Richard B. Katz · Igor Kleyner · Peter Liiva · Adam Matuszeski · Erwan M. Mazarico · Jan F. McGarry · Anne-Marie Novo-Gradac · Melanie N. Ott · Carlton Peters · Luis A. Ramos-Izquierdo · Lawrence Ramsey · David D. Rowlands · Stephen Schmidt · V. Stanley Scott III · George B. Shaw · James C. Smith · Joseph-Paul Swinski · Mark H. Torrence · Glenn Unger · Anthony W. Yu · Thomas W. Zagwodzki

Received: 5 January 2009 / Accepted: 15 April 2009 / Published online: 16 May 2009

Abstract The Lunar Orbiter Laser Altimeter (LOLA) is an instrument on the payload of NASA's Lunar Reconnaissance Orbiter spacecraft (LRO) (Chin et al., in Space Sci. Rev. 129:391–419, 2007). The instrument is designed to measure the shape of the Moon by measuring precisely the range from the spacecraft to the lunar surface, and incorporating precision orbit determination of LRO, referencing surface ranges to the Moon's center of mass. LOLA has 5 beams and operates at 28 Hz, with a nominal accuracy of 10 cm. Its primary objective is to produce a global geodetic grid for the Moon to which all other observations can be precisely referenced.

Keywords Moon · Shape · Space instrumentation · Topography

D.E. Smith · G.A. Neumann · H. Riris · X. Sun · E.M. Mazarico · J.F. McGarry · D.D. Rowlands · V.S. Scott III · J.-P. Swinski · T.W. Zagwodzki
Solar System Exploration Division, NASA Goddard Space Flight Center, Greenbelt, MD 20771, USA

M.T. Zuber (✉)
Department of Earth, Atmospheric and Planetary Sciences, Massachusetts Institute of Technology, Cambridge, MA 02139-4307, USA
e-mail: zuber@mit.edu

G.B. Jackson · J.F. Cavanaugh · R.S. Zellar · C. Coltharp · J. Connelly · R.B. Katz · I. Kleyner · A. Matuszeski · A.-M. Novo-Gradac · M.N. Ott · C. Peters · L.A. Ramos-Izquierdo · L. Ramsey · S. Schmidt · G.B. Shaw · J.C. Smith · G. Unger · A.W. Yu
Advanced Engineering Technology Directorate, NASA Goddard Space Flight Center, Greenbelt, MD 20771, USA

M.H. Torrence
Stinger Ghaffarian Technologies, Greenbelt, MD 20770, USA

P. Liiva
Sigma Space Corporation, Lanham, MD 20706, USA

1 Introduction

The Lunar Orbiter Laser Altimeter (LOLA) is an instrument designed to assist in the selection of landing sites on the Moon for future robotic and human exploration. The primary motivation is that any future mission proposed to land on the surface of the Moon would need to be certain from the best available information that the location would be safe for robotic and/or human endeavors. In this context, LOLA was proposed for the payload of the Lunar Reconnaissance Orbiter (LRO) mission (Chin et al. 2007) to provide the geodetic location, the direction and magnitude of surface slopes, and the elevation variation at scales relevant to landing and surface mobility, the latter generally referred to as surface roughness. In addition, since one of the objectives of NASA's Exploration Program is to attempt to detect the presence of water ice on or near the surface, LOLA was designed to measure surface reflectance at the laser wavelength as any significant water ice crystals on the lunar surface would provide a measurable increase in reflectance.

The present knowledge of the topography of the Moon (cf. Wieczorek 2007) is not adequate for ensuring a safe landing in any but a few locations, such as those visited by the Apollo flights. Based on all previous missions to the Moon, knowledge of the position of features on the surface is at approximately the kilometer level and in many areas, such as the lunar far side, that knowledge is of order many kilometers. In addition, the slope of the surface at the majority of locations is completely unknown on scales of 10's to 100's of kilometers. For topography, and more generally for the shape of the Moon, the major source of data for the last 15 years has been the Clementine mission (Nozette et al. 1994), launched in 1994, which carried a laser altimeter that provided the first near-global topographic mapping of the Moon (Smith et al. 1997; Zuber et al. 1994); the exception being the regions within about 10 to 15 degrees of each pole. Clementine provided 100-m quality radial information and similar horizontal quality position information with a spatial resolution of order 1 degree (~30 km) at the equator. The lunar shape measured by Clementine is shown in Fig. 1 (Zuber et al. 1994). A more recent effort combined Clementine altimetry (Smith et al. 1997) with elevation models of the lunar polar regions from stereo-photogrammetry (e.g. Cook et al. 2000). This model (U.S. Geological Survey 2002) has improved spatial resolution at the poles but lacks geodetic control.

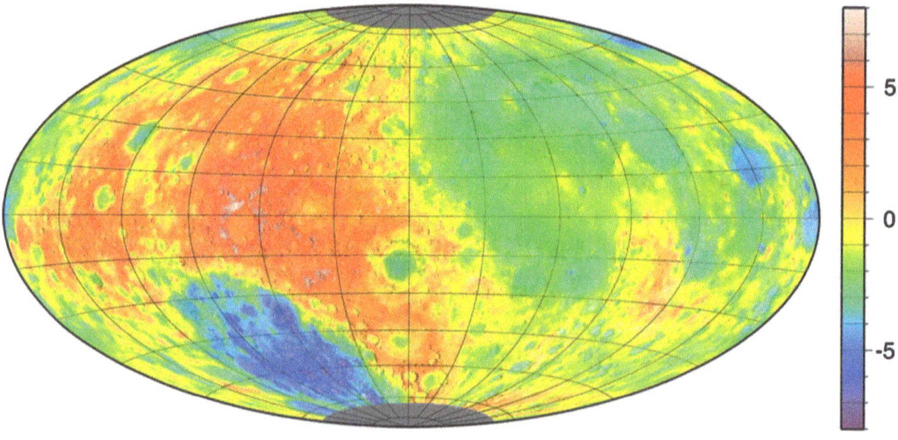

Fig. 1 Clementine lunar topography (Smith et al. 1997). Clementine altimetry data did not extend to the poles; estimated overall radial accuracy is 100 m. The topography scale is in kilometers

A new USGS Unified Lunar Control Network (ULCN 2005) based upon most available data was released in 2006 (Archinal et al. 2006).

Recently, laser altimeters have flown to the Moon on missions from the Chinese, Japanese and Indian space agencies. The Chinese (Qian 2008) and Japanese (Araki et al. 2009) instruments, both single-beam 1-Hz ranging systems, operated successfully in lunar orbit. The Indian instrument (Kamalakar et al. 2005), a single-beam 10-Hz ranging system similar to MOLA at Mars (Smith et al. 1999; Zuber et al. 1992), began mapping in fall 2008. While so far the Chinese and Japanese instruments have improved lunar topographic knowledge, their reported range accuracy of 5 to 10 m, and measurement spacing along track of 150 m to 1.5 km dictate that none of these systems meet the requirements identified by the NASA Exploration Program, which included 30-m along-track measurement spacing, meter-level ranging, and surface roughness and slope measurements.

In order to improve significantly our knowledge of the lunar topography, it was necessary employ a greater accuracy altimetric ranging system and be able to determine the position of the LRO spacecraft to significantly greater accuracy than had been achieved on earlier missions. In addition, the spacecraft altitude and the timing of the observations would be needed to commensurate accuracy.

2 Investigation Description

2.1 Exploration

The LOLA investigation is designed to assist in the selection of future landing sites by providing topographic, surface roughness, surface slope, and surface reflectance measurements, and a global lunar coordinate system that provides information on the positions of lunar surface features. In the design and development of the instrument, certain assumptions were made about the geodetic requirements for landing on the lunar surface, particularly spatial resolution and positional accuracy. The assumptions used to guide the instrument design are shown in Table 1. These assumptions are in three categories: landing site knowledge, instrument performance, and spacecraft positional and orbital knowledge.

In order for positional accuracies to be achieved globally, it is necessary to develop an improved lunar reference system or grid that has an internal consistency at the ~50-m level. This grid could then be compared with the known positions of the Apollo landing sites so that all LRO and previous lunar data could be incorporated into the LRO/LOLA-derived

Table 1 Assumptions in LOLA instrument investigation that guided design of the system and were not being met by other lunar laser altimeters

Lander scale	~5 m	Approx. size of Apollo landed module
Landing area	~50 to 100 m	
Slope accuracy in 2 directions	±1°, 10-m baseline	
Elevation precision	±0.1 m	
Radial accuracy	<1 m	Includes orbit and instrumental errors
Horizontal accuracy	<50 m	Landing site size
Surface roughness	0.3 m	Typical height clearance of Mars lander
Spacecraft position	H: 50 m; V: <1 m	Landing site positional knowledge
Spacecraft timing	3 ms	Spacecraft velocity is ~1.6 m/s

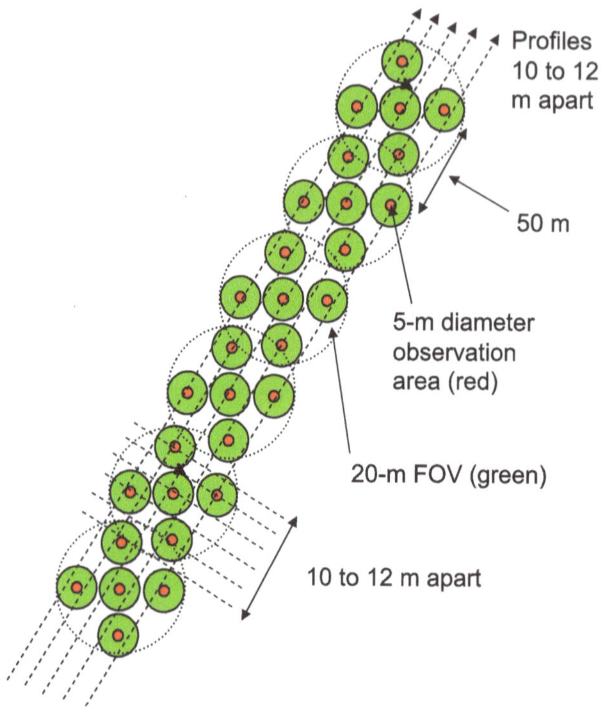

Fig. 2 LOLA's five laser spot pattern on the lunar surface

coordinate frame. The development of an accurate (∼50-m) global geodetic grid was considered to be the highest priority objective of the LOLA investigation and will ultimately provide the control at regional and local scales. An internal consistency at the 50-m level also requires spatial positioning of the LRO spacecraft of similar quality, which in itself requires improvements be made to the lunar gravity model. This improvement is expected to come from the global coverage of laser altimeter data in combination with tracking of the LRO spacecraft. Nominal LRO tracking is by the Universal Space Network (USN) at S-band with approximately ±1 mm/s Doppler per 10 s augmented by laser ranging (LR) (Zuber et al. 2009), which is able to provide approximately ±10 cm range measurements at 28 Hz.

The determination of the full surface slope requires the measurement of slope in two directions and hence the altimeter must make measurements across track as well as along track. This requirement is accomplished in LOLA by a multi-beam approach that provides 5 measurements in an X-pattern from which the slopes along track and across track can be derived on scales of approximately 25 m. This X-formation also has the advantage that, suitably rotated, will provide, 5 near-equidistant parallel profiles spaced at approximately 12 m. Figure 2 shows the 5-spot pattern on the lunar surface that LOLA will implement.

The 5 parallel profiles, providing 5 measurements of altitude, surface roughness, and surface brightness for each laser pulse, characterize a continuous strip of the lunar surface 50- to 60-m wide and will provide an assessment of the suitability of the area as a "lunar landing strip". Graphical displays of the data from LOLA, provided within hours of the data being returned will show the elevation of the surface, an estimate of the surface roughness and surface reflectance, and the derived measurements of the slope of the surface. Figure 3 shows the large number of slopes on various baselines derivable from 10 observations obtainable in less than 0.1 seconds of operation.

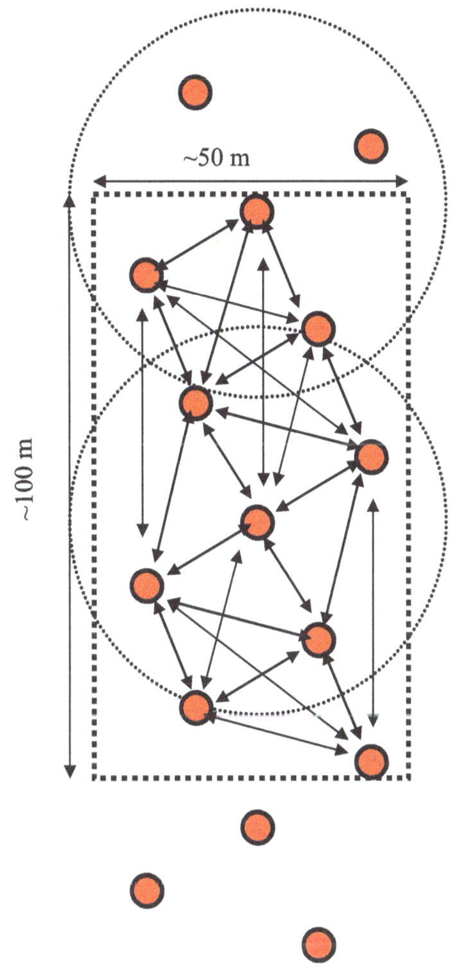

Fig. 3 Surface slope measurements from LOLA profiles showing along and across-track capability on baselines of 25 m to 50 m

The surface reflectance represents a measure of the albedo of the surface at the laser wavelength of 1064 nm and its measurement has been designed into the instrument with the aim of being able to detect the presence of surface ice at small areal density. The measurement is accomplished by estimating the energy of the return laser signal and comparing it with the output energy. Note that this measurement is calibrated in a relative sense, with respect to pre-launch testing, as the instrument lacks a source with known brightness in flight. Nonetheless, experience has shown the measurement to be extremely stable based on experience in ranging at Mars (Smith et al. 2001; Sun et al. 2001). The brightness of ice crystals on the lunar surface is expected to have an albedo of >0.8 (Abshire et al. 2005), and is approximately 4 times the average albedo (∼0.2) of the regolith. Thus an albedo measurement of 10 to 15% in each of the 5-m spots on the lunar surface would enable ice crystal areal densities of 4 or 5% to be detected. The possible detection of surface ice by LOLA, although not the instrument's prime measurement, is considered a potential high-priority opportunity to address one of LRO's primary mission objectives.

The resolution across track depends on the number of orbits of the spacecraft. After one year the across-track resolution is approximately 0.04°, providing an average separation of

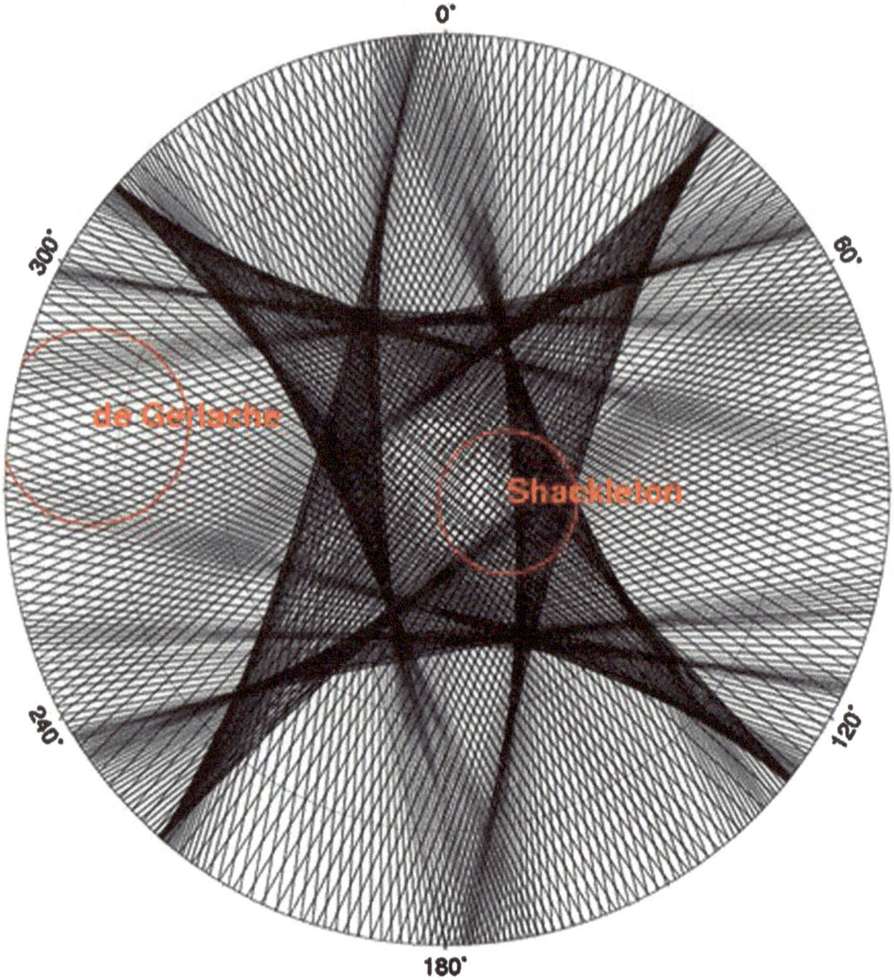

(a)

Fig. 4 (**a**) Example of south polar coverage, lat 88° S to the pole. (**b**) Example of coverage lat 80S to the pole showing the range of spacecraft heights. The height scale is in kilometers

the LOLA swaths at the equator of order 1.2 km, and at latitudes 80N and S a separation of order 200 m. Figures 4a and 4b show the coverage from 88S to the pole for a typical month and 80S to the pole; each month is very similar to the next although the ground tracks do not repeat.

2.2 Science

In addition to providing measurements for exploration, the LOLA Investigation Team will also conduct a variety of scientific studies to address a wide variety of outstanding questions in lunar science.

(a) LOLA will improve the current understanding of impact flux on the Moon, especially during its early history, by helping to identify and characterize topographic expressions of

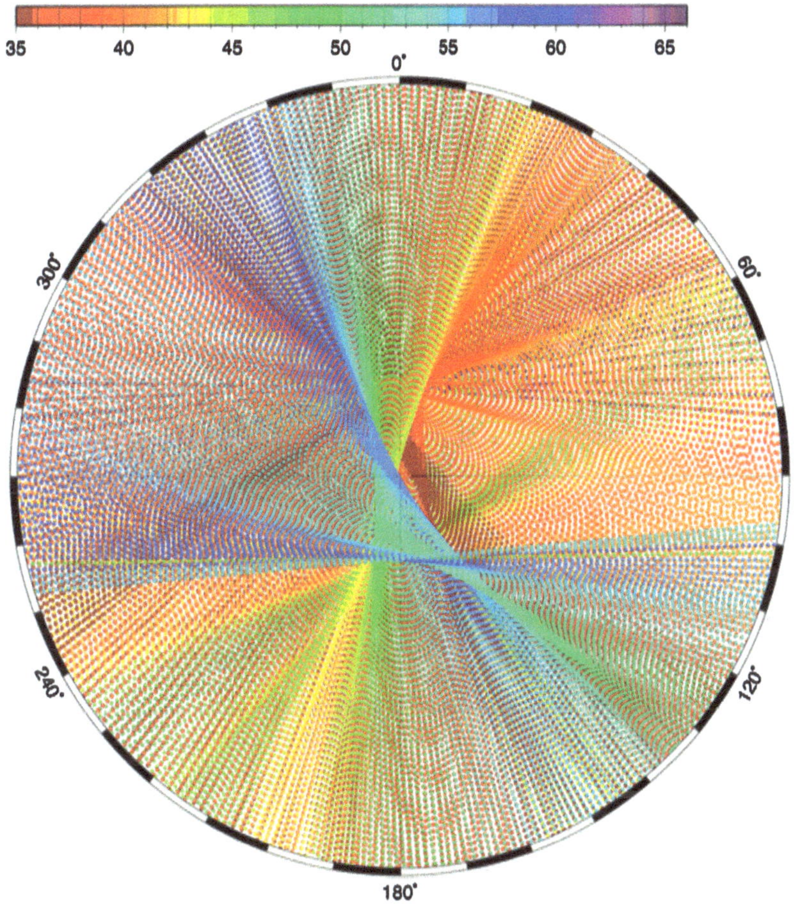

(b)

Fig. 4 (*Continued*)

crater-like circular structures otherwise undetectable in visible images (Frey et al. 2002). Their size-frequency distribution will provide new constraints on the cratering history subsequent to the late heavy bombardment. In addition, LOLA will provide greater insights on the factors that contribute to the formation of craters, including improvements to the connection between an observed crater and the projectile size/impact velocity which generated it (cf. Melosh 1989), yielding an improved understanding of the size-frequency distribution of the primary impactor population striking the Moon (Hartmann 1970).

(b) Topography and gravity are fundamental measurements that provide information on the lunar interior structure. LOLA topography in combination with high-resolution gravity soon to be available from the Gravity Recovery and Interior Laboratory (GRAIL) mission (Zuber et al. 2008a) will be used to map the Moon's global crustal and lithosphere thickness as well as crustal and mantle density anomalies. These observations will be applied, *inter alia*, to study the structure of impact basins as a measure of the influence of impactors on the shallow interior of the Moon.

Models of the planetary interior depend explicitly on the estimate of the polar moment of inertia, which measures the radial distribution of mass. The k_2 Love number of the Moon, which measures the tidal response of the potential, has been determined from analysis of satellite tracking data (e.g., Konopliv et al. 2001), with $k_2 = 0.026 \pm 0.003$ (Goossens and Matsumoto 2008), from a re-analysis of Lunar Prospector (LP) and historical data, found $k_2 = 0.0213 \pm 0.0075$. The LRO data, LOLA altimetry, S-band tracking, and LR ranging, will lead to an independent estimate of this important physical parameter. In addition, from altimetry and LRO laser ranging (LR) (Zuber et al. 2009), we will be able to estimate the global h_2 Love number—which reflects the global geometric radial response of the Moon to tidal deformation and whose uncertainty is currently $\sim$25%.

(c) Lunar albedo measurements, independent of illumination conditions, can be made by LOLA. To characterize possible volatiles in permanently shadowed polar regions, LOLA provides a bidirectional reflectance (albedo) at zero phase angle by means of the ratio of transmitted and returned energy measurements, and is capable of revealing concentrations of a few percent of surface water ice (see Sect. 2.1).

(d) LOLA data will be used to make quantitative assessments of volcanic landforms and source regions, and the characterization of lava flows and their stratigraphic relationships. Magmatic intrusions (dikes and sills) result in unique patterns of uplift and deformation at the surface that can be used to invert for the geometry of the underlying magma bodies. LOLA topographic data will be used to discriminate between magmatic and non-magmatic tectonism, and to constrain the volumes of the underlying magma bodies. Similar analyses have been successfully applied to the study of dikes on the Earth and Mars (e.g., Schultz et al. 2004).

(e) Ridges and rilles are tectonic structures that reflect basin-scale loading processes (Quaide 1965). The high spatial and vertical resolution of LOLA data will allow significant improvements in the estimation of strain associated loading of maria units, and will improve understanding of both the thermomechanical state of the underlying lithosphere and the history of volcanic extrusion within major impact basins.

(f) LOLA data will be used to improve knowledge of impact cratering and the processes in shaping planetary crusts, particularly, early in their history (Melosh 1989). New information will be obtained at all scales including multi-ring basins (e.g., Hartmann 1971), which play a significant role in shaping the topography of the terrestrial planets (cf. Spudis et al. 1994). The topographic signature of basin ring structures on the Moon is presently poorly resolved in Clementine topography (Zuber et al. 1994). LOLA topography is expected to enable the first very high-resolution study of well-preserved multi-ring structures, thereby providing important information regarding their underlying structure and formation.

(g) Ejecta generated by impact processes constitute one of the primary sources of regolith and influences its subsequent evolution (Melosh 1989). High-resolution topographic analyses of ejecta at lunar craters using LOLA will provide new constraints on the thickness, run out, roughness, and block-size distributions of such ejecta as a function of crater and basin size (Soderblom 1970). These measurements will collectively provide a quantitative means to assess the role of ejecta in the formation and evolution of regolith.

In addition, many science objectives are enabled or supported by the knowledge of topography, roughness, and surface slopes, which depend on precision laser tracking and orbit determination of LRO. Thus, LOLA will re-analyze its data using the GRAIL gravity model that is expected in 2012 and thereby improve geodetic framework for all other lunar data. In addition, we plan to combine the Kaguya (Araki et al. 2009) and Chandrayaan (Kamalakar et al. 2005) altimetry with LRO data for an extended, fully compatible and inter-calibrated dataset. We will also combine data from LOLA and the Lunar Reconnaissance Orbiter Camera (LROC) (Robinson et al. 2009) in a joint LOLA/LROC activity, for geodetically-controlled stereo topography to obtain the highest-resolution Digital Elevation Models (DEMs). The LOLA global data products, working with LROC (Robinson et al. 2009), will be cartographically-tied to the Lunar Laser Retro Reflectors (Dickey et al. 1994) and the Apollo and Lunakhod sites and thus incorporate the most precise lunar orientation/libration properties.

2.3 Orbit Determination for LOLA

The quality of the topographic measurements and models depends upon the quality of the altimeter and its supporting data, and also on the knowledge of the spacecraft orbit. The routine operational tracking data of LRO will be S-band Doppler acquired for approximately 20 hours per day and an accuracy of order 1 mm/s every 10 seconds (1σ). The altimeter has decimeter accuracy but the orbital error is expected to be significantly larger. Orbits for LRO will be developed by the LOLA team using a combinations of the S-band Doppler and laser tracking (Zuber et al. 2009) in conjunction with the altimetry.

An important tool of the LOLA orbit determination activity is the altimeter used in the orbital crossover mode. At an orbital crossover the lunar radius is the same by definition, as is the surface roughness, and the implied surface slopes. The equivalence of these measurements at the same location on different orbits enables the location of the crossover to be determined and the altimeter measurements to be used as an observable in determining the orbit of LRO, and also for estimating other parameters, such as pointing and timing, and gravity coefficients of the lunar gravity field (Rowlands et al. 1999). In addition, the very precise laser ranging observations (Zuber et al. 2009) will constrain the LRO orbit to the decimeter level over the lunar near side and will significantly improve the orbit accuracy beyond that which will be achievable with S-band Doppler tracking alone, and also enable the estimation of additional force model parameters. At the conclusion of the analysis of all LOLA data and the simultaneous LRO orbit determination we expect the global radial positional accuracy at the 50-cm level and at the 25-m level in the horizontal. Further details on the crossover analysis are given in Zuber et al. (2009).

3 Instrument Design

3.1 Overview

The LOLA instrument configuration is shown in Fig. 5 and the key instrument parameters are shown in Table 2. LOLA uses a Q-switched Nd:YAG laser at 1064 nm and a silicon avalanche photodiode detector (SiAPD) to measure the time of flight (TOF) to the lunar surface from a nominal 50-km orbit. The transmitted laser beam is split in five different beams by a diffractive optical element (DOE), each beam having a 0.5-mrad spacing. The receiver telescope focuses the reflected beams into a fiber optic array, placed at the focal plane of the telescope. The array consists of five fibers and each fiber in the array is aligned with a laser spot on the ground. The fibers direct the reflected beams into five separate detectors. The detector electronics amplify the signal and time stamp the received pulses relative to the spacecraft mission elapsed time (MET) using a set of time-to-digital converters (TDC) with 0.5-ns resolution. The transmitted pulse is also time stamped and the TOF to the lunar surface can be determined. The LRO spacecraft carries an ultra-stable quartz oscillator and distributes the timing signal to LOLA and other instruments. A signal-processing algorithm running on an embedded processor continually adjusts the receiver gain and threshold levels

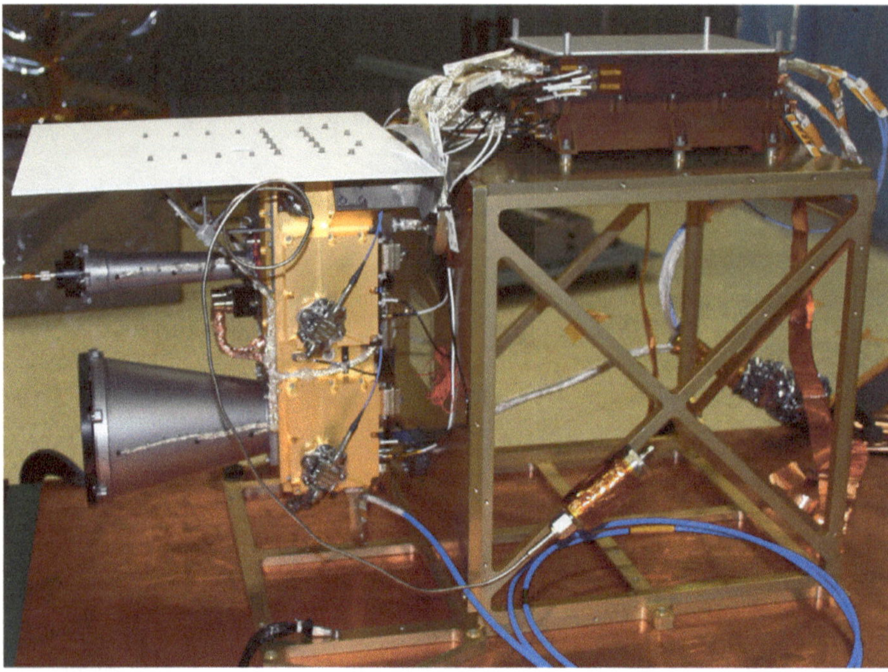

Fig. 5 LOLA instrument configuration prior to thermal vacuum testing. The frame is a holding mechanism during the test. The electronic box is on the *top side of the frame*, the *white rectangle* is the radiator; the grey cone on the left is the receiver telescope; the laser beam expander is the small cone above the receiver telescope

Table 2 Key LOLA instrument parameters

Parameter	Value
Laser wavelength	1064.4 nm
Pulse energy	2.7/3.2 mJ (laser1/laser2)
Pulse width	∼5 ns
Pulse rate	28 ± 0.1 Hz
Beam divergence	100 ± 10 μrad
Beam separation	500 ± 20 μrad
Receiver aperture diameter	0.14 m
Receiver field of view	400 ± 20 μrad
Receiver bandpass filter	0.8 nm
Detector responsivity (nominal)	300 kV/W
Detector active area diameter	0.7 mm
Detector electrical bandwidth	46 ± 5 MHz
Timing resolution	0.5 ns

and maintains the range window centered on the lunar surface return. A functional block diagram of LOLA is shown in Fig. 6.

The LRO spacecraft also carries a unique laser ranging (LR) system (Zuber et al. 2009) for precise orbit determination. The laser ranging system consists of a 30-mm aperture op-

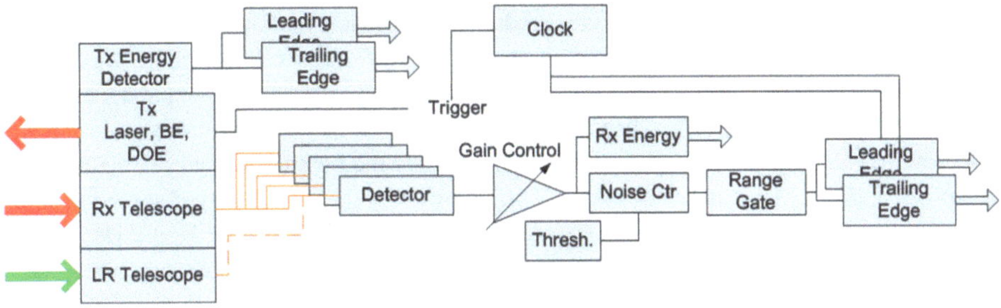

Fig. 6 LOLA functional block diagram

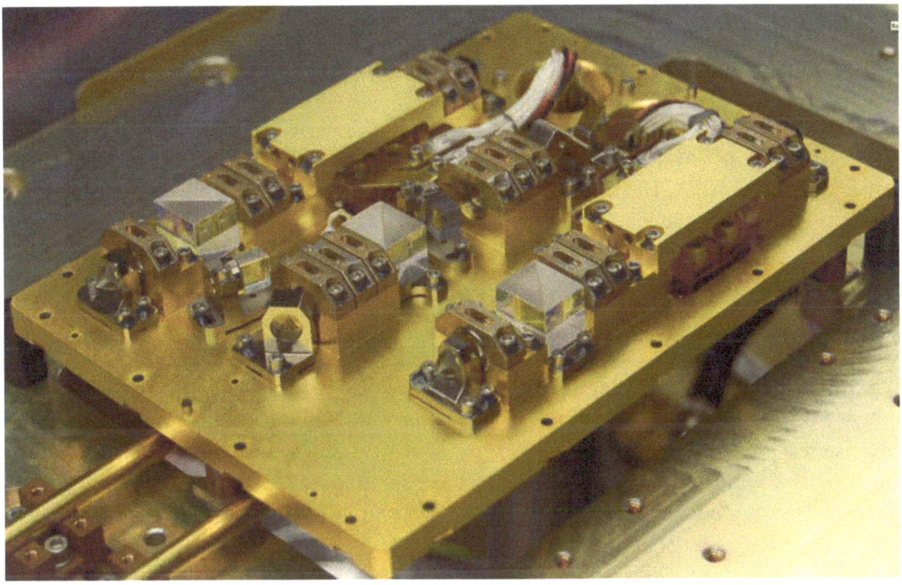

Fig. 7 Laser transmitter showing two lasers

tical receiver mounted on the LRO spacecraft high gain antenna used for communications and data transmission. The ranging precision of the LR is nominally 10 cm at 28 Hz. The receiver is pointed to a ground (Earth) based laser satellite ranging station that sends a 532-nm laser pulse to LRO. The receiver focuses the incoming 532-nm beam into a fiber bundle and the laser pulses are then directed onto one of the five LOLA detectors modified to accommodate both signals at both 532 and 1064 nm. The combination of LOLA altimetry and LR on the lunar near side is expected to provide the position of LRO at the decimeter level and constrain the orbit of LRO on the near side in the orbit determination process. The LR system is described in a companion paper (Zuber et al. 2009).

3.2 LOLA Transmitter

The LOLA transmitter consists of two virtually identical, diode pumped, Q-switched Nd:YAG oscillators operating at 1064.4 nm (Fig. 7). The diode pump lasers are derated to increase their lifetime. The laser beams are combined with polarizing optics and only one laser is operating at a time; the other one is redundant. The laser repetition rate is 28 ± 0.1 Hz, the energy per pulse is ~2.7 mJ for Laser 1 and ~3.0 mJ for Laser 2. The pulse width is approximately 5 ns. The output of the laser is directed through an ×18 beam expander and then through the DOE to produce five beams separated by 500 ± 20 μrad. The laser beam prior to the beam expander has a divergence of 1.8 mrad. After the beam expander and DOE, each beam has a 100 ± 10 μrad divergence and approximately the same energy; some variation in the energy between beams is discernible due to imperfections in the DOE. The LOLA laser is designed to operate in vacuum but a significant amount of the subsystem and system testing was done in air. The laser cavity was integrated in a clean room and two filters in the laser housing are designed to prevent particulate and molecular contamination of the lasers prior to launch.

The laser beam pattern is rotated 26° relative to the spacecraft velocity vector. From the nominal orbit of 50 km, each laser spot is ~5 m in diameter on the ground. The five-spot pattern will allow LOLA to measure both the slope and roughness of the lunar surface. The five beam spot pattern on the ground (including the field of view) relative to the spacecraft velocity vector is shown in Fig. 2.

A small kick-off mirror placed inside the laser box directs a small fraction of the laser power onto the SiAPD. The detector monitors the energy of the transmitted laser pulse and it also used as a fire acquisition signal that turns off the drive to the pump diode lasers. In addition, the outgoing pulse is time tagged and provides timing information for the time of flight measurement.

Pulse width LOLA monitors the laser-pulse energy and the pulsewidth at threshold crossing, which may be used to infer the transmitted laser-pulse shape. Note that the pulse shape output from the LOLA SPOT 4 detector is the convolution of the actual pulse shape and the impulse response of the transmit energy monitor. The relationship between the LOLA measured pulse width and the threshold value for Laser 1 and 2 was measured and the result is given in Fig. 10. The nominal transmit threshold setting is 116 mV. The average transmitted laser pulse energy is not expected to decrease significantly over the mission lifetime, thus the threshold value for the transmitted pulse should not have to be adjusted. If the laser or the transmitter common optics were to degrade, the threshold could be lowered to continue detecting transmitted laser pulses. There is approximately a factor of six adjustment that can be made on the transmit threshold (from 116 to 19 mV).

Life Testing of Transmitter As part of the development of the transmitter, a parallel life test program was conducted in order to verify the basic design, identify areas requiring modification, and provide an estimate of the lifetime that can be expected on orbit. This was particularly important for LOLA because of the use of a double transmitter system that would enable extended life of the instrument. The goal of the instrument was to acquire 1 billion pulses from each laser thus providing approximately 9 billion altimetric, pulse spreading and surface albedo measurements.

Three life tests at various stages were conducted. The first was on an early design that operated continuously for approximately 560 million shots that verified the basic design. The second test on a pre-engineering model (pre-EM) of the Laser 1 operated for approximately 520 million shots, and test 3 (presently in process) on an EM of Laser 2, has accumulated over 500 million shots without degradation. The requirement of the LRO mission for exploration is approximately 880 million total shots.

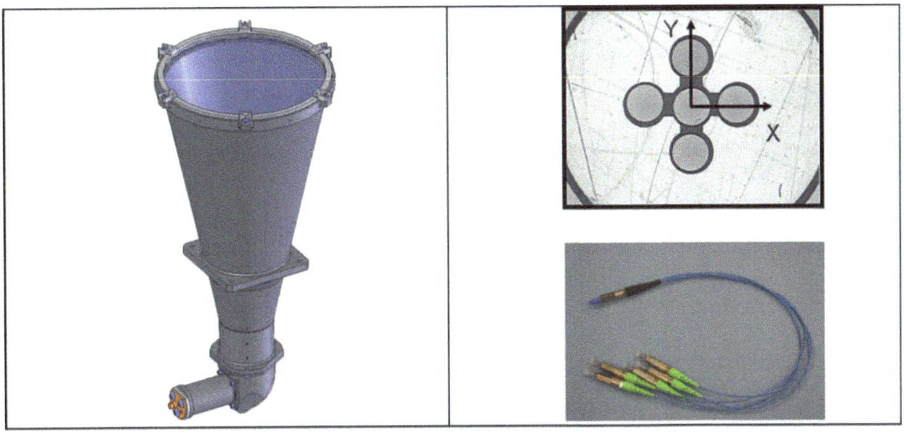

Fig. 8 Receiver and fiber bundle. The fiber bundle takes the photons from the receiver telescope to the detectors

3.3 LOLA Receiver and Detector

The LOLA receiver consists of a 14-cm-diameter clear aperture refractive telescope that focuses the received photons on to a fiber optic bundle (FOB). The effective focal length of the telescope is 500 mm. The design and materials of the telescope were chosen to minimize the thermal fluctuations expected on orbit. Since LOLA is not an imaging system, the concern is not to maintain the image quality but to collect the maximum number of photons with the fewest possible losses and to minimize the background radiation. The telescope assembly includes a dielectric fold mirror, which lets all radiation other than 1064 nm pass through and reflects the laser radiation onto the FOB. This minimizes the amount of background solar radiation incident on the detectors. The fiber bundle consists of five identical fibers. Each fiber in the bundle directs the reflected energy into the aft-optics assembly for each detector. The receiver and fiber bundle are shown in Fig. 8.

The aft optics assemblies for detectors (channels) 2–5 are identical (Fig. 9). Detector 1 houses the LR aft optics. The aft-optics assembly mounts directly on a flange that is an integral part of the detector housing assembly. The housing assembly or detector plate includes the detector and all the associated electronics. The aft optics consists of collimating and focusing optics to collimate the output of the fiber and send it on to the detector. A band-pass filter tuned to the laser wavelength is also included in the optical train to minimize the background solar radiation. In addition, the aft optics assembly includes a separate test port with an FC connector. The test port is intended for calibration purposes during instrument integration and testing. Signals from optical test sources can be injected into the test port and exercise the signal processing algorithms and electronics of the instrument. It is also possible to back-illuminate the detectors through the test port and monitor the field of view (FOV) and boresight alignment during the integration process.

The LOLA detectors are part of the detector housing assembly. The housing assembly includes the aft optics, the detector hybrid and the LOLA detector board. The detectors and detector board electronics have low noise and sufficient bandwidth to allow detection of the reflected laser pulses from the lunar surface and measure the time of flight (TOF). Each detector has its own independent gain control and threshold setting. The detector hybrids are the same silicon avalanche photodiodes that were used on the Mars Orbiter Laser Altimeter

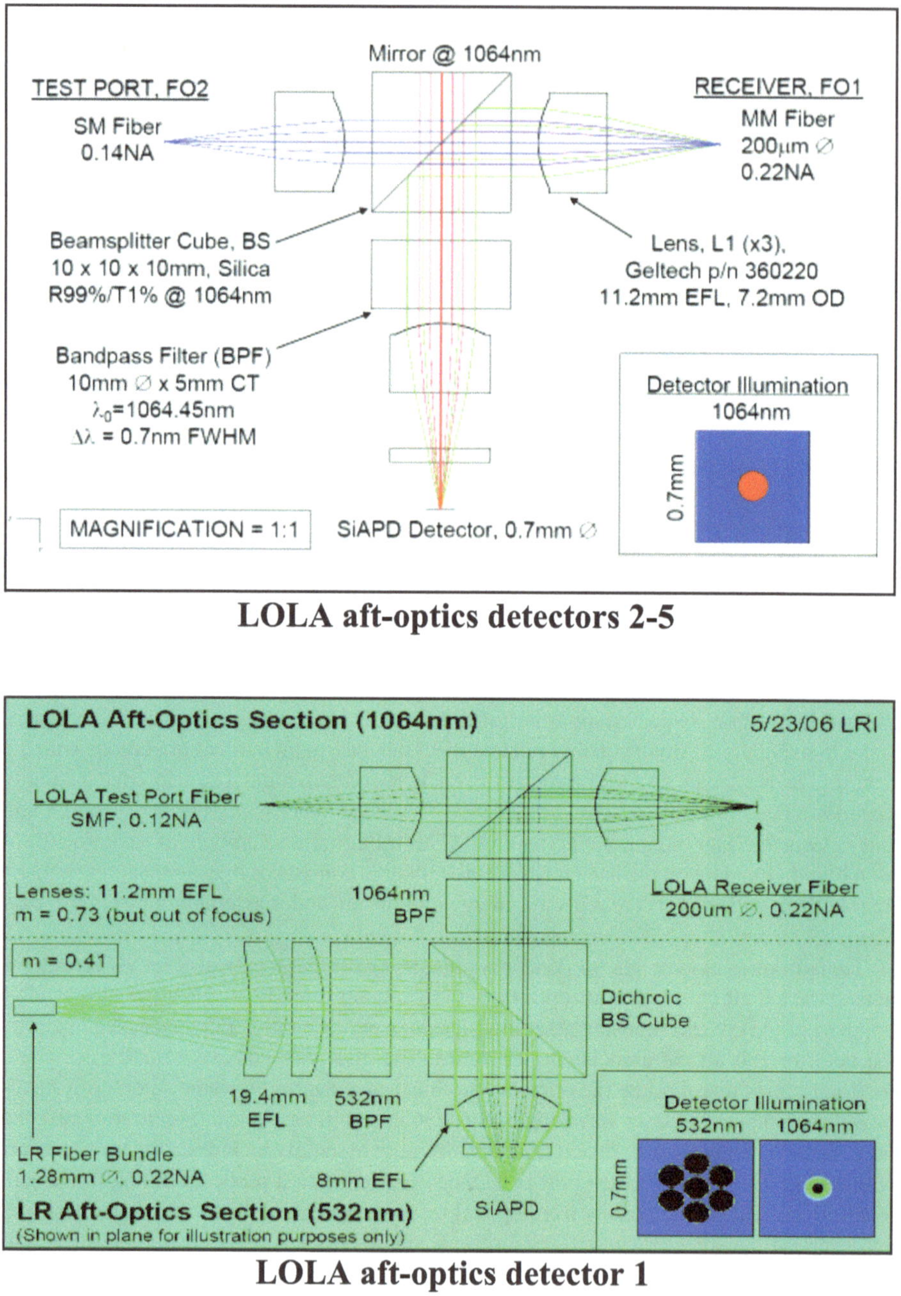

Fig. 9 Aft optics and test port

(MOLA) (Zuber et al. 1992) and the Geoscience Laser Altimeter (GLAS) (Abshire et al. 2005; Schutz 2001).

The LOLA detector board amplifies the detector hybrid output, and performs two separate functions: a discrimination function to provide timing information to the Digital Unit

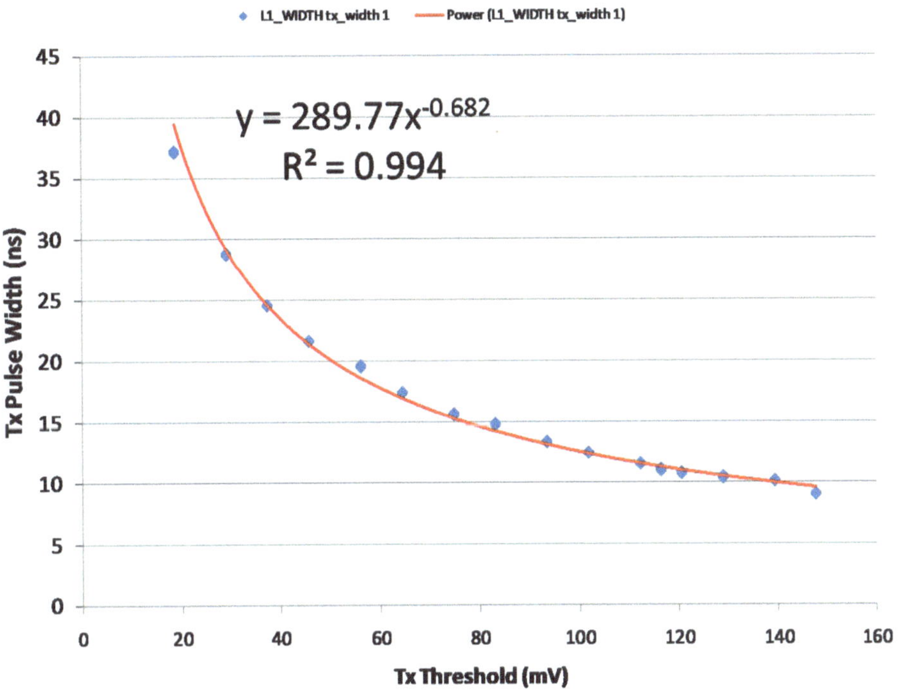

Laser 1 Tx pulse threshold sweep

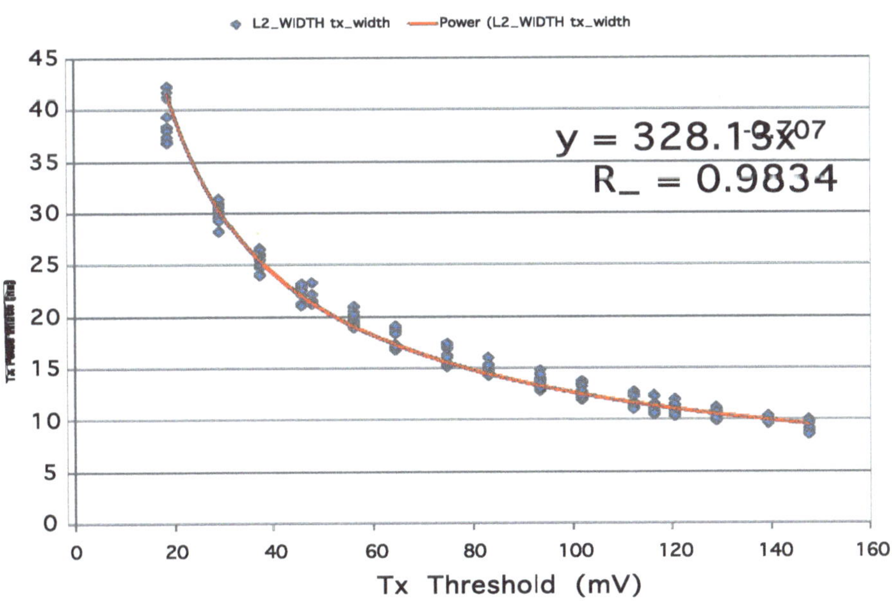

Laser 2 Tx pulse threshold sweep

Fig. 10 Detector thresholds

(DU), and an energy measurement function that integrates and samples the peak of the amplified hybrid output.

The amplification of the hybrid output is performed by a variable gain amplifier (VGA) and a fixed gain buffer with a gain of 5. The VGA gain is variable over a range of <0.5 to 10, controlled by an externally generated d.c. voltage from the DU. The gain control voltage is scaled and level-shifted on the detector board. The discrimination function for the timing information is performed by a high-speed comparator. The threshold level is set by the DU and fed directly to the comparator input.

The energy measurement function is performed by an integrator and a peak detector. It is not dependent on the threshold set by the DU but it is affected by the gain setting. The integrator stage is combined with a track-and-hold function, which is controlled by a latch. The latch is set by a peak detector, and later reset by a signal from the DU. The peak detector responds to the integrator output so as to set the latch at the time of the maximum output from the integrator. The latch then puts the integrator into its HOLD mode until its output is digitized. The peak detector is an a.c.-coupled, d.c. offset constant-fraction discriminator. The output is proportional to the received energy.

3.4 LOLA Digital Unit

The LOLA Digital Unit provides two fundamental categories of functionality for the instrument. The first category—range measurement—includes firing the laser and acquiring LR data. The other category—command and telemetry (C&T)—comprises receiving and distributing spacecraft commands, collecting, formatting and transmitting telemetry, as well as monitoring and processing ranging information for real-time control of the altimeter's parameters such as the range gate, amplifier gain and detector threshold.

Two different time measurements are made by the electronics. The primary is the time-stamping of the laser transmit pulse and the five lunar return pulses, and the LR pulse. The timing reference is based on redundant 20 MHz ultra stable oscillators (USO) located in the spacecraft, which is divided to 5 MHz and used as the internal timing reference. The spacecraft provides a non-redundant 1 Hz pulse, which is used to synchronize all LOLA activities. The fault-tolerant design allows for the spacecraft timing signals to replace the function of the internal clock oscillator, while the internal oscillator is sufficient to operate the instrument in the absence of any spacecraft timing reference. Also, circuits will use the spacecraft's USO to measure the frequency of the oscillator, for purposes of calibration and trending.

LOLA's pulse repetition frequency (PRF) of 28 Hz is phase locked to the spacecraft's 1-Hz timing signal. Each 1-s interval is referred to as a major frame, which consists of 28 approximately 35.7 ms periods referred to as a minor frames. The time within a minor frame is allocated for various functions, which include an 8 ms window for receiving the Earth laser signal (LR), a window for receiving the reflections from the Moon, and time for transmitting science and engineering data to the C&T Electronics. The Range Measurement Electronics also synchronizes the operation of the Analog Electronics and controls resetting of the energy measurement circuits for Detector boards.

The Range Measurement Electronics control is implemented in a single RTAX2000S Field Programmable Gate Array (FPGA). This FPGA also contains a variety of counters which implement the 200-ns coarse timer for each channel, noise counters, event counters, and calibration counters for measuring the margin in the digital phase lock loops in each of the twelve fine timing microcircuits. There are six physical channels in the Range Measurement Electronics. One channel is utilized for the timing laser firing pulse, and the other five are for the lunar return pulses, with one of the return channels also time-shared for the Earth laser time measurement. The time measurement method utilizes a coarse timer implemented in the FPGA along with fine timing supplied by the digital ASIC time-to-digital converter (TDC) originally designed to support the European Space Agency's (ESA's) Automated Transfer Vehicle (ATD). The FPGA and TDC circuits together allow for time-tagging over planetary distances of the time that the laser is fired, the time the Earth-based LR signal arrives at the spacecraft, and the time of each of 5 returns are received from the Moon; the LOLA altimeter has a time resolution of approximately 28 ps.

3.5 Laser Transmitter Telescope

The LOLA laser transmitter telescope has two functions: it reduces the divergence of the LOLA laser oscillator input beam, and it splits the output beam into five to generate a cross pattern in the far field. The telescope is an 18× magnification, afocal, Galilean beam expander that expands and collimates the LOLA Laser 1 mm and 1.8 mrad $1/e^2$ diameter input beam to generate the required 100 μrad $1/e^2$ diameter output beam divergence. The telescope has a single Corning 7980 fused silica negative lens, a BK7G18 positive lens group. It also contains the DOE that splits the output beam. The DOE optic is a 40-mm diameter by 6-mm thick fused silica substrate etched on one side with a transmission phase grating and anti-reflection (AR) coated on both sides to maximize transmission at 1064 nm. The LOLA DOE was manufactured by MEMS Optical from a master pattern that is replicated via a proprietary lithographic process onto a fused silica wafer from which multiple parts can be cored out. The flight DOE had an overall efficiency of 80% with the center beam having nearly twice the energy of the outer beams.

The telescope tube is beryllium with a titanium flexure pre-loaded to 30 lbs in order to securely hold the positive lenses and DOE over all expected mechanical and thermal loads. The exit optics contain a 36-mm-diameter exit clear aperture. The LOLA assembly is 165 mm long and weighs 120 g, and has an operational thermal range of 20 ± 20°C, and a survival thermal range of −30°C to +70°C. Two flight model (FM) Laser Transmitter Telescopes were integrated, functionally tested in air and vacuum, thermally qualified for survival, and delivered to the LOLA laser team for integration to the Laser Bench. Further details on the transmitter telescope are given in Ramos-Izquierdo et al. (2009).

3.6 LOLA Receiver Telescope

The LOLA Receiver Telescope (LRT) has a 150-mm-diameter objective lens and an unfolded path length to provide an $f/2$ objective lens speed. The LRT opto-mechanical design is a beryllium structure with titanium flexures, tolerances, optic mounting techniques, and alignment compensators for focus and boresight. The receiver FOV is determined by the FOB located at the telescope focal plane. The telescope end of the FOB has five 200-μm core diameter, 0.22 NA, multimode, step-index fiber-optics spaced 250 μm apart in a cross pattern; these fiber-optics define five 400-μrad diameter FOV's with a center-to-center spacing of 500 μrad. The FOB telescope connector is clocked 26° to align the five receiver FOV's with the five transmitted laser beams. Behind the telescope connector the FOB fans out into five independent fiber-optic cables that attach to the five Aft-Optics/Detector Assemblies. The LOLA receiver system FOV was selected to provide sufficient boresight alignment margin for each laser/receiver pair while having enough angular distance between neighboring channels to prevent crosstalk. Further details regarding the LRT are given in Ramos-Izquierdo et al. (2009).

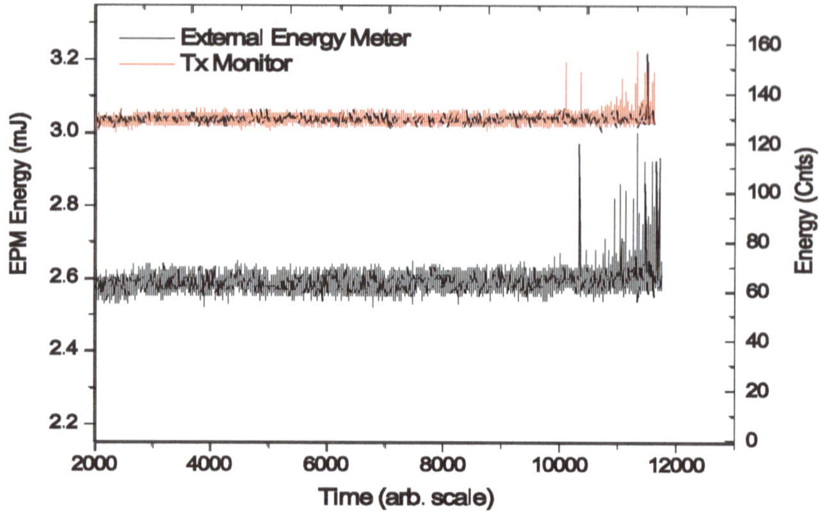

Fig. 11 Pulse energy calibration

4 System Calibration

4.1 Transmitter Energy Calibration

The LOLA transmitter has an on-board laser energy sensor (detector), which also serves as the fire acquisition detector. The detector is commonly referred as the SPOT4 or Transmit energy (Tx Energy) monitor. The Tx Energy monitor was correlated with measurements made with an external energy meter to assess its performance during pre-launch system-level testing (Fig. 11).

By correlating and normalizing the Tx Energy monitor with the external energy monitor it is possible to measure the transmitted energy with approximately 1.3% accuracy. There was no degradation in the performance of the Tx Energy monitor as a function of temperature. Using data for Laser 1 in the hot and cold plateaus we determined that in the hot case the accuracy was 0.5–1% and in the cold case it was approximately 1.0%. Thus, we can expect the Tx energy accuracy requirement of 2% to be met at all temperatures.

The equation that converts Tx Energy monitor from counts to laser energy is (Riris et al. 2008)

$$Energy(\text{mjoules}) \approx 0.0207 \times (Tx\, Energy(\text{counts}) - 12(\text{counts})). \tag{1}$$

The equation was obtained using ambient data (where the full energy of the laser beam could be measured) and it is an average approximation from different data sets.

The responsivity of the Tx Energy monitor changes significantly as function of temperature, therefore a temperature correction must be applied to (1). The temperature correction was derived from the breadboard laser test data and the EM laser TVAC data, where a stable external monitor was present, and it is

$$E(T) = E(T_0) \times [0.82702 \times e^{0.0081816 \times T}], \tag{2}$$

where T_0 is the energy at room temperature ($T_{room} = 23.2°\text{C}$). There is no temperature sensor on the SPOT4 detector itself. However, the detector should follow the laser bench temperature and the laser electronics assembly (with some lag). It does not get as cold as the laser bench temperature or as hot as the laser electronics. The average of the two temperatures should be a good indicator of the SPOT4 temperature.

4.2 Receiver Energy Calibration

Each LOLA detector has an energy monitor circuit that estimates the received energy (Rx Energy) incident on each detector. The received energy per pulse is a direct measure of the lunar surface reflectivity and is given by Gardner (1982)

$$E_{Rx} = E_{Tx} \times \varepsilon_{Rx} \times \frac{A_{Rx} \times \rho}{\pi \times r^2} \tag{3}$$

where E_{Rx} is the received energy, E_{Tx} is the transmitter energy, ε_{Rx} is the efficiency of the receiver optics (from the telescope to the detector active area), A_{Rx} is the receiver aperture area, ρ is the surface reflectivity and r is the range to the target (lunar surface). By normalizing the received energy to the transmitter energy and knowing the range, we can determine the surface reflectivity of the lunar surface.

The energy monitor is essentially a peak sample and hold circuit with an integrator. Its circuitry is after the variable gain amplifier (VGA) and voltage buffer but before the timing circuit which means it is affected only by the gain setting but not the threshold setting. The circuit is optimized for detection of energies from 0.1 to 3.0 fJ.

The specification for the energy monitor precision is to measure the relative energy within 12% from 0.1 to 3.0 fJ. This specification is met for energies above 0.2 fJ. At 0.1 fJ the precision will can be as large as 30% depending on the detector and temperature. The impact on ice detection has been modeled and its impact should be minimal.

4.3 Counts to Energy Calibration

The energy monitor was characterized at subsystem testing and system-level integration. The energy monitor is to a first approximation, linear with energy, but has a gain-dependent slope and offset. The following equations can be used to convert counts to energy, E, (in fJ)

$$E = slope(G) \times Counts + offset(G) \tag{4}$$

where, $slope(G)$ and $offset(G)$ are the gain-dependent slope and offset. The equation that describes the gain dependent slope is given by

$$slope(G) = A + B \times e^{-\frac{Gain}{\varepsilon}}. \tag{5}$$

The offset is best described by a quadratic fit, although a linear fit also gives adequate results:

$$offset(G) = A + B \times Gain + C \times Gain^2. \tag{6}$$

In the above equations *Gain* is the gain read back.

Table 3 gives the coefficient values for all 5 detectors. These equations are valid for energies $\geq$0.2 fJ. At 0.1 fJ the equations do not give an accurate energy value. Note that *absolute* energy accuracy of Rx energy monitor (i.e. the counts to fJ conversion) is no better than 20–30%. However, the *relative* energy values (in counts) will remain within the specification of 12% except at the lower energies (0.1 fJ) as indicated in Table 4.

Table 3 Laser output beam properties

	Laser 1 (μrad)	Circularity	Laser 2 (μrad)	Circularity
Ambient 1 Atm	153.1	Out of focus	147.9	Out of focus
Ambient 0 Atm (Start)	100.0	1.08	94. 9	1.19
Hot Qual.	106.7	0.96	97. 5	1.18
Cold Qual./Op.	103.6	1.30	110.7	1.24
Hot Operational	104. 5	1.29	102.5	1.17
Ambient 0 Atm (End)	109.9	1.14	101.7	1.30

Table 4 Energy monitor counts to energy calibration

	Slope coefficients			Offset coefficients		
	A	B	C	A	B	C
Detector 1	0.00914	0.07437	13.4092	−0.84429	0.03313	−0.00032235
Detector 2	0.00829	0.07399	12.9082	−0.82769	0.03376	−0.00034980
Detector 3	0.01103	0.0813	11.44682	−0.63522	0.02518	−0.00025046
Detector 4	0.00985	0.08125	11.10461	−0.8598	0.03452	−0.00033139
Detector 5	0.00852	0.06865	14.19706	−0.74418	0.02174	−0.00010699

4.4 Transmitter Calibration

The laser far field pattern was measured at the subsystem (laser) level with a 4-m focal length off-axis parabola collimator and a charge-coupled device CCD camera at the focal point. The data were analyzed using commercial beam analysis software (BeamView© by Coherent). The laser beam profiles after the beam expander integration are shown in Fig. 12.

The laser divergence prior to the beam expander integration was (Riris et al. 2008):

Laser 1
FF Divergence 1.76 mrad (=97.8 μrad after ×18 beam expander)
99% energy enclosed within aperture diameter of 5.7 mm—1.6 × $1/e^2$ dia, or <1% of energy lies outside of 2 × $1/e^2$ diam.
Circularity = 0.91

Laser 2
FF Divergence 1.72 mrad (=95.6 μrad after ×18 beam expander)
99% energy enclosed within aperture diameter of 5.7 mm—1.7 × $1/e^2$ dia, or <1% of energy lies outside of 2 × $1/e^2$ diam.
Circularity = 0.97

During system level environmental testing the laser far field pattern was measured using an off-axis parabola and a different CCD camera. The data were analyzed using custom-developed software that fit the laser images to a two-dimensional Gaussian surface. Sample of laser images (along with reference cube images and a fiduciary image) are shown in Fig. 13.

The laser far field pattern remained nearly in the fundamental or TEM_{00} mode over the entire operating temperature range. The divergence was calculated as the average of the x

Fig. 12 Laser transmitter profiles showing transmit pulse shape

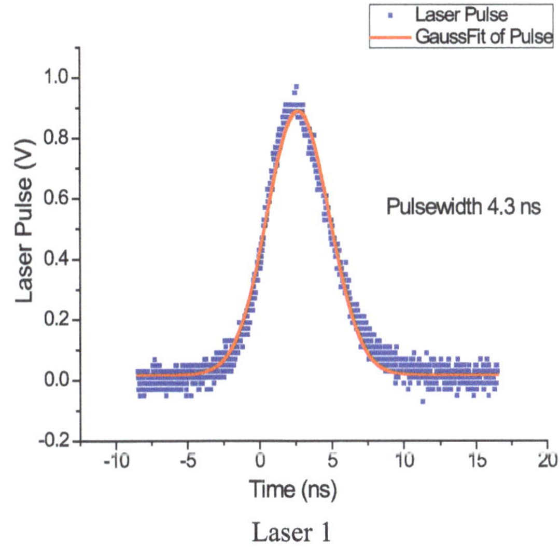

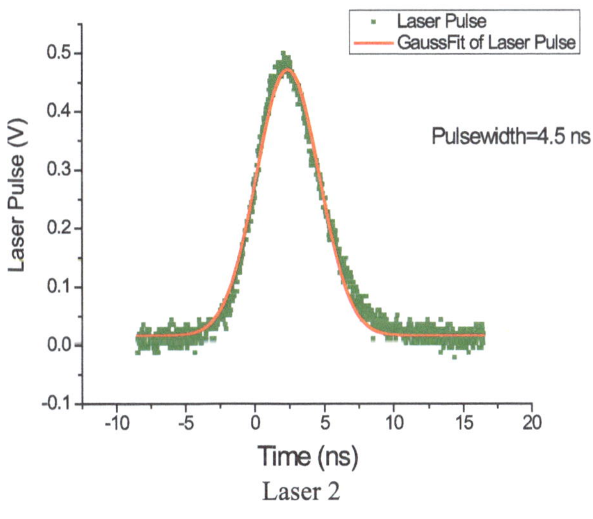

and y Gaussian $1/e^2$ widths. The laser divergence and circularity (defined as the ratio of major to minor axes, in this case, the ratio of the x and y Gaussian widths obtained by the fit) as a function of temperature are shown in Table 3. Note that at ambient pressure the transmitter optics is out of focus, which results in high divergence.

4.5 Laser Pointing Jitter

The laser pointing jitter was monitored at the subsystem level and also during the system level environmental testing. Figure 14 shows the Laser 1 pointing jitter after analyzing 450 images taken under vacuum and at ambient temperature over al temperatures. The images were analyzed for the relative separation between the central laser spot and a reference

Fig. 13 Laser far field beam patterns showing 5 spots and energy profiles. The center spot has slightly more energy than the outer 4 spots

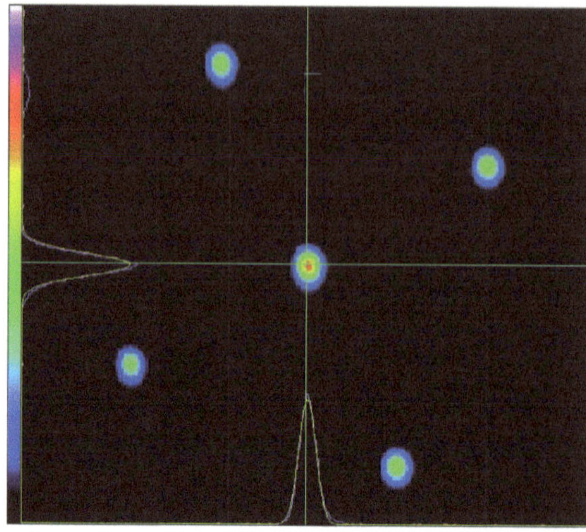

Laser 1 profile with CCD at simulated vacuum Focus (18 mm from OAP focal plane)

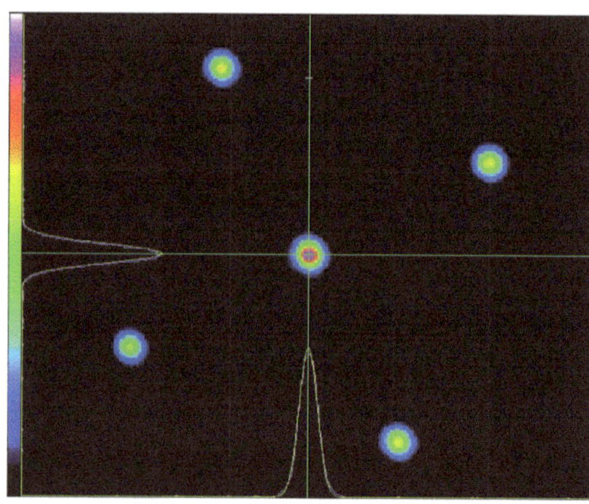

Laser 2 profile with CCD at simulated vacuum Focus (18 mm from OAP focal plane)

cube. The jitter in both subsystem and system level environmental testing was approximately 5 mrad in the x and y directions (Fig. 15).

4.6 Laser Wavelength

The laser wavelength at the subsystem delivery was 1064.4 nm. During system-level environmental testing the laser wavelength was monitored by a wavemeter (operating in continuous wave—CW—mode). The overall change in the laser wavelength was ~0.1–0.13 nm. Discontinuous changes in the laser wavelength were 0.04–0.05 nm. To first order the change

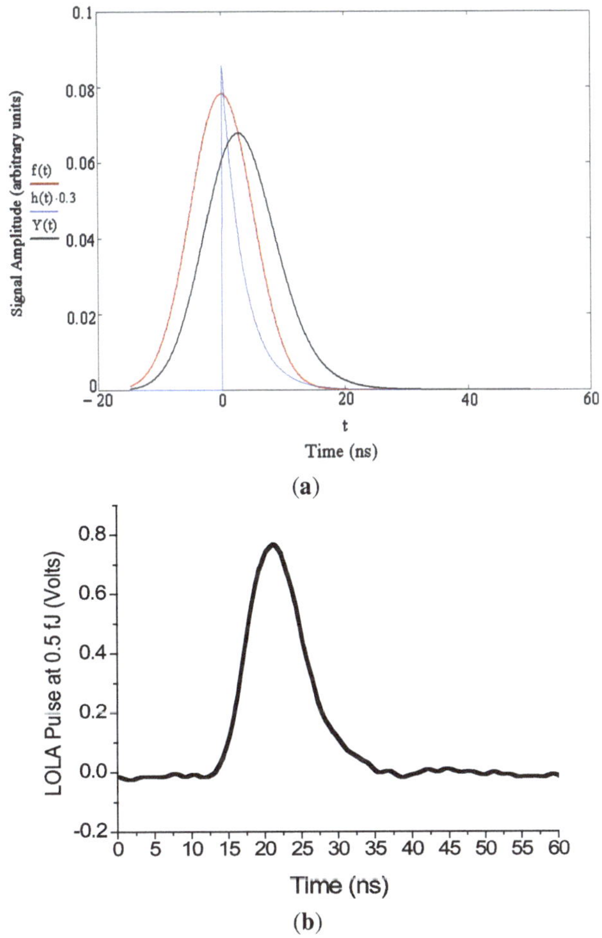

Fig. 14 (**a**) Convolution of Gaussian pulse with exponential delay function. (**b**) LOLA pulse at the output of the VGA

in wavelength (without the discontinuous changes) was ~0.06–0.07 nm/°C. The discontinuous changes in the wavelength could be attributed to changes in the mode structure of the laser and/or the wavemeter operation, which will report the wavelength of only one mode.

4.7 Laser Pulse Shape and Mode-Beating

The laser pulse shape was measured at subsystem (laser) integration using a high-speed photo detector and oscilloscope and during system environmental testing. Sample waveforms from the system level environmental tests are shown in Fig. 12. The pulse width varies from 4.3 to 4.9 ns full width at half maximum (FWHM) amplitude.

The LOLA laser is not a single mode laser. It will occasionally operate in more than one mode and those modes may beat against each other producing a series of very short pulses (shown in Fig. 16 as seen by a high speed photo detector and a fast oscilloscope). The impulse response of the transmit energy monitor will "smear out" the short mode-beating pulses. However, when the laser mode beats there will be an impact on the time of flight (range) measurement. Laser 1 rarely exhibits multi-mode behavior (1–2% of the total number of shots in steady state), whereas Laser 2 exhibits multi-mode behavior much more

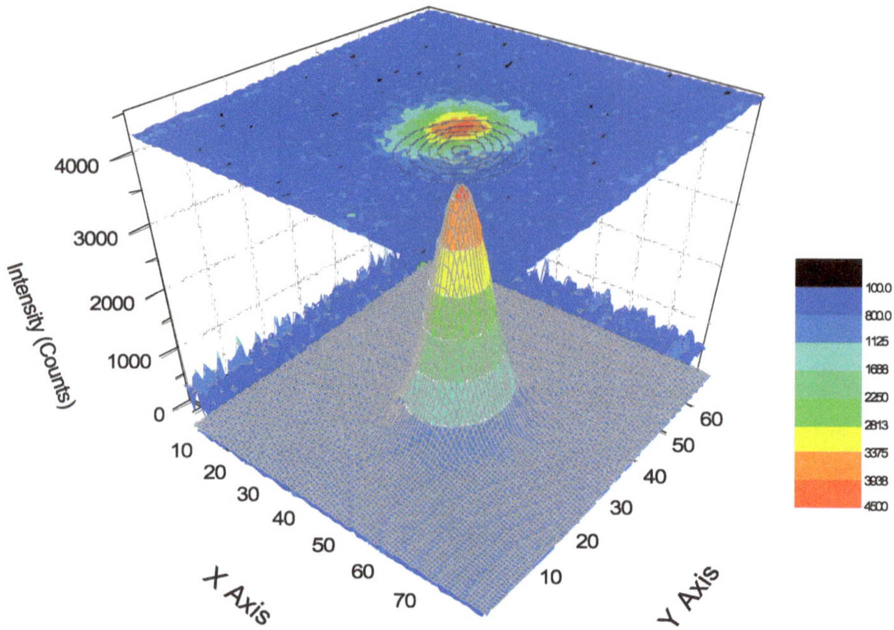

Fig. 15 Spatial pattern of pulse energy

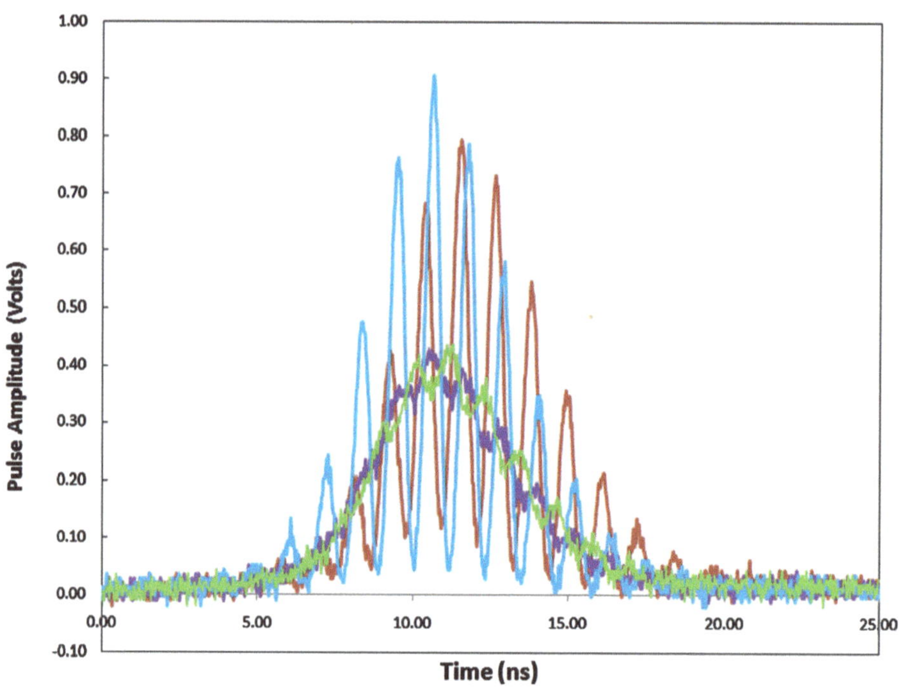

Fig. 16 Examples of laser mode beating. Mode beating is observed occasionally in laser 1 and more frequently in laser 2

frequently (about 5–25% of the total number of shots) depending on the temperature. The Tx Energy monitor can be used to detect the multi-mode behavior of the laser since the peak energy shows a large increase, The multi-mode behavior of the laser will impact the time of flight calculation since the transmit pulse will be distorted. However, the Tx Energy monitor can be used to flag and filter the returns that show multi-mode behavior. An approximate correction can be applied which will be some fraction of the pulse width.

4.8 LOLA Ranging (TOF) Calibration

The primary function of LOLA is to provide time of flight (TOF) or altimetry measurements of the laser pulse which, along with the spacecraft position, can be used for determining the topography of the lunar surface. The LOLA TOF is determined by the following measurements: the Tx laser pulse leading edge time stamp (*LETx*), the Tx laser pulse trailing edge time stamp (*TETx*), the Rx laser pulse leading edge time stamp (*LERx*), the Rx laser pulse trailing edge time stamp (*TERx*). The distance from the LRO spacecraft to the illuminated spot on the lunar surface, D, is related to the laser-pulse TOF by:

$$D = \frac{c}{2} \times (T_x - R_x) \tag{7}$$

where c is the speed of light in vacuum, and Tx and Rx are the times of the transmitted and echo laser-pulses. If the pulse shapes were perfectly symmetric, the pulse arrival times are simply taken as the average value of the signal threshold crossing times at the leading and trailing edges, or

$$T_x = \frac{1}{2} \times (LET_x + TET_x) \tag{8}$$

and

$$R_x = \frac{1}{2} \times (LER_x + TER_x). \tag{9}$$

The distance traveled by the laser-pulse and the spacecraft orbit position and pointing angle at the moment the laser-pulse was transmitted defines a vector in space, which can be used to solve for the coordinates and the height of the laser footprint on the planet surface given the position of the planet center of mass. A topographic map of the planet can be constructed with a great number of such measurements along different ground tracks. For a perfectly symmetric Gaussian pulse waveform, the leading and trailing edge points also provide a measurement of the Tx and Rx pulsewidth.

4.8.1 LOLA TOF to Range

In an ideal altimeter where the pulses are perfectly symmetric there would also be no inherent biases in the time of flight. Of course, LOLA is not an ideal altimeter and we need to correct for several biases which are different for each detector. These corrections can be divided in three broad categories. The electronics TOF bias (TDC trailing edge bias), mainly the time to digital converter biases for each channel, the impulse response bias which makes the pulses asymmetric and introduces a range bias (or "time walk") that is a function of the received signal amplitude and finally fiber delays due to the different fiber lengths for each channel. The range is also reference to a physical point on the instrument (zero range bias).

These corrections will be refined on orbit.

Table 5 TDC bias

	Phase A		Phase B	
	Average (ns)	Stdev (ns)	Average (ns)	Stdev (ns)
tx	2.22	0.21	1.83	0.15
rx1	1.83	0.18	1.71	0.21
rx2	1.89	0.18	1.83	0.18
rx3	1.77	0.18	1.38	0.18
rx4	1.56	0.18	1.5	0.21
rx5	1.86	0.18	2.04	0.15

4.8.2 TDC Trailing EdgeTime Bias

Each TDC in the digital unit has its own time delay and bias. The trailing edge biases for each TDC were measured during subsystem testing and are listed in the Table 5. The TDC trailing edge biases varied very little over temperature.

The trailing edge bias for each TDC can then be accounted for when computing the midpoint for each pulse using the equations below

$$T_{x_i} = \frac{1}{2} \times [LET_{x_i} + (TET_{x_i} - TDC\,Bias_i)] \tag{10}$$

and

$$R_{x_i} = \frac{1}{2} \times [LER_{x_i} + (TER_{x_i} - TDC\,Bias_i)] \tag{11}$$

where $i = 1, 2, 3, 4, 5$ for each channel/detector.

4.8.3 Pulsewidth and Impulse Response Bias

The impulse response of the detectors may be modeled, to first order, by a simple RC low-pass filter. The convolution of a Gaussian pulse with the impulse response of an RC low pass filter is the error function. Figure 14a shows a simulation of a Gaussian pulse (red trace) with the impulse response of an exponential decay function from an RC filter (blue trace) and the resulting convolution (black trace) and Fig. 14b shows an actual pulse from the LOLA detector from subsystem testing. The impulse response of the detector will smear out any high frequency components in the laser pulse resulting in slightly wider pulsewidth and a "time-walk" as the energy and pulsewidth increase.

Since LOLA does not use a constant fraction discriminator it is necessary to correct for this "time-walk", which is dependent on the gain, threshold, and return pulse energy. For a perfectly symmetric pulse the "time-walk" could have been compensated for by taking the mid-point of the pulse. The "time-walk" correction due to the impulse response of all channels may be modeled in three different ways:

1. A simple exponential (a lowpass RC filter with $R = 50$ W and $C = 68$ pF).
2. A single-pole butterworth filter based on the actual frequency response of the subsystem testing (cf. Riris et al. 2008).
3. An empirical formula derived from test data that accounts for the pulse broadening and "time walk" effect based on the gain and threshold readback and the received energy.

Table 6 Minimum detectable energy (in counts) in each LOLA detector

Detector 1	12
Detector 2	13
Detector 3	10
Detector 4	13
Detector 5	13

We are currently investigating and modeling all 3 options. At this time we have preliminary results for the third method which seems to yield the satisfactory results and is the one we will present in this paper. The final time-walk corrections will be presented in the final calibration report to the LRO Project and will be verified on orbit.

In order to account for the different thresholds, gains and energies we form a new variable:

$$x \equiv \frac{\left(\frac{Threshold\,Readback}{Gain\,Readback}\right)}{Energy - Energy_{\min}} \tag{12}$$

where, *Energy* is the energy in counts and $Energy_{\min}$ is the minimum energy (in counts) for each channel, which is listed in Table 6. Then the range offset due to the time walk correction is given by

$$\begin{aligned} Range\_Offset &= slope\_n(gain\_readback) * \ln(Threshold\,Readback/energy) \\ &\quad + offset\_n(gain\_readback). \end{aligned} \tag{13}$$

Using the following polynomials

Slopes

$$\begin{aligned} slope1 &= -0.0009x^2 + 0.0605x - 1.8952, \\ slope2 &= -0.0008x^2 + 0.05x - 1.5695, \\ slope3 &= -0.0007x^2 + 0.0442x - 1.5382, \\ slope4 &= -0.0008x^2 + 0.0548x - 1.6802, \\ slope5 &= -0.0008x^2 + 0.051x - 1.7203. \end{aligned}$$

Offsets

$$\begin{aligned} offset1 &= -8E\text{-}05x^3 + 0.0053x^2 - 0.0945x + 4.770, \\ offset2 &= -1E\text{-}04x^3 + 0.0067x^2 - 0.1449x + 4.277, \\ offset3 &= -7E\text{-}05x^3 + 0.0047x^2 - 0.1124x + 3.135, \\ offset4 &= -9E\text{-}05x^3 + 0.006x^2 - 0.1158x + 0.000, \\ offset5 &= -3E\text{-}05x^3 + 0.0012x^2 - 0.0133x + 1.120. \end{aligned}$$

If $Energy - Energy_{\min} = 0$, then only the constant term in *offset_n* above is used.

Table 7 LOLA detector timing biases

Detector 1 (ns)	Detector 2 (ns)	Detector 3 (ns)	Detector 4 (ns)	Detector 5 (ns)
10.9236 ± 0.227	9.4533 ± 0.226	8.1327 ± 0.289	5.3142 ± 0.369	7.221 ± 0.240

4.8.4 Fiber Time Bias and Zero Reference (Zero Range)

The range offset bias due to the different length of the fiber optic cables in the 5 channels was measured at the system level. Table 7 gives the timing bias for each channel. The range bias is referenced to the front of the LOLA receiver. It can also be referenced to the LOLA cube. The distance between the LOLA receiver and the LOLA reference cube is 15.174 cm.

5 Observation Strategy

In terms of coverage, LOLA will under sample the lunar surface in cross-track coverage and for this reason the basic observational strategy is to operate continuously and obtain as much coverage of the Moon as is possible within operational constraints. Further, the instrument is a nadir-viewing and therefore primarily acquires observations along the spacecraft ground track and provides the "landing strip" information described earlier. However, certain exploration and science investigations can require LOLA observations away from the ground track and therefore LOLA is required to operate off-nadir. Off-nadir observations are possible with LOLA and other laser altimeters. Notably, MOLA (Zuber et al. 1992) at Mars and MLA (Cavanaugh et al. 2007) at Mercury, have both obtained off-nadir observations; MOLA on a regular basis as the MGS spacecraft changed attitude to enable the imaging system to acquire specific targets (Smith et al. 2001), and MLA during two flybys in which the emission angle exceeded 60° (Zuber et al. 2008b).

The observation of potential landing sites and special locations, such as inside permanently shadowed regions (where LOLA is one of the few instruments capable of making observations), may require dense observations with coverage as uniform as possible. Off-nadir pointing by only 1 or 2 mrad from 50-km altitude can move the LOLA swath on the surface by 50 or 100 m and thereby improve the distribution of LOLA tracks across a region.

6 Data Analysis, Interpretation, and Modeling

6.1 LOLA Data

LOLA data packets consist of time-ordered, round-trip, time-of-flight ranges to the lunar surface, preceded by housekeeping and ancillary data. These 3424-byte packets are output over the spacecraft IEEE-1553 data bus at 1-s intervals and aggregated into ~1 megabyte-sized files on the LRO Solid State Recorder. The raw, uncalibrated data comprise the experiment data record (EDR) product. After range calibration and orbital processing, the position of each laser spot is located on the surface using a spacecraft trajectory, attitude history, and a lunar orientation model. The reduced data records (RDR) contain calibrated, geolocated pulse returns, altitudes and reflectivities. Higher-level gridded and transformed data products are produced from the cumulative RDR product.

The LOLA science data from each downlink tracking pass will be aggregated by the LRO Ground Data System and transmitted daily to the GSFC LOLA Science Operations Center (SOC) computer, together with spacecraft event, housekeeping and attitude data. The raw Deep Space Network (DSN), Universal Space Network (USN), and White Sands 1 (WS1) tracking data and supporting products will also be forwarded to the SOC. Each EDR file, ~300 per day, is processed in a pipeline. The geolocation processing cycle will collect the data records into half-orbits, 25–26 per day, with manageable file sizes (~25 megabyte per product). The northern half-orbit begins at the spacecraft ascending node and the southern half-orbit begins at the descending node, following the LRO Project orbital numbering convention.

The timing of LOLA instrument events is derived from the LRO USO, which is monitored by ground stations. Time systems aboard LRO employ Coordinated Universal Time (UTC) to correlate spacecraft Mission Elapsed Time (MET) to ground time with an accuracy of ±3 ms. The correlation of MET time to Barycentric Dynamical Time (TDB) is maintained at much higher precision by the LR system and orbital theories. The LOLA data analysis uses TDB as its primary time system (see http://tycho.usno.navy.mil/systime.html for details regarding the difference). Spacecraft states relative to the Solar System Barycenter (SSB) at the laser transmit and detector receive times are projected along the instrument boresight and return path vectors to match the observed time of flight, correcting for the aberration of light and general-relativistic time delays. SSB states are determined in the Earth Mean Equator of 2000 (J2000) inertial reference frame using lunar spacecraft trajectories and the DE421 planetary ephemeris (Folkner and Williams 2008). The speed of light is applied to the boresight time-of-flight to locate the laser bounce point in barycentric coordinates. The bounce points are transformed to a selenodetic coordinate frame about the center of mass of the Moon using the DE421 Moon Mean-Earth and Rotation Pole (MOON_ME) lunar orientation model. The DE421 model incorporates 37 years of Lunar Laser Ranging (LLR) data, thereby providing the geodetic framework for LOLA. The lunar radius and position are then subjected to orbital crossover analysis (Neumann et al. 2001; Rowlands et al. 1999) to minimize the terrain mismatch at the intersections of ground tracks. Empirical adjustment of short-term pointing biases is combined with orbital estimation from tracking data to bring the ground tracks into agreement, and produce a dynamically-consistent, geodetically-precise lunar coordinate framework, as well as an improved lunar potential solution, as discussed earlier. The crossover adjustment is performed at monthly intervals following the propulsive orbital adjustment maneuvers that disrupt the dynamical orbit solutions.

The cumulative altimetric data are binned and interpolated at appropriate resolution to generate the gridded data record (GDR) products described in the next section. These products are then transformed to Spherical Harmonic Analysis Data Records (SHADR) of shape, time-averaged reflectivity, and related geodetic parameters.

Initial quick-look profiles are generated using orbital and attitude data provided by the GSFC Flight Dynamics Facility, and can be displayed in a web browser. Automated editing flags the noise returns and occasional Earth ranges received by Detector 1. After error checks, labeling, manual inspection of edited products, and registration with images obtained by the co-boresighted LRO cameras (Robinson et al. 2009), the EDR and RDR products are validated for delivery to the Planetary Data System (PDS). Refined altimetry is generated in the course of orbital processing by the GEODYN software system (Pavlis et al. 2001), and is released at quarterly intervals as specified in the LRO-LOLA Data Management and Archive Plan. Releases to the PDS Geosciences Node occur at specific intervals, while the current best solutions are made accessible to LRO scientists via a Data Node of the PDS hosted by the LOLA SOC.

Table 8 Standard data product sizes and delivery rates

Product	Product size	Production sate	Expected number of products for nominal mission	Expected total data volume for nominal mission
LOLA_EDR	1 MByte	296 per day, average	108000	108 GB
LOLA_RDR	24 MByte	25–26 per day	9400	225 GB
LOLA_GDR	~2 GByte	monthly refinements of ~100 data products	100	200 GB
LOLA SHADR	5 MByte	release at quarterly intervals	4	~1 GB

6.2 Quick-Look Data Analysis

A quick-look web-based interface has been developed for near-real-time assessment of LOLA data. Data are displayed for each of the five LOLA profiles, and also for the surface reflectivity and roughness. These data are uncorrected and are not suitable for scientific analysis but together with the global map of data distribution can be used for assessing adequacy of coverage and general functioning of the instrument, and in particular for any missing data or changes in the probability of data return from the lunar surface.

7 Science Data Products

Table 8 summarizes expected sizes and production rates for the LOLA Standard Data Products. These products, with detached PDS labels, will be organized into daily and monthly directories. The RDR observations from multiple orbits will be transformed to useful map projections, binned (median-averaged) and interpolated to fill regions lacking data.

A control network for the Moon, consisting of images and control points, will be tied to a topographic reference surface, for which the LOLA GDR is the primary datum. The GDRs are raster Digital Elevation Models of the lunar radius with respect to a spherical datum about the center of mass, and are generated at multiple resolutions (Table 9) as accumulating coverage permits to match the characteristics of image mosaics and other instrument products. Surface slope, roughness, and albedo gridded products will likely be noisier and less amenable to interpolation than altimetry, so will be generated only at a subset of resolutions given in Table 9 unless a specific need arises, in which case special products will be produced. The size of the products at highest resolution may require that they be aggregated in subsets of global coverage. Global products use the equi-rectangular map projection, while the higher-resolution products afforded by dense polar coverage use the Polar Stereographic projection. Tiling will be employed to limit the size of individual products to less than 2 GB to facilitate electronic data transfer. The GDRs are formatted as binary images.

Table 9 LOLA equi-rectangular map-projected Digital Elevation Models

Product	Product size	Pixel size (m in lat)	Number/size of tiles	Bits per pixel
LDEM_4	4 MByte	7580.8	Global, 0-360	32
LDEM_16	64 MByte	1895	Global, 0-360	32
LDEM_32	256 MByte	947.6	Global, 0-360	32
LDEM_64	1 GByte	473.8	Global, 0-360	32
LDEM_128	2 GByte	236.9	Global, 0-360	16
LDEM_256	4 × 2 GB	118.45	4 tiles, longitudes 0:180:360 by N/S	16
LDEM_512	16 × 2 GB	59.225	16 tiles, longitudes 0:45:90:135:180:225:270:315:360 by N/S	16
LDEM_1024	64 × 2 GB	29.612	64 tiles, longitudes as above, in 22.5° latitude bands	16

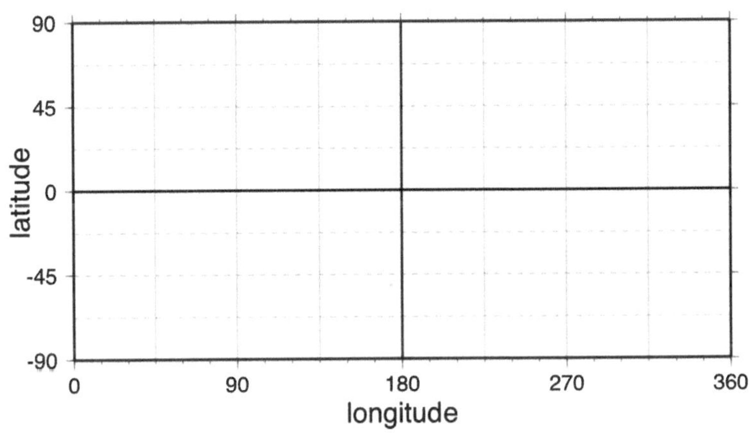

Fig. 17 Equirectangular map projection. Lines show tiling subdivisions of LDEM_256 (*bold*), LDEM_512 (*fine*), and LDEM_1024 (*dashed*)

Binary elevation data may consist of 16-bit integers scaled to a dynamic range of 1728 to 1748 km, or as 32-bit integers scaled to millimeters. The resolution of equi-rectangular pixels will be powers of two pixels per degree of longitude and latitude. The resolution of polar projected pixels will be integral numbers of meters, scaled by a radius of 1737.4 km. Anticipated tiles are shown in Figs. 17 and 18. Where feasible, products may also be provided in a geo-referenced JPEG-2000 format.

8 Summary

The LOLA instrument was delivered to the LRO spacecraft in April 2008 and along with the other instruments integrated with the spacecraft. As of January 2009, the LRO spacecraft

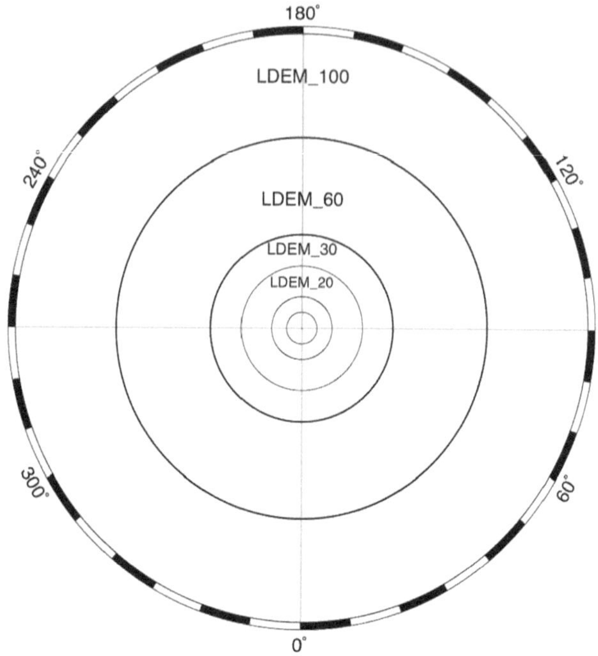

Fig. 18 Polar stereographic map projection from 45N to pole. Circles show outlines of LDEMs at various resolutions. Innermost circles enclose 10- and 5-m DEMs

Table 10 Polar stereographic map-projected Digital Elevation Models

Product	Dimension	Pixel size	Latitude range	Bits/pixel
LDEM_100	28786 × 28786	100 × 100 m	+/−45° to pole	16
LDEM_60	31036 × 31036	60 × 60 m	+/−60° to pole	16
LDEM_30	30496 × 30496	30 × 30 m	+/−75° to pole	16
LDEM_20	30400 × 30400	20 × 20 m	+/−80° to pole	16
LDEM_10	30342 × 30342	10 × 10 m	+/−85° to pole	16
LDEM_5	30328 × 30328	5 × 5 m	+/−87.5° to pole	16

has completed thermal vacuum testing and is being readied for shipment to the launch site. The LOLA instrument is a significant advance in capability over previous planetary laser altimeters because of its multiple beams and laser ranging component, and is expected to contribute a unique dataset for lunar science and exploration.

Acknowledgements We are pleased to acknowledge the support and assistance of the LRO Project, particularly the Payload Manager, Arlin Bartels and Project Manager Craig Tooley, without whom the LOLA instrument could not have been built. We also thank the Review Boards that provided sanity and constructive thought during the LOLA development process.

References

J.B. Abshire et al., Geoscience Laser Altimeter System (GLAS) in the ICESat mission: On-orbit measurement performance. Geophys. Res. Lett. **43**, L21S02 (2005)

H. Araki et al., Lunar global shape and polar topography derived from Kaguya-LALT laser altimetry. Science **323**, 897–900 (2009)

B.A. Archinal, et al., U.S. Geological Survey Open File Report 2006-1367, 21 pp., 2006. pubs.usgs.gov/of/2006/1367/

J.F. Cavanaugh et al., The Mercury Laser Altimeter instrument for the MESSENGER mission. Space Sci. Rev. **131**, 451–480 (2007)

G. Chin et al., Lunar Reconnaissance Orbiter overview: The instrument suite and mission. Space Sci. Rev. **129**, 391–419 (2007). doi:10.10007/s11214-007-9153-y

A.C. Cook et al., Lunar polar topography derived from Clementine stereoimages. J. Geophys. Res. **105**, 12,023–12,033 (2000)

J.O. Dickey et al., Lunar laser ranging: A continuing legacy of the Apollo program. Science **265**, 482–490 (1994)

W.M. Folkner, J.G. Williams, *Planetary Ephemeris DE421 for Phoenix Navigation* (Jet Propulsion Laboratory, Pasadena, 2008)

H. Frey et al., Ancient lowlands on Mars. Geophys. Res. Lett. **29** (2002). doi:10.1029/2001GL013832

C.S. Gardner, Target signatures for laser altimeters: an analysis. Appl. Opt. **21**, 448–453 (1982)

S. Goossens, K. Matsumoto, Lunar degree 2 potential Love number determination from satellite tracking data. Geophys. Res. Lett. **35** (2008). doi:10.1029/2007GL031960

W.M. Hartmann, Lunar cratering chronology. Icarus **13**, 299–301 (1970)

W.M. Hartmann, Moon: Origin and evolution of multi ring basins. Moon **3**, 3 78 (1971)

J.A. Kamalakar et al., Lunar ranging instrument for Chandrayaan-1. J. Earth Syst. Sci. **114**, 725–731 (2005)

A.S. Konopliv et al., Recent gravity models as a result of the Lunar Prospector mission. Icarus **150**, 1–18 (2001)

H.J. Melosh, *Impact Cratering: A Geologic Process* (Oxford University Press, New York, 1989). 245 pp.

G.A. Neumann et al., The crossover analysis of MOLA altimetric data. J. Geophys. Res. **106**, 23,753–23,768 (2001)

S. Nozette et al., The Clementine mission to the Moon: Scientific overview. Science **266**, 1835–1839 (1994)

D.E. Pavlis et al., GEODYN Operations Manuals. Raytheon ITTS Contractor Report, Lanham, MD, 2001

H. Qian, Topography of the Moon from the Chang'e Laser Altimetry Data, 2008

W. Quaide, Rilles, ridges and domes—Clues to maria history. Icarus **4**, 374–389 (1965)

L. Ramos-Izquierdo et al., The Lunar Orbiter Laser Altimeter (LOLA) optical subsystem, 2009

H. Riris et al., LOLA Calibration Report. NASA/Goddard Space Flight Center, Greenbelt, MD, 2008

M.S. Robinson et al., The Lunar Reconnaissance Orbiter Camera (LROC). Space Sci. Rev. (2009, this issue)

D.D. Rowlands et al., The use of laser altimetry in the orbit and attitude determination of Mars Global Surveyor. Geophys. Res. Lett. **26**, 1191–1194 (1999)

R.A. Schultz et al., Igneous dikes on Mars revealed by Mars Orbiter Laser Altimeter topography. Geol. Soc. Am. Bull. **32**, 889–892 (2004)

B.E. Schutz, Laser altimetry and LIDAR from ICESat/GLAS. IEEE Geosci. Remote Sens. **3**, 1016 1019 (2001)

D.E. Smith et al., Topography of the Moon from the Clementine LIDAR. J. Geophys. Res. **102**, 1591–1611 (1997)

D.E. Smith et al., The global topography of Mars and implications for surface evolution. Science **284**, 1495–1503 (1999)

D.E. Smith et al., Mars Orbiter Laser Altimeter: Experiment summary after the first year of global mapping of Mars. J. Geophys. Res. **106**, 23,689–23,722 (2001)

L. Soderblom, A model for small-impact erosion applied to the lunar surface. J. Geophys. Res. **75**, 2655–2661 (1970)

P.D. Spudis et al., Ancient multiring basins on the Moon revealed by Clementine laser altimetry. Science **266**, 1848–1851 (1994)

X. Sun et al., Radiometry measurements of Mars at 1064 nm using the Mars Orbiter Laser Altimeter. EOS Trans. Am. Geophys. Un. 82, 2001

U.S. Geological Survey, Color-coded topography and shaded relief map of the lunar near side and far side hemispheres. Flagstaff, AZ, 2002, pp. I-2769

M.A. Wieczorek, Gravity and topography of the terrestrial planets. Treatise Geophys. **10**, 165–206 (2007)

M.T. Zuber et al., The Mars Observer Laser Altimeter investigation. J. Geophys. Res. **97**, 7781–7797 (1992)

M.T. Zuber et al., The shape and internal structure of the Moon from the Clementine mission. Science **266**, 1839–1843 (1994)

M.T. Zuber et al., Outstanding questions on the internal structure and thermal evolution of the Moon and future prospects from the GRAIL mission. Lunar Planet. Sci. Conf. **XXXIX**, #1074 (2008a)

M.T. Zuber et al., Laser altimeter observations from MESSENGER's first Mercury flyby. Science **321**, 77–79 (2008b)

M.T. Zuber et al., The Lunar Reconnaissance Orbiter laser ranging investigation. Space Sci. Rev. (2009, this issue)

Space Sci Rev (2010) 150: 243–284
DOI 10.1007/s11214-009-9584-8

CRaTER: The Cosmic Ray Telescope for the Effects of Radiation Experiment on the Lunar Reconnaissance Orbiter Mission

H.E. Spence · A.W. Case · M.J. Golightly · T. Heine · B.A. Larsen · J.B. Blake · P. Caranza · W.R. Crain · J. George · M. Lalic · A. Lin · M.D. Looper · J.E. Mazur · D. Salvaggio · J.C. Kasper · T.J. Stubbs · M. Doucette · P. Ford · R. Foster · R. Goeke · D. Gordon · B. Klatt · J. O'Connor · M. Smith · T. Onsager · C. Zeitlin · L.W. Townsend · Y. Charara

Received: 10 April 2009 / Accepted: 23 October 2009 / Published online: 6 January 2010

Abstract The Cosmic Ray Telescope for the Effects of Radiation (CRaTER) on the Lunar Reconnaissance Orbiter (LRO) characterizes the radiation environment to be experienced by humans during future lunar missions. CRaTER measures the effects of ionizing energy loss in matter due to penetrating solar energetic protons (SEP) and galactic cosmic rays (GCR), specifically in silicon solid-state detectors and after interactions with tissue-equivalent plastic (TEP), a synthetic analog of human tissue. The CRaTER investigation

H.E. Spence (✉) · A.W. Case · M.J. Golightly · T. Heine · B.A. Larsen
Boston University, Boston, MA, USA
e-mail: spence@bu.edu

J.B. Blake · P. Caranza · W.R. Crain · J. George · M. Lalic · A. Lin · M.D. Looper · J.E. Mazur · D. Salvaggio
The Aerospace Corporation, El Segundo, CA, USA

J.C. Kasper
Harvard-Smithsonian Center for Astrophysics, Cambridge, MA, USA

T.J. Stubbs
University of Maryland, Baltimore County, Baltimore, MD, USA

M. Doucette · P. Ford · R. Foster · R. Goeke · D. Gordon · B. Klatt · J. O'Connor · M. Smith
Massachusetts Institute of Technology, Cambridge, MA, USA

T. Onsager
NOAA Spaceweather Prediction Center, Boulder, CO, USA

C. Zeitlin
Southwest Research Institute, San Antonio, TX, USA

L.W. Townsend · Y. Charara
University of Tennessee, Knoxville, TN, USA

quantifies the linear energy transfer (LET) spectrum in these materials through direct measurements with the lunar space radiation environment, particularly the interactions of ions with energies above 10 MeV, which penetrate and are detected by CRaTER. Combined with models of radiation transport through materials, CRaTER LET measurements constrain models of the biological effects of ionizing radiation in the lunar environment as well as provide valuable information on radiation effects on electronic systems in deep space. In addition to these human exploration goals, CRaTER measurements also provide new insights on the spatial and temporal variability of the SEP and GCR populations and their interactions with the lunar surface. We present here an overview of the CRaTER science goals and investigation, including: an instrument description; observation strategies; instrument testing, characterization, and calibration; and data analysis, interpretation, and modeling plans.

Keywords LRO · Galactic cosmic rays · Solar energetic protons · Radiation effects · LET spectrum · Lunar science

1 Introduction

The Lunar Reconnaissance Orbiter (LRO) mission obtains exploration-enabling observations needed to facilitate returning humans safely to the Moon as part of NASA's longer-term Exploration Initiative. A NASA-established LRO Objectives/Requirements Definition Team (ORDT) identifies the following high-priority objective for this initial robotic mission in the Lunar Exploration Program (LEP): "Characterization of the global lunar radiation environment and its biological impacts and potential mitigation, as well as investigation of shielding capabilities and validation of other deep space radiation mitigation strategies involving materials." This critically-important LRO objective selected by the ORDT traces to earlier NASA planning documents, including the 1998 "Strategic Program Plan for Space Radiation Health Research" (Life Sciences Division, Office of Life and Microgravity Sciences and Applications, National Aeronautics and Space Administration). Appendix H of that document summarizes the most critical questions concerning space radiation and its broad impact on all NASA activities. Strategic questions in the categories of space radiation environment, nuclear interactions, and human effects include:

- For a given mission, what are the fluxes of galactic cosmic rays (GCR) in interplanetary space as a function of particle energy, linear energy transfer (LET), and solar cycle?
- What are the doses related to heavy ions in deep space?
- How is a radiation field transformed as a function of depth in different materials?
- What are the optimal ways of calculating the transport of radiation through materials?
- What will the radiation environment be within the space vehicle, and what factors influence the flux, energy, and linear energy transfer spectrum of radiation?

The importance of this LRO objective and related outstanding questions is underscored in more recent NASA planning documents, such as the 2005 "Bioastronautics Roadmap: A Risk Reduction Strategy for Human Space Exploration". That document outlines several key goals including the development of new measurement approaches required for understanding the spectral properties of penetrating space radiation, and validating radiation shielding designs needed to define risk levels on exploratory-class missions and for risk mitigation. To date, these vitally important goals and questions remain open, in part because of limits in our ability to understand and simulate the evolution of radiation in space as it passes through human tissue. We are limited not only by scarce information about the cross sections for nuclear fragmentations at high energies but also by the lack of targeted observational data required to constrain radiation transport models. While much is known and understood about radiation exposure in low earth orbit (cf., Badhwar et al. 1996; Golightly et al. 1994), comparatively little is known at the Moon and in deep space.

The Cosmic Ray Telescope for the Effects of Radiation (CRaTER) instrument is designed *to characterize the global lunar radiation environment and its biological impacts*. CRaTER investigates the effects of galactic cosmic rays (GCR), solar energetic protons (SEP), and any secondary radiation produced by interactions between GCRs and SEPs at the lunar surface on tissue equivalent plastic (TEP) as a constraint for models of biological response to radiation in the lunar environment. CRaTER accomplishes this by producing linear energy transfer (LET) spectra over a wide dynamic range behind different areal densities of TEP and under different levels of solar activity and GCR flux. Functionally, CRaTER consists of a stack of circular silicon semiconductor detectors and cylindrical sections of TEP arranged so the ends of the stack have unobscured views of deep space and the lunar surface. Radiation passing through the stack, including ions and electrons, and to a lesser extent neutrons and gamma-rays, loses energy while passing through the silicon detectors. When ionizing radiation passes through a detector a signal is produced that is proportional to the total energy ΔE lost in the detector. Combined with the thickness Δx of the detector, an approximate LET is determined for the single particle as $LET = \Delta F/\Delta x$. Detectors are in pairs, with a thin detector connected to a low gain amplifier and a thick detector connected to a high gain amplifier. Together, the detector pairs and associated amplifiers make CRaTER sensitive to a broad range of LET from approximately 0.1 keV μm^{-1} to 2.2 MeV μm^{-1}. By combining signals from different detectors to identify the path of individual particles, CRaTER can be used to understand how radiation loss evolves in human tissue and how dose rates change during periods of heightened solar activity and ultimately over the course of the solar cycle. Key unique capabilities of CRaTER are the inclusion of the TEP to make biologically relevant radiation measurements and a telemetry rate sufficient to capture high resolution LET values for up to 1,200 events per second. Whereas previous instruments were hampered by the need to reduce LET resolution or the number of events recorded owing to telemetry constraints, CRaTER will be able to produce spectra with high resolution in both LET and time.

This paper is structured as follows. Section 2 describes the CRaTER investigation, from science goals to measurement requirements and data products. We motivate the measurements needed and the resulting instrument design and discuss our anticipated results. Section 3 describes the instrument, from the telescope stack of TEP and silicon detectors that measure LET through the electrical design and the measurement process, our observing strategy, and science and housekeeping data products. Section 4 describes the instrument testing, characterization, and calibration. Section 5 describes the science operations center, as well as the data analysis, interpretation, and modeling. Finally, in Sect. 6, we summarize briefly several representative science studies made possible by CRaTER's design—a design intended initially for exploration-enabling measurements, but also capable of high quality science measurements of energetic particle phenomena at and near the Moon.

2 Investigation Description

2.1 Linear Energy Transfer and the LET Spectrometer

The measurement goal of CRaTER is to produce linear energy transfer (LET) spectra. $LET(\Delta E/\Delta x)$ is the mean energy ΔE lost locally, per unit path length Δx, when a charged particle traverses material (LET is expressed in units of MeV $g^{-1}cm^2$, or, in a given system with known areal density, in MeV μm^{-1}). An LET spectrometer measures the amount of energy lost in a detector of some known thickness and material property as a high energy particle passes through it, usually without stopping. While LET spectrometers do not (necessarily) resolve mass, LET measurements do include all the species, with the possible exception of neutrons, that are relevant to the energy deposited behind a known amount of spacecraft shielding. Relevant LET spectra are a missing link, currently derived largely by models, and require experimental measurements to provide critical ground truth—CRaTER will provide this essential quantity needed for model closure. Before proceeding further, we comment on the distinction between CRaTER LET measurements and the instantaneous energy loss rate, dE/dx. We recognize that instruments such as CRaTER typically cannot measure dE/dx directly because the detector has a finite thickness and the path length is not fully well-known for each particle detected. However, CRaTER is designed such that pathlength uncertainty is small ($<10\%$). Additionally, by using sufficiently thin detectors, dE/dx is approximately constant within each detector. Consequently, as a large number of events are accumulated a spectrum of $\Delta E/\Delta x$ converges statistically to dE/dx (typically to within $\sim 10\%$, the pathlength uncertainty).

As the particles comprising the solar and galactic cosmic ray environment penetrate shielding, whether the shielding is provided by spacecraft or habitat structure, or by an overlying atmosphere (such as on the Martian surface), or even by the body's own tissue that overlays a critical body organ, they lose energy due to the myriad of Coulomb interactions between the charged ions penetrating the material and the atomic electrons attached to the atoms and molecules of the target materials, and their nuclei. As these penetrating ions slow, their rates of energy loss to the target medium begin to increase. Hence their LET increases. On occasion, the nuclei of these penetrating ions undergo nuclear collisions with the nuclei of the target atoms. Occasionally the collisions are elastic, but more often than not, they are highly inelastic resulting in fragmentation (breakup) of the colliding nuclei. These lighter fragments, having lower charge and mass numbers than the parent nucleus, have reduced energy losses per distance traveled in the target medium, i.e. their LET decreases. Thus, the overall effects of the shield materials on the LET spectra are a complicated mixture of competing processes, which tend to significantly alter the LET distribution, making measured LET spectra dependent on the type, quantity, and geometric configuration of shielding that the incident particle spectrum has penetrated as well as on variations in the incident spectrum itself. In general, material shielding tends to reduce the overall fluence of high LET particles and increase the fluence of lower LET particles, for incident solar and GCR spectra.

A variety of LET measurements behind various thicknesses and types of material is of great importance to spacecraft engineers, radiation health specialists, and to modelers who estimate impacts of the penetrating radiation. LET is one of the most important quantitative inputs to models for predicting human health risks and radiation effects in electronic devices. By relaxing the demand to measure the entire parent cosmic ray spectrum to one where only that part of the energy spectrum deposited in a certain thickness of material is needed, the challenging requirements of measuring total incident cosmic ray particle energy is removed. This change in focus greatly simplifies the complexity, cost, and volume of the required instrument. And, in addition to these savings, an LET spectrometer essentially provides the key direct measurement needed to bridge the gap between well measured cosmic ray intensities that will be available from other spacecraft and specific energy deposition behind shielding materials, vital exploration-enabling knowledge needed for the safety of future astronauts working in the harsh space radiation environment.

2.2 Measurement Objectives

The first high level LRO measurement objective relevant to CRaTER is that "*The LRO shall characterize the deep space radiation environment at energies in excess of 10 MeV in lunar orbit, including neutron albedo*". CRaTER is designed to measure the LET spectra of energetic particles across a range in charge and energy of biological relevance. One significant gap in knowledge of the deep space radiation environment is the uncertain LET spectra produced by the highest-energy, high-Z particles from both the Sun and the galaxy. While composition and variability of galactic cosmic rays above 10 MeV are fairly well defined, the spectra of the energetic component during intense solar energetic particle events is not well understood, demonstrates significant variability, and needs to be better specified. The current gap in understanding of this energetic component (>10 MeV) is critically important. Thus, we aim to target >10 MeV energies by virtue of their known damaging effects to man and machine and by virtue of their unpredictable time variability. CRaTER will provide measurements of the deep space radiation environment above 10 MeV with a primary focus on charged particles. The LEND instrument on LRO is designed specifically to characterize the radiation effects of neutrons, the remaining aspect of the first LRO measurement objective (Mitrofanov et al. 2009, this issue).

The second LRO objective relevant to CRaTER is: "*The LRO shall measure the deposition of deep space radiation on human equivalent tissue while in the lunar orbit environment*". CRaTER will provide LET spectra behind different amounts and types of areal density, including tissue equivalent plastic. TEP is an inert solid substance that has radiation absorption characteristics similar to human tissue and has been used extensively in laboratory and space-based studies of radiation effects on humans. Understanding the evolution of the LET spectrum as it passes through human tissue is an important step in predicting, and possibly mitigating the effects of radiation on human bodies and spacecraft components. The pathlength through CRaTER's TEP represents the characteristic depth of radiation penetration to blood forming organs.

2.3 Anticipated Results

The exploration goal of CRaTER is to provide the first deep space validation of radiation transport models. CRaTER will do so by producing LET spectra for comparison with simulation predictions. This validation is critical for transport models that are used to design shielding and estimate astronaut radiation risks.

Radiation transport models that take solar and galactic energetic particle spectra and produce predictions of evolving LET spectra within materials serve a critical role in risk reduction for the manned space program. For example, Kim et al. (1994) evaluated shield designs by calculating LET spectra using the 1977 solar minimum GCR spectrum as input into the "HZETRN" code. This simulation showed that LET curves decreased with increased shield thickness for high LET values but that for low LET values, the LET distributions increased, rather than decreased, with increasing shield thickness. Predicted LET spectra are strong functions of both the materials used and uncertainties in nuclear processes. These uncertainties are due to a lack of experimental data and to simplifications introduced into the models to speed up calculations. Simplified code is necessary to make the transport problem tractable and to permit exploration of a large number of possible configurations, but the simplifications also potentially limit model accuracy. The ability of these codes to accurately handle, for instance, spallation of heavy ions into lighter fragments and propagation through complex geometries and materials awaits experimental confirmation.

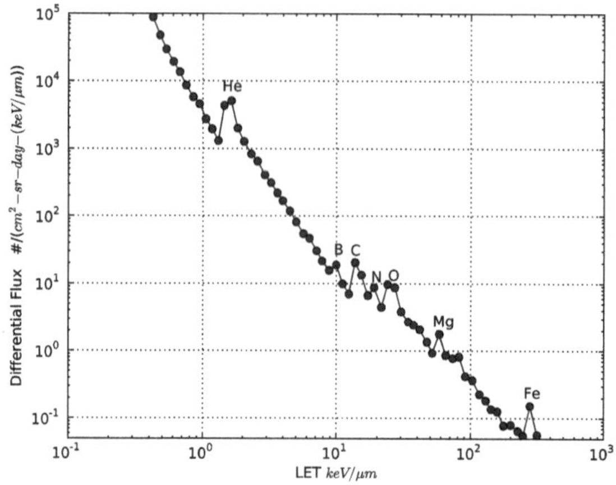

Fig. 1 Modeled LET spectrum in silicon behind 1 mm of aluminum using the CREME86 model and a galactic cosmic ray spectrum appropriate to the 1996 solar minimum, similar conditions to that expected during LRO's prime mission

Once LET spectra data begin to be taken by CRaTER, the data can be used to aid in transport code validation for the lunar space radiation environment. Instruments on several NASA spacecraft such as ACE (Stone et al. 1998) and STEREO (Luhmann et al. 2007) as well as spacecraft from other agencies (e.g., NOAA–GOES) will be used to record the primary galactic and solar particle spectra contemporaneously. The populations measured at these high-altitude locations will serve as excellent proxies for the population of cosmic rays present at the lunar orbit at the same time (or at some easily calculated propagation time delay in the case of SEP events). Galactic cosmic rays are not considerably localized in space, so that measurements anywhere within several hundred Earth radii of the Moon will be sufficient. Transport models such as the HZETRN code developed at NASA Langley (J. W. Wilson et al. 1995) or the Monte Carlo code HETC-HEDS (Townsend et al. 2005), will then be used to predict the LET distributions seen by CRaTER. The comparison of our observed LET spectra with these predictions will allow us to provide closure between key measurements and models for understanding radiation effects.

To illustrate how CRaTER observations will be used to validate radiation transport models, Fig. 1 shows a representative example of an LET spectrum. We created this spectrum using the Cosmic Ray Effects on MicroElectronics (CREME86) model with an input GCR spectrum based on the last solar minimum in 1996. The figure shows the LET spectrum in silicon expected behind 1 mm of aluminum plotted in the same flux and LET units as planned for the CRaTER data products. Peaks in the spectrum are the result of major ion species (e.g., He) or ion groups (e.g., CNO). With such GCR rates, particularly for the heaviest ions that produce the largest and rarest response, LET spectra must be integrated over days, weeks, or months in order to resolve spectral features, intervals consistent with the one-year prime mapping duration of LRO. In later sections, we will show how CRaTER's design, in particular the ability to process and return unprecedented volumes of particle energy deposits, enables the construction of accurate LET spectra also over short time intervals, including during intense SEP events.

3 Instrument Description

Figure 2 illustrates the overall mechanical design of CRaTER, which physically consists of an electronics box and a telescope assembly. The instrument consists of a rectangular

Fig. 2 CRaTER flight model: telescope assembly is the rightmost portion while the signal processing section and spacecraft interface is in the slanted enclosure to the left

electronics box (approximately 24.1 cm long by 23.0 cm wide by 15.9 cm tall) with a slanted top cover. The electronics box bolts to the spacecraft through six feet along the base. The CRaTER telescope assembly attaches to the electronics box (see right portion of Fig. 2). The telescope assembly and the electronics box are assembled separately and then integrated into the flight configuration as shown. The nadir, or lunar-viewing, aperture of the telescope is delineated by the circular area on top of the telescope assembly in Fig. 2. A normal to the circular aperture defines the telescope axis, which remains aligned along the nadir zenith line during regular in-flight operations while LRO is in lunar polar orbit.

The electronics box houses a digital processing board (DPB) and an analog processing board (APB). The DPB interfaces with the spacecraft through the four connectors seen near the lower edge of the electronics box in Fig. 2. These connectors serve to relay commands and telemetry, to provide power, and to provide a timing signal to CRaTER.

Figure 3 illustrates in cross-section the elements internal to the telescope assembly. The telescope assembly holds the telescope stack and the telescope electronics board (TEB). The TEB (green area) connects the telescope to the electronics box, delivers bias voltages to the detectors, and sends detector signals and calibration pulses through preamplifiers back to the APB. The preamplifiers are highly sensitive and so it is desirable to limit electrical interference near the detectors. Therefore the telescope assembly is electrically isolated from the electronics box, and grounded on the same path as the signals from the preamplifiers to the APB. The telescope stack consists of aluminum caps on the nadir (top end cap in Fig. 3) and zenith (bottom end cap) sides to shield the telescope detectors from low energy particles, followed by three pairs of thin and thick silicon detector pairs surrounding two sections of A-150 Tissue Equivalent Plastic (TEP) of different length (dark grey areas).

Figure 4 shows a functional block diagram of the entire CRaTER instrument, from detectors to the spacecraft bus; data flows from left to right in this block diagram. It shows the most critical components necessary for reliable functioning of the instrument. As described above, the telescope assembly houses the telescope stack of detectors and TEP, and

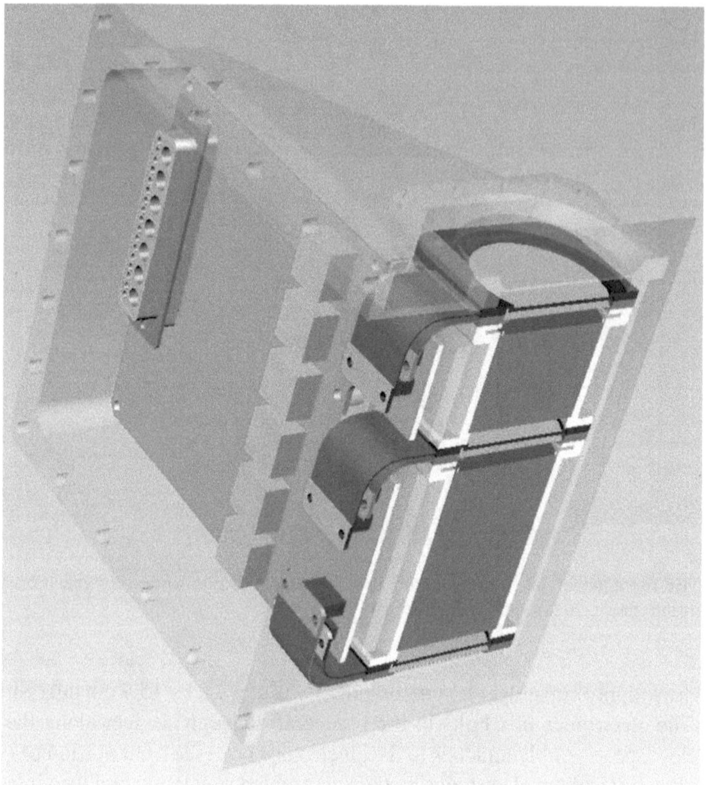

Fig. 3 (Color online) Cross-section schematic of the telescope assembly, showing the two sections of TEP (*gray*) sandwiched between three detector pairs, connected via flexible cables (*brown*) to the analog signal processing board (*green*)

the telescope board. The electronics box assembly houses the APB and the DPB. The APB shapes the pulses from each of the detector preamplifiers, further amplifies the signals, and generates calibration pulses for testing the response of each signal path. The DPB identifies and processes particle events and generates scientific measurements, controls power distribution within the instrument, records housekeeping data, and receives commands and sends telemetry to the spacecraft.

Table 1 summarizes the main properties of the CRaTER instrument, including physical resources, key design values, and measurement performance metrics. These quantities of the as-delivered flight model represent the optimal values resulting from science and engineering trade studies completed throughout the instrument development. The remainder of Sect. 3 describes details of the design which drive and determine the high-level values shown in Table 1.

3.1 Telescope Design

3.1.1 Overview

In this section, we summarize the physical construction of CRaTER, with a focus on the sensing portion of the instrument, or the telescope assembly. As described in the instrument

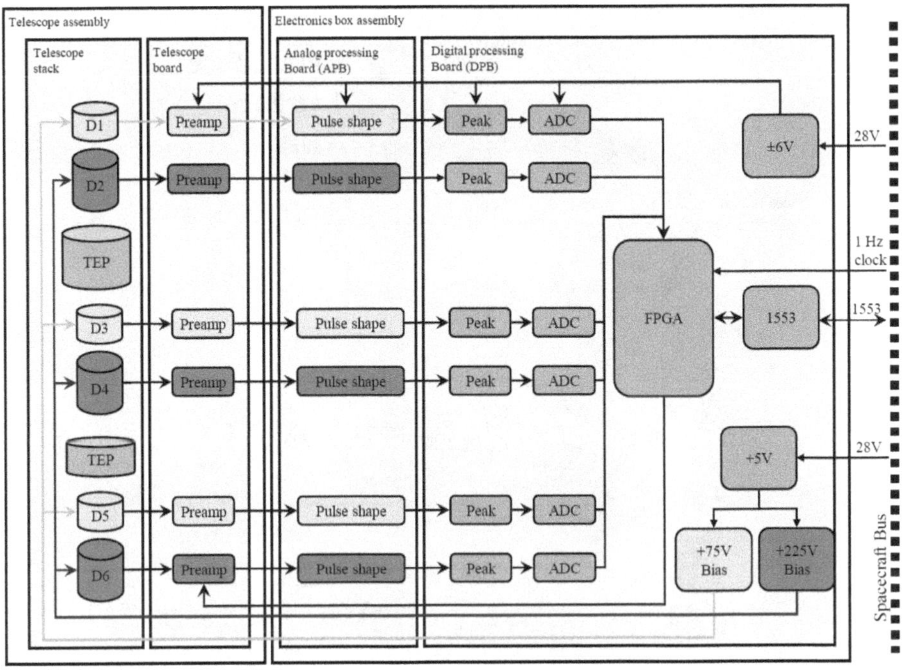

Fig. 4 CRaTER functional block diagram showing the critical components of the instrument. Particle detection, signal processing, and communication with the LRO spacecraft flow from left to right

Table 1 Summary of as-delivered CRaTER flight unit physical properties and measurement performance

Property	Value	Comments
Mass	5.53 kg	6.36 kg allocation
Power	6.66 W	9.00 W allocation
Maximum telemetry rate	89.1 Kbps	Sized for largest historic solar proton event
Maximum event transmission rate	1200 events/sec	Event defined as pulse height analysis on all 6 detectors for any valid detection
Low LET range	0.09 keV/μm to 85 keV/μm	Determined with thick detectors (D2, D4, D6)
High LET range	2.3 keV/μm to 2.2 MeV/μm	Determined with thin detector (D1, D3, D5)
LET resolution	<0.3% of maximum LET in each range	Net RSS value including detector and electronics noise, and gain uncertainty
Minimum geometric factor	0.57 cm^2 sr	Defined by D1–D6 geometry
Zenith full angle field of view	31.4°	Defined by D1–D6 geometry
Nadir full angle field of view	65.9°	Defined by D4–D6 geometry

overview, CRaTER consists of two physical parts, the telescope assembly and the electronics box. The telescope is mechanically mounted to the electronics box, but the two structures are electrically isolated from one another. The entire telescope assembly is instead grounded

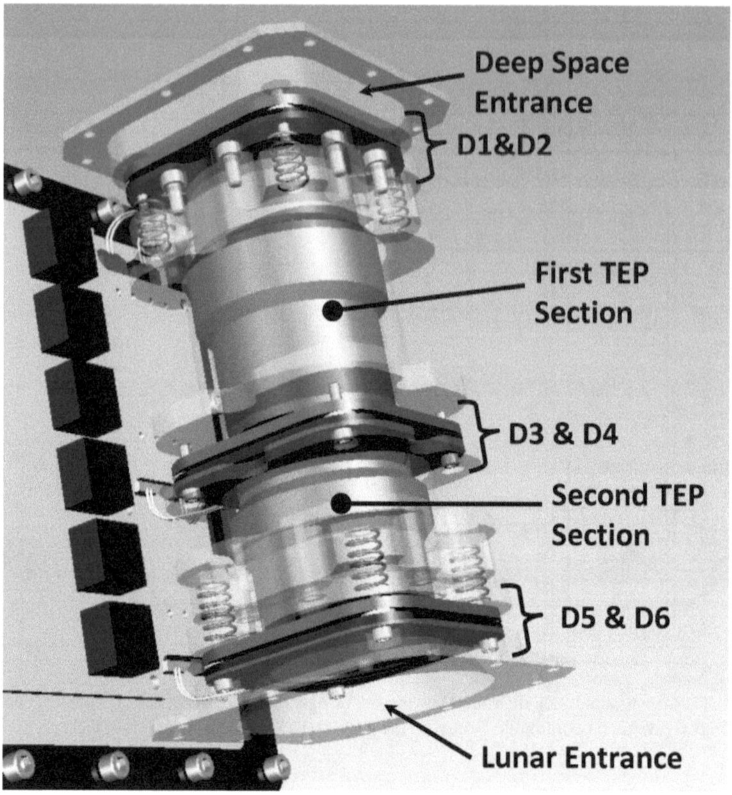

Fig. 5 Transparent view of the telescope assembly, showing the pairs of thin and thick detector sandwiching the two TEP volumes in the telescope stack, and the associated telescope electronics board. A cable with signal and ground runs from each of the detectors to one of six preamplifiers on the telescope electronics board

to the analog signal ground. This is done to reduce noise on the spacecraft chassis ground reaching the sensitive detectors and their preamplifiers.

As illustrated and identified in Fig. 5, the CRaTER telescope stack consists of three pairs of thin and thick detectors surrounding two pieces of TEP. From the zenith side of the stack the components are the zenith shield (S1), the first pair of thin (D1) and thick (D2) detectors, the first TEP absorber (A1), the second pair of thin (D3) and thick (D4) detectors, the second TEP absorber (A2), the third pair of thin (D5) and thick (D6) detectors, and the final nadir shield (S2). We shall use these shorthand designations henceforth. With this configuration, paired thin and thick silicon detectors thereby measure the LET spectrum at three depths through the telescope stack. Section 3.1.2 reviews the detectors selected for CRaTER and discusses the need for the pairs of thin and thick detectors to cover the full range of LET possible in silicon. The three pairs of detectors cover the range in LET expected by SEP and GCR ions in silicon and after evolving through the telescope assembly, including passage through the dominant component, the TEP. Section 3.1.3 describes the A-150 Tissue Equivalent Plastic used as the human tissue analog and primary absorber. Section 3.1.4 describes the optimization of the location of components in the telescope stack.

Key factors guiding our design of the telescope stack were the total pathlength through TEP desired, the sizes of the different pieces of TEP, and the geometric factor and resulting sensitivity of the entire telescope. By requiring that the minimum energy of protons that can just exit the TEP be ~100 MeV and that the TEP rather than the silicon in the detectors dominates the areal density of the telescope stack, we identified a goal pathlength through the total amount of TEP in the telescope of ~60 mm. Our design divides the total TEP pathlength into two components of different length, 1/3 and 2/3 the total length of the TEP. If the total TEP is 71 mm in length, then the TEP section closest to deep space will have a length of approximately 54 mm and the second section of TEP will have a length of approximately 27 mm. A variety of LET measurements behind various thicknesses and types of material is of great importance to spacecraft engineers, radiation health specialists, and to modelers who estimate impacts of the penetrating radiation. Simulations suggested that splitting the TEP into two asymmetric components that are 1/3 and 2/3 the total length provides a useful combination of lengths, also similar to typical thicknesses through human tissue to depths of blood forming organs.

Statistically significant LET spectra should be accumulated over short enough time intervals to resolve dynamical features in the galactic cosmic ray (GCR) and solar energetic particle (SEP) flux. During quiescent intervals, the counting rate will be dominated by the slowly varying GCR light ions. With typical GCR fluxes expected for solar minimum conditions (Webber and Lezniak 1974), a minimum geometric factor of 0.57 cm^2 sr (shown later to be the as-designed value) will yield LET rates of ~0.3 events/second from protons with incident energy sufficient to pass through the entire telescope. In one hour, a statistically significant sampling of ~1000 events would permit construction of longer-term average spectra; this one-hour interval is still short compared to typical GCR modulation timescales.

The geometric factor of a cylindrical telescope of this type is a function of both the distance between the detectors and the diameter of the detectors. We conducted a trade study to select the diameter of the CRaTER detectors. Our study focused on the benefit of a larger diameter detector in the form of larger collecting areas and geometric factors, versus the risk of unacceptably high single particle fluxes during major solar energetic particle (SEP) events. While both protons and heavy ions are enhanced during SEP events, the flux of hydrogen is many orders of magnitude larger and their detection dominates the counting rates. The hydrogen therefore was the focus of this trade study.

The first thick detector in the zenith (deep space) direction, D2, is designed to detect protons and is therefore most susceptible to enhanced fluxes during SEP events. The other thick detectors lie behind shielding in the form of the TEP and telescope mass and ultimately the Moon below, and are therefore expected to have lower rates. The concern is not that the D2 detector could be damaged by the high flux, but rather that it could experience electronic pulse pileup, in which multiple protons pass through the detector within a detection interval and corrupt the signal. The predicted worst-case singles count rate for D2 is shown in Fig. 6, estimated from the >10 MeV proton channel on the GOES spacecraft during the 225 most extreme SEP events over the last 33 years. Peak, instantaneous (5 minute average) proton rates are combined with the D2 single-detector geometric factor (~30 cm^2-sr) for this analysis We find that CRaTER's normal operation mode is sufficient to robustly measure all particles for ~90% of the peak rates encountered during extreme SEP events. To recover the remaining 10%, CRaTER is deliberately designed to reconfigure in response to such especially high rates. If the D2 signal saturates and thereby would otherwise overwhelm event processing with low energy protons, CRaTER has additional features to avoid any decrease in the utility of the measurements returned to the ground; this rate event mode switching is described fully at the end of Sect. 3.3.6.

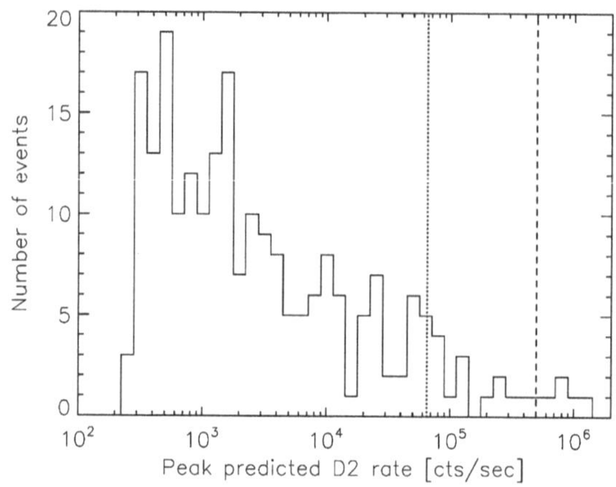

Fig. 6 Predicted number of >10 MeV solar proton events versus 5-minute-average D2 peak rates, based on 33 years of GOES proton data. Approximately 90% of these 225 representative >10 MeV SEP events will be cleanly measured in CRaTER's normal operating mode; the 10% with rates higher than that delineated by the left-most vertical line will be measured cleanly in CRaTER's elevated rate model (see Sect. 3.3.6 for details). *The vertical dotted line* represents the instrumental limit for recording true rates; *the vertical dashed line* is the ideal analog electronic limit for event processing

3.1.2 Detectors

CRaTER uses silicon semiconductor detectors to characterize the rate of energy loss of charged particles passing through the telescope. In this section we motivate our choice of detector thicknesses and the LET ranges the detectors are used to measure. We begin by reviewing the operating principles of semiconductor detectors. Figure 7 depicts a simplified detector cross-section illustrating the key components of the CRaTER detectors. At energies of interest to CRaTER, silicon semiconductor detectors are useful for characterizing the energy loss of ionizing radiation as it passes through or stops in it. In the CRaTER detectors, silicon is doped to make it a semiconductor, with electron-hole pairs in valance levels with energy levels much smaller than the first ionizing potential of an atom, making the detector much more sensitive than a proportional counter. The silicon is composed of both an N-type and a P-type substrate, making it a large diode that can be biased at a relatively high voltage with little leakage current.

When an energetic charged particle passes through the silicon, it liberates electron-hole pairs, which rapidly drift to either end of the detector in the electric field established by the bias voltage. The resulting signal from the drifting electron-hole pairs is linearly proportional to the total energy deposited. An analog circuit is then used to amplify these signals when they occur, and to stably and accurately determine the height of each pulse in multiple detectors. Both detector types are ion-implanted, totally-depleted structures formed from an N-type substrate. The Phosphorous-implanted N-type substrate is the ohmic side of the detector and the Boron-implanted P-side is the junction. These impurities generally require lower implantation energies and thus result in low implant depths of about 0.3 μm. Both detector areas are circular, have thin junction and ohmic windows, and have relatively fast timing capability, meaning that the electron hole pairs rapidly diffuse to the surfaces and provide a sharp signal to the electronics. Although fast timing is not critical for CRaTER, it is desired to have the metalization made in such a fashion as to reduce surface resistivity. There is a guard ring (indicated by the region labeled Guard) around the active junction to

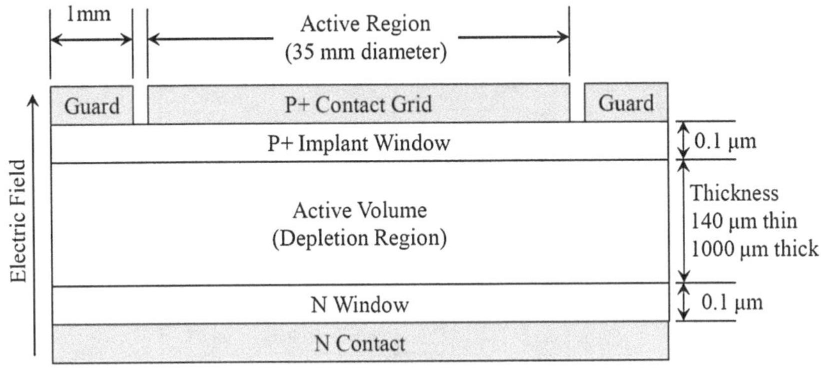

Fig. 7 Schematic cross section of the silicon detectors in the CRaTER instrument

improve edge uniformity and a neighboring field plate ring to aid discharge of oxide stray charge. Each thin and thick detector mounts to its own small passive printed circuit board and connects to the telescope electronics board by a flexible cable.

Thin and thick detector pairs are used to span the entire range of LET and to provide a system that is less sensitive to low energy protons in the event of intense SEPs associated with solar flares and coronal mass ejections. The thickness of the thin detectors is approximately 150 microns while the thick detectors are approximately 1000 microns. The need for pairs of detectors arises from the large range in LET that ions can produce in silicon through ionizing radiation and from the different amplitudes of energy loss expected from light and heavy cosmic ray ions. In order to understand the range of LET expected in the detectors, we review the common modes of energy loss by a moderately relativistic particle.

A moderately relativistic proton or heavier charged ion passing through a medium such as human tissue, TEP, or the silicon detectors will lose energy through electromagnetic interactions with electrons in the medium and through weak and strong interactions with nucleons. Generally the nuclear interactions are less common, but they become more important for the heavier ions above 20 MeV/nuc and are a major source of uncertainty in predicting LET. Nuclear interactions are also a mode whereby neutrons can produce a signal in the CRaTER detectors, although this process is also uncommon. For the purposes of optimizing CRaTER, we focus on electromagnetic energy loss. The average rate of energy loss, or stopping power, of energetic but not ultra-relativistic charged particles other than electrons is given by the Bethe-Bloch equation. The energy loss dE/dx of a particle with charge z, energy E, and speed $\beta = v/c$ passing through a medium with atomic number Z and atomic mass A is given by,

$$-\frac{dE}{dx} = Kz^2\frac{Z}{A}\frac{1}{\beta^2}\left[\frac{1}{2}\ln\left(\frac{2m_e c^2\beta^2\gamma^2 T_{\max}}{I^2}\right) - \beta^2\right] \text{MeV g}^{-1}\text{ cm}^2$$

where $K = 0.307075$ MeV g^{-1} cm^2, m_e is the electron mass, c is the speed of light, $\gamma^2 = 1/(1-\beta^2)$, $T_{\max}$ is the maximum energy that the ionizing particle can impart in a single collision, and is the minimum energy required to excite the medium (Yao et al. 2006). There is a strong dependence of the stopping power on the charge of the incident particle, and a weaker dependence on the mass of the particle through $T_{\max}$. Corrections exist at both high energies, where polarization and shielding effects become more difficult to treat, and at low energies where assumptions about the adiabaticity of the collisions become invalid.

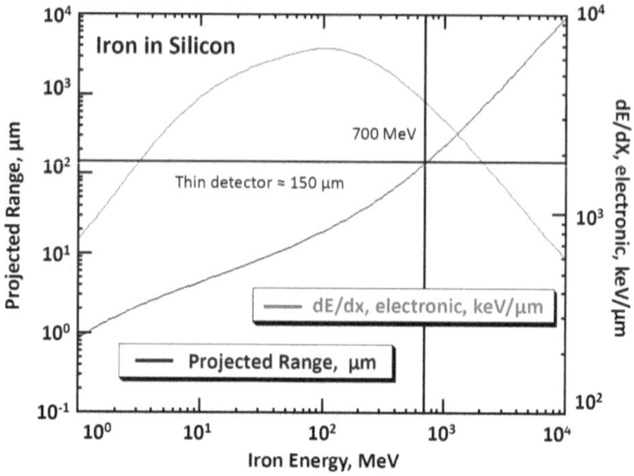

Fig. 8 The upper energy range of the thin detectors is determined by the maximum energy expected from an iron nucleus as determined using SRIM (Ziegler et al., 1984). This is a plot of the projected range and LET of iron in silicon showing a peak energy of approximately 700 MeV

Additionally, an ion will experience multiple Coulomb scatterings, leading to a trajectory that deviates from a straight line. This leads to a statistical distribution in the path length that a given species of particles will traverse in passing through or stopping in the detector, even with identical initial energies. This results in the phenomenon of "straggling", which will manifest itself as a finite width in the range of energy deposited by a single type and energy of particle, no matter how fine the energy resolution of the detector or how low the noise level. This "noise" in the signal due to straggling places a lower limit on the energy resolution needed to distinguish events. Three implications of the Bethe-Bloch equation are that the energy deposited in the detectors increases with the square of the charge of the ionizing particle (so iron will deposit much more energy than hydrogen), that the peak of the energy deposition occurs as the particle is just stopping in the detector, and that the minimum-ionizing radiation rate occurs at ∼2 GeV/nuc. Consequently, we set the minimum measurable LET to that of minimum-ionizing protons in silicon. The maximum LET is set by the energy that an iron nucleus will deposit when it just stops in a thin detector. Figure 8 shows projected range and LET of iron in silicon using SRIM (Ziegler et al. 1984). SRIM (The Stopping and Range of Ions in Matter) is a group of programs which calculates the stopping and range of ions (10 eV–2 GeV/nuc) into matter using a full quantum mechanical treatment of ion-atom collisions.

The CRaTER telescope uses pairs of thin (∼150 micron) and thick (∼1000 micron) detectors. As discussed in the electronics section, the thin detectors are connected to a low gain amplifier, while the thick detectors are operated with a high gain. This design makes the thick detectors sensitive to particles with very low LET, while a very high LET is required to produce a measurable signal in the thin detectors. This use of high and low gain thin and thick detectors is motivated by the range of LET we need to measure at each point in the telescope stack and in the infeasibility of doing that broad an LET range with a single detector. Within the limits of the noise level of the detectors, it is desirable to detect high energy particles with the minimum ionizing deposition rate, so each point in the telescope where the LET spectrum is to be observed, the minimum LET measured shall be no greater than 0.25 keV/micron in the silicon. Practical considerations effectively constrain the high end of the LET energy range. Slow moving, high-Z ions that give up much of their energy upon interaction will by definition yield large LET events. Therefore, the instrument should be able to measure such high-Z particles. On the other hand, the probability of a high energy heavy ion just stopping in a detector and depositing the maximum signal is small. Models and observations of heavy ions with prototype instruments show that these particles will produce signals commensurate with a deposition of 2 MeV/micron, so at each point in the telescope where the LET spectrum is to be observed, the maximum LET measured shall be no less than 2 MeV/micron in the silicon.

3.1.3 Tissue Equivalent Plastic and Composition of CRaTER materials

As discussed in the previous section, the evolution of LET spectra is determined both by the composition of the incident high energy radiation and by the composition of the material being encountered. This is especially significant when considering biological effects. For example, one study has shown that as radiation travels through organic matter, the contribution to biological change (gauged by the neoplastic transformation of C3H101T/2 mouse cells, a well studied cell line) from light ions increases, while the importance of heavy ions (such as iron) decreases (Kim et al. 1994). At equivalent depths of approximately 5 g/cm^2, hydrogen and helium are the dominant contributors to the measured biological change. Since the LET spectrum is such a sensitive function of the absorbing medium, the bulk of the material within CRaTER must be similar to human tissue in order to produce results related to biological effects. Furthermore, the volume of TEP employed must be sufficient to cover the range in depths where the dominant species is expected to change from heavy to light ions.

The muscle equivalent tissue equivalent plastic (TEP) A-150 is the most popular material used today in the construction of instruments that measure the evolution of ionizing radiation in the presence of materials similar to human tissue. A150 is an Exradin proprietary plastic blend designed to be tissue equivalent. It is also known as Shonka tissue equivalent plastic after its original creator, Francis Shonka. By material, A-150 TEP is a combination of Polyethylene, Polyamide, Carbon black, and Calcium fluoride. TEP has the appearance of a black crayon but is very stiff and is easily machined. The TEP is slightly hydroscopic, and will absorb up to ~3% water by weight in a 50% humidity environment. Our tests showed that other than the water, little material outgassed from TEP samples after they were baked out in a vacuum. The composition and thicknesses of the TEP used in CRaTER is given in Table 2; the detailed TEP composition and dimensions are important as inputs to radiation transport code models.

The two pieces of TEP used in the telescope stack have a 2 : 1 length ratio, with the longer component on the deep space side of the telescope. The pieces are right cylinders and are held firmly in place at each end. The three thick-thin detector pairs sandwich the two pieces of TEP. The different lengths produce an asymmetric geometry. This configuration serves

Table 2 Composition of A-150 TEP used in CRaTER by mass

A-150 tissue equivalent plastic (TEP)		
Element	Mass composition (%)	Use for CRaTER (%)
H	10.33 ± 0.07	10.33
C	76.93 ± 0.09	76.93
N	3.30 ± 0.08	3.30
O	6.94 ± 0.51	6.93[a]
F	1.14 ± 0.60	1.14
Ca	1.37 ± 0.06	1.37
Total		*93.07*
Density	1.127 ± 0.005 g/cm^3	1.127 g/cm^3

TEP on-axis linear dimensions: 53.992 mm (zenith section) and 26.972 (nadir section)

[a]O composition not measured-assumed to be the balance of the elemental composition

Table 3 Composition and thickness of Al 6061-T6 aluminum endcaps used in CRaTER by mass

Al 6061-T6		
Element	Mass composition (%)	Use for CRaTER (%)
Al	95.8–98.6	97.200
Cr	0.0400–0.350	0.195
Cu	0.150–0.400	0.275
Fe	$\leq$ 0.700	0.455
Mg	0.800–1.20	1.000
Mn	$\leq$ 0.150	0.075
Si	0.400–0.800	0.600
Ti	$\leq$ 0.150	0.075
Zn	$\leq$ 0.250	0.125
Other, each	$\leq$ 0.0500	
Other, total	$\leq$ 0.150	
Total		*100.00*
Density	2.70 g/cm^3	2.70 g/cm^3

Aluminum end cap thickness: 812.8 μm (zenith cap) and 810.3 μm (nadir cap)

several purposes. The initial LET spectrum is measured at D1/D2 after first passing through the thin aluminum zenith cap. (Since the aluminum end caps are part of the material traversed by detected particles, and because the material properties of this matter is important also for a robust model comparison, we include Table 3 which shows the detailed composition and thicknesses of the end cap aluminum. In addition, a thin single layer of thermal blanket material covers each end cap on the exterior of the telescope.) After passing through D2 the LET spectrum is representative of the corresponding spectrum that would be present just on the inside of a manned space vehicle. The particles then traverse the longer piece of TEP whose length of several inches is similar to the depth to blood-forming organs in a human body. D3/D4 measure and quantify the evolution of the LET spectrum after passing through this first length of TEP. Those particles with enough energy then traverse the second, shorter TEP segment simulating a longer pathlength to a blood-forming organ. D5/D6 measures the fully evolved LET spectrum resulting from passage of particles through the entire length of the telescope.

3.1.4 Fields of View and Geometric Factors

Motivated by the need for sensor simplicity and in keeping with the primary measurement requirement, CRaTER does not determine the vector trajectory of individual particles (through position sensing strips, for example). Additionally, no means are employed to identify particles that pass through the side of the instrument and are detected (using a scintillator shroud surrounding the entire telescope stack coupled to a photo-multiplier tube, for example). Such techniques are often used to constrain the pathlength traversed by particles in a cosmic ray telescope and to reject events consisting of multiple particles or which scatter out of the instrument volume. Since we are interested in studying the spectrum of all particles within the telescope, including those that scatter at large angles, and because the techniques described above introduce a great deal of complexity, telemetry, and mass, we instead define the fields of view of the instrument (and hence statistically well-constrained path lengths) through n-fold detector coincidences.

The CRaTER measurement requirements place constraints on the field of view of the instrument only indirectly, through restrictions on the uncertainty in the pathlength traversed through the TEP and the detectors, and through lower limits on geometric factors, for example. While large fields of view are not a requirement, they are desirable in that they increase the geometric factor and therefore the statistical accuracy of the LET spectral measurements, particularly of the low intensity GCR population. In addition, since it is possible that SEPs and GCRs reflecting off the lunar surface—especially at small grazing angles of incidence—could provide significant diagnostics of the composition of the Moon and of secondary radiation from the surface, it is desirable that the nadir end of the telescope have as large a field of view as possible (owing to the presumed lower flux of cosmic ray albedo). This is another driver for having the shorter piece of TEP on the nadir end of the telescope as the D3/D4 and D5/D6 coincidences define a relatively larger opening angle than do D1/D2 and D3/D4.

The optimization of the CRaTER fields of view is discussed in the following section. Here we note the fields of regard that the spacecraft is able to support for the instrument. It is essential that the fields of view of CRaTER are not obstructed by components of the spacecraft, as CRaTER will not be able to distinguish between the primary spectrum of GCRs and SEPs and the secondary spectrum produced by interactions with spacecraft components.

The final spacing between the components of the telescope stack was determined by conducting a trade study that identified the configuration which optimized the parameters captured in the design requirements. The instrument requirements that are affected by the telescope stack are the following: 'the minimum pathlength through the TEP should be >60 mm' and 'the two pieces of TEP should be different lengths (herein chosen to be 1/3 and 2/3 of the total length)'. In order to have sufficient counting rates of the most energetic particles during normal times CRaTER should have a minimum, full telescope geometric factor of $\sim 0.1\ \mathrm{cm}^2\ \mathrm{sr}$. In order to yield a sufficiently small uncertainty on the inferred LET, a geometry which produces $<10\%$ variation in path length between D1 and D6 is required. We constrained both the zenith and nadir field of view to be within their fields-of-regard (35 and 85 degrees, respectively). Owing to the proximity of CRaTER to the spacecraft, some fields-of-view defined by detector coincidences intercept part of the spacecraft. However, the required fields-of-regard assure unobstructed views of deep space for particles producing six-fold detector coincidences and for those producing a D4–D6 coincidence from the nadir direction. Furthermore, these two important view geometries are also unobstructed by the instrument itself.

We found that the telescope stack optimization could be expressed by varying two free parameters, which we called L1 and L2. L1 is the length between the top of detector D2 and the top of detector D4, and L2 is the length between the top of detector D2 and the top of detector D6. We considered other heights less important. For example, the spacing between the thin and thick detectors in each pair is set by the natural value of the printed circuit boards the detectors are mounted on. We found that placing the zenith and nadir shields further away from the closest detector pair only increased the necessary diameter of the shield needed to provide a uniform thickness over the field of view. Therefore the shields are simply best put close to the adjacent detectors, as limited solely by sensible design practices. Since the TEP sections have a sufficiently larger diameter than the detectors, it did not make a practical difference where the TEP was mounted relative to the detector pairs. Thus, L1 and L2 summarize the two unique parameters varied in the trade study.

From this trade study the optimal values set a D1–D6 geometric factor of $0.57\ \mathrm{cm}^2\ \mathrm{sr}$ with a zenith or nadir field of view of $31.4°$, well within the zenith and nadir fields of regard. In addition, the D4–D6 coincidence provides a geometric factor of $2.564\ \mathrm{cm}^2\ \mathrm{sr}$ and a field

Table 4 Linear dimension of telescope components

Location	Material	Thickness (mm)	Cumulative distance (mm)
S1	Aluminum	0.8128	0.8128
S1 (back) → D1 (front)	*(void)*	*2.363*	*3.1758*
D1	Silicon	0.148	3.3238
D1 (back) → D2 (front)	*(void)*	*3.360*	*6.6383*
D2	Silicon	1.000	7.6838
D2 (back) → T1 (front)	*(void)*	*3.943*	*11.6268*
A1	A-150	53.9670	65.5938
A1 (back) →D3 (front)	*(void)*	*3.554*	*69.1478*
D3	Silicon	0.149	69.2968
D3 (back) → D4 (front)	*(void)*	*3.568*	*72.8648*
D4	Silicon	1.000	73.8648
D4 (back) → A2 (front)	*(void)*	*11.900*	*85.7648*
A2	A-150	26.9800	112.7448
A2 (back) → D5 (front)	*(void)*	*11.900*	*124.6448*
D5	Silicon	0.149	124.7938
D5 (back) → D6 (front)	*(void)*	*3.060*	*127.8538*
D6	Silicon	1.000	128.8538
D6 (back) → S2 (back)	*(void)*	*2.362*	*131.2158*
S2	Aluminum	0.81026	132.0260

Component designations are described fully in Sect. 3.1.1; "S" refers to end cap shields, "D" refers to detectors, and "A" refers to A-150 TEP

Table 5 Fields-of-view, geometric factors, and threshold energies of representative coincidences

Representative coincidences	Field of view (full angle)	Geometric factor (cm^2-sr)	Proton threshold energy (MeV)
D1·D2	169.0	24.152	12.7 (z)
D1·D4	53.4	1.679	90.8 (z)
D1·D6	31.4	0.569	114.5 (z)
D6·D5	170.0	24.566	17.7 (n)
D6·D4	65.9	2.564	63.9 (n)

of view of 65.9°, again within the nadir field of regard. In the nominal LRO orbit (polar inclination, 30–50 km lunar altitude) this nadir field of view is completely filled by the lunar surface, thus maximizing the probability of detecting albedo particles.

Table 4 contains the critical linear dimensions of telescope elements along the resulting optical path. The detector diameters are each 35 mm and all other components have diameters slightly greater than 35 mm. Table 5 contains the resultant fields of view for key representative detector coincidences along with their geometric factors. The final column shows the threshold proton energy needed to create the coincidence from the indicated direction of incidence ("z" for zenith and "n" for nadir). Finally, Fig. 9 shows the minimum energy required of ions (hydrogen to iron, using each ion's most common isotope) normally-incident from zenith to reach D2, D4, and D6.

The geometric factors presented in Table 5 are based purely on geometry and coincidence requirements. For particles passing through the sides of the instrument, detectors will have a complex and energy-dependent geometric factor. It is important to understand and limit the

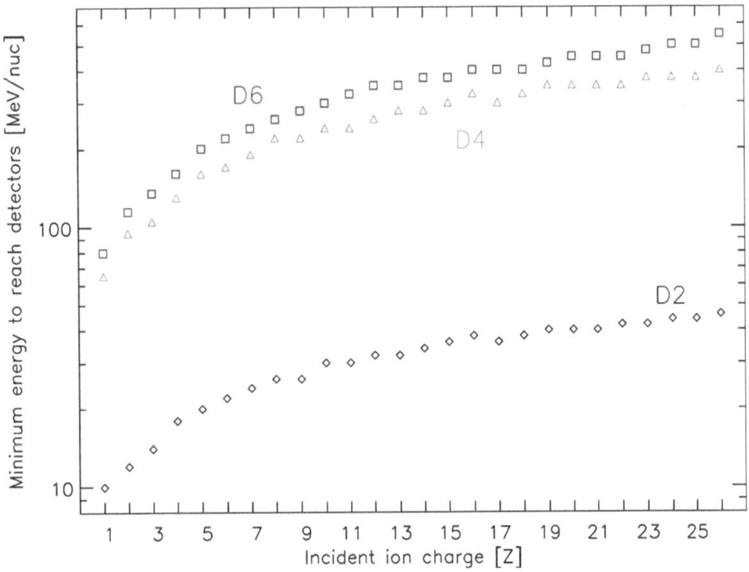

Fig. 9 (Color online) Minimum energy required of an incident ion, from hydrogen to iron (using each ion's most common isotope) normally-incident from the zenith direction to reach detectors D2 (*lower blue curve*), D4 (*middle green curve*), and D6 (*upper red curve*)

sensitivity of the detectors to such side penetrators, in order to limit false coincidences and to maximize performance during high flux intervals. To this end, reserve mass was placed in a thick ring around each detector pair (see Fig. 2). We used GEANT4 to show that this extra shielding mass, along with the instrument housing, provides effective shielding of side penetrating protons up to typically ∼50 MeV.

3.2 Electrical Design

3.2.1 Overview

In this section we cover the flow of electrical signals from the detectors through the telescope board to the analog processing board and finally the digital processing board. A key electrical aspect of the overall design of CRaTER is that the telescope housing is connected to the analog ground inside the telescope assembly, and nowhere else. The telescope housing is electrically isolated from the electronics box in order to prevent noise coupling from the spacecraft into the detector front-end. The electronics box housing is connected to the digital ground inside the electronics box.

A high-resolution measurement of the energy deposited is required to characterize the LET spectrum and to distinguish between the effects of the primary radiation and secondaries produced through interactions. The pulse height analysis of the energy deposited in each detector shall have an energy resolution better than $1/200$ the maximum energy measured by that detector.

3.2.2 Telescope Board

The purpose of the telescope board is to bias each of the detectors, and to provide for an initial pre-amplification of the signals in close proximity to the detectors in order to reduce

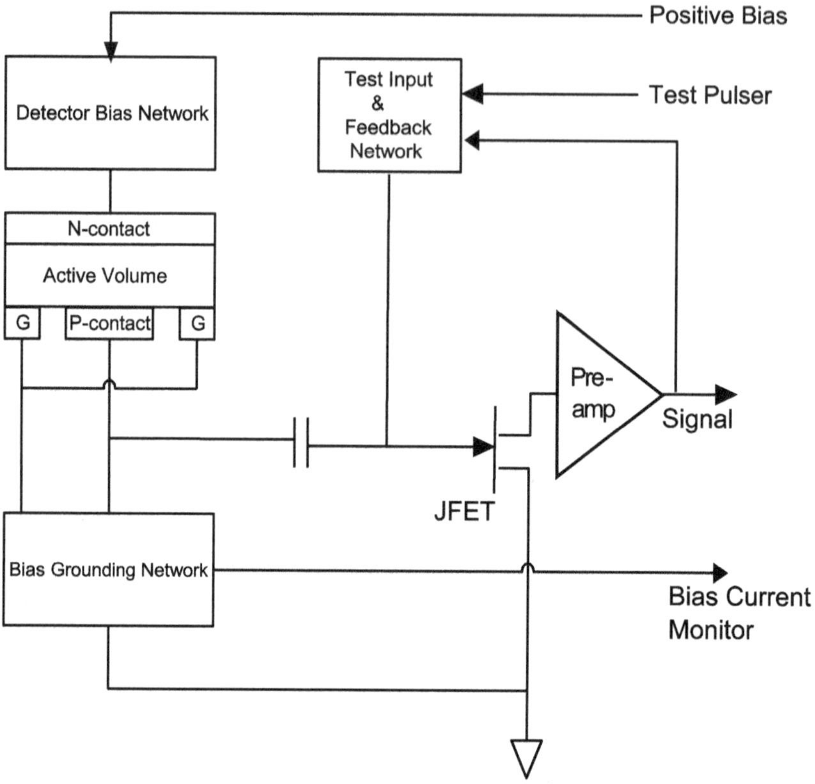

Fig. 10 Schematic illustrating the components of the telescope board for a single detector. A positive voltage biases the detector on the N-contact with the signal coming off the P-contact. Bias grounding networks connect the P-contact and G-contact to ground. Current from the detector is AC coupled to a jFET and preamplifier and the amplified output signal is sent to the electronics box along with a bias current monitor. A feedback network dissipates the collected signal from the detector and accepts external test pulses from the electronics box for calibration purposes

noise as the small initial signal is transported. Figure 10 is a block diagram of the components of the telescope board for a single detector.

3.2.3 Analog Processing Board

The Analog Processing Board (APB) provides a linear transfer function of output signal amplitude to detector energy deposit for three thin detectors and three thick detectors. A functional block diagram of a single amplifier string is shown in Fig. 11. The signal from the preamplifier is passed through a system to remove constant offsets and give the signal a symmetric Gaussian profile. The thin and thick detector signals are Gaussian shaped pulses with a peaking time of ~1 μs. This pulse shaping time was optimized by a consideration of the noise level of the thick detectors at 20°C as a function of the shaping time. The design architecture for the APB is drawn from significant flight heritage such as in the Imaging Proton Spectrometer instrument on the NASA Polar spacecraft (Blake et al. 1995).

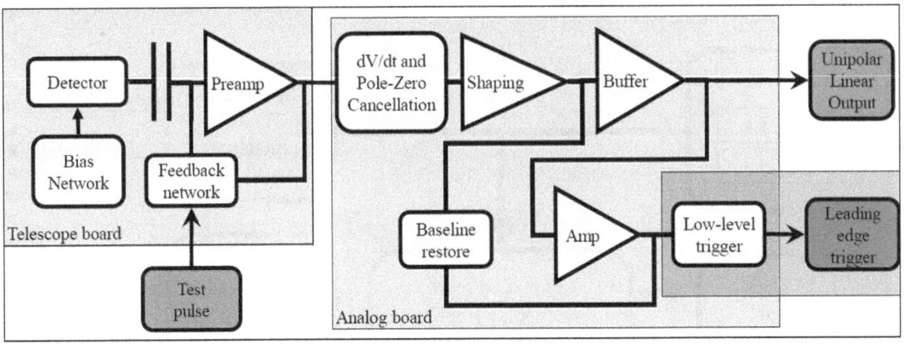

Fig. 11 Block flow diagram illustrating the signal pulse shaping process. The preamp processes the signals from either the detector or the test pulser. The signal is then amplified and given an approximately 1 us pulse profile. A baseline restorer is used to help maintain the baseline voltage with larger events. The baseline restorer can also be used to generate a low level trigger—this feature is not used in the current design in flight but is used on the ground while testing the system

3.2.4 Digital Processing Board

The purpose of the Digital Processing Board (DPB) is to identify events that are valid for pulse height analysis, digitize the pulse-heights of the APB output signals, provide control of APB test pulser, supply power and detector bias, and provide the command and telemetry interface to the spacecraft. Figure 12 is a block diagram of the functions of the field-programmable gate array (FPGA) and the DPB.

Test Pulse Generator An internal pulse generator is used during ground test phases and in orbit to monitor the stability of the pulse forming network with time and to perform instrument aliveness tests. The test pulse generator injects a known charge into the front of each preamplifier with two gain settings, an 8-bit resolution amplitude, and low (8 Hz) and high (2 kHz) frequency rates. A connection is available for each detector for the injection of externally generated pulses. This external option was used on the ground to distinguish between noise and variation in the APB and in the test pulse generator and is not available in flight.

Low Level Discriminator Two Low Level Discriminator (LLD) threshold voltages are generated by converting Pulse Width Modulated FPGA outputs to DC voltages. The discriminator amplitude, set via the 8-bit command interface registers, ranges from −0.047 to 0.141 V-DC. The thin detector discriminator setting drives the D1, D3 and D5 comparators; the thick detector discriminator setting drives the D2, D4 and D6 comparators. The instrument does not perform a pulse height analysis for an event unless the energy deposited in at least one detector produces a signal stronger than its corresponding LLD threshold.

3.2.5 Power

The DPB monitors and enables low voltage DC-DC converters that changes the 28 VDC provided by the spacecraft into +/−5 V for use by the analog electronics, and a separate +5 V for use by the digital electronics and bias supplies. The bias voltages were designed to be about 30 volts higher than the full depletion voltage of the detectors over a large range of leakage currents. The thin detector bias is approximately +75 V and the thick detector bias is approximately +225 V.

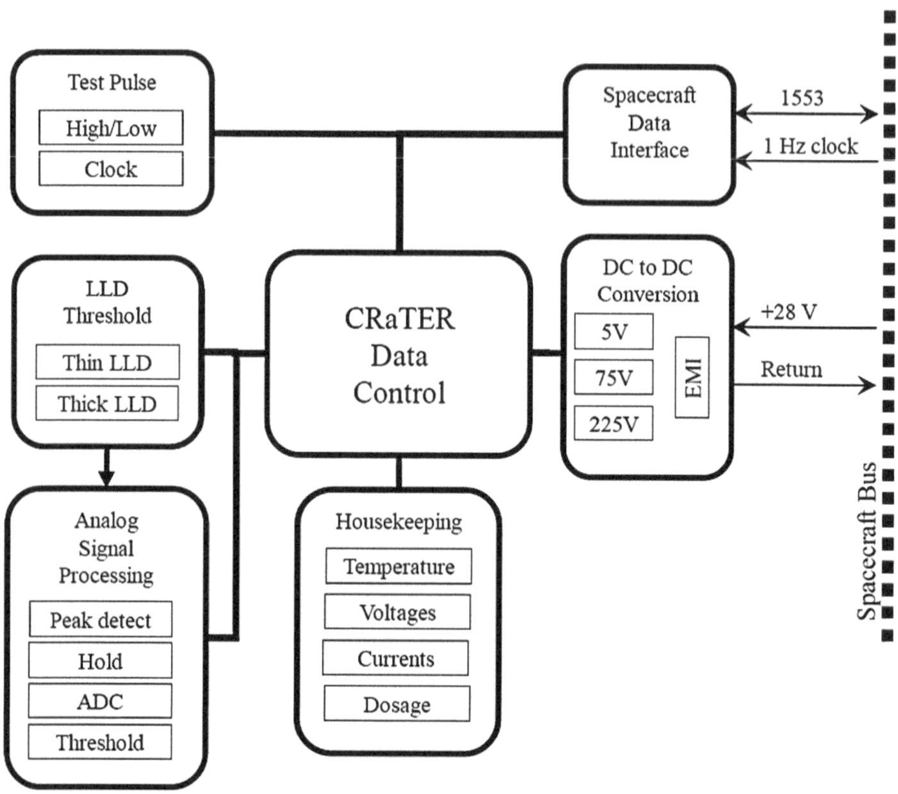

Fig. 12 Block diagram of the digital system. The FPGA generates test pulses for calibration and broadcasts the LLD for the thin and thick detectors, performs the PHA, enables bias voltages, and monitors housekeeping data such as temperatures, voltages, currents, and total dose within the telescope. The FPGA communicates to the spacecraft through the 1553 interface and receives the 1 Hz clock from the spacecraft

3.3 CRaTER Measurements

3.3.1 Overview

If a particle deposits enough energy in a detector, then a primary science measurement will be triggered. The measurement is of the energy deposited separately in each of the six detectors. It should be noted that regardless of which detector(s) cross their respective LLD thresholds, all six detectors are analyzed and reported in the data.

3.3.2 Measurement Sequence

The primary science measurements consist of the amplitudes of pulses detected from each of the detectors. The six signals from the pulse-forming and amplification networks on the APB are sent to the digital board, where the signals from the thin and thick detectors are separately compared to a low level discriminator (LLD) reference voltage. If any of the six signals crosses the corresponding LLD threshold, then a pulse height analysis (PHA) of all six signals is triggered.

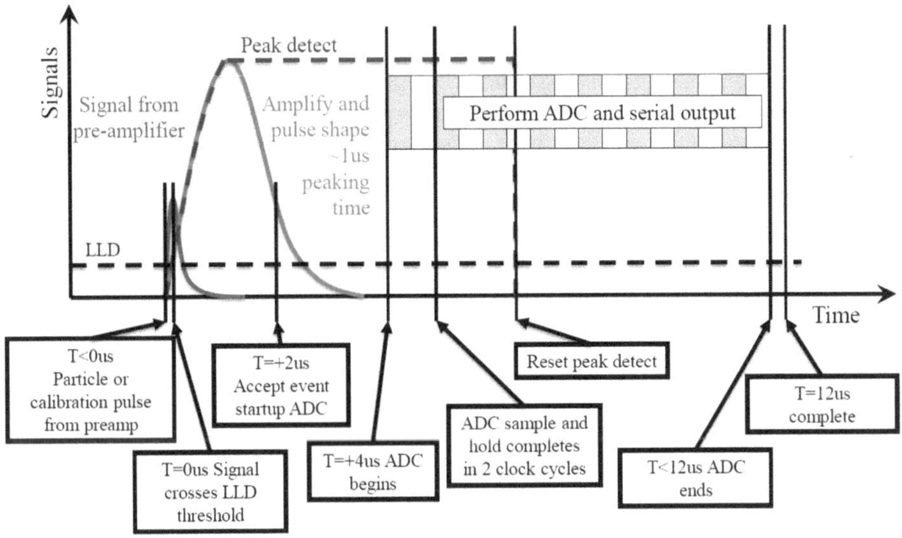

Fig. 13 Schematic illustration of the pulse height analysis (PHA) process on the analog and digital boards in the electronics box

After the PHA is complete, the list of triggered detectors is compared with a programmable mask that encodes the sets of detector coincidences that correspond to events of interest for subsequent analysis on the ground. If an event falls within the mask set, it is added to a telemetry buffer within the FPGA (a "good" event), otherwise it is discarded (a "reject" event). With the exception of extreme high rate events, CRaTER shall operate with a coincidence mask that accepts all possible detector coincidences (i.e., virtually all particle detections are telemetered to the ground).

Figure 13 provides a schematic illustration of the pulse height analysis. At some time $T < 0$ μs a current pulse produced by either migrating electron-hole pairs in the detector or a test pulse from the digital board passes through the preamplifier in the telescope board and arrives at the analog board. At $T = 0$ μs the signal from the preamplifier crosses the low level discriminator (LLD) and triggers the PHA. At $T = 2$ μs the peak detect is set to hold and the analog-to-digital converter (ADC) is commanded to power up. At $T = 4$ μs the ADC is powered up and does a sample and hold in the following two clock cycles. The ADC then performs a serial conversion over the next 12 clock cycles, clocking the results out to the digital board. About 1/3 of the way through the ADC the peak detector is reset. The ADC and the measurement process is complete at $T = 12$ μs, defining the fixed measurement dead time.

3.3.3 Science Data Products

Primary Science Data Products The event data for a single event in the primary science packet consists of a 9-byte block containing the six 12-bit numbers corresponding to the pulse height analysis of the six detectors. The FPGA contains two buffers. While primary science data are being written to one buffer during a given one second interval, the FPGA reads out the data stored in the previous second from the other buffer.

Secondary Science Data Products Secondary science telemetry packets are read during every 1-second data interval in which primary science data are being collected. These secondary science values set the time resolution that the state of the instrument (calibration commands, bias supply status, and disabling of detector processing) can be tracked on the ground. Parameters included in the secondary science data stream include: whether or not the calibration system is enabled, the calibration system rate, whether or not the thin and thick detectors are biased, whether or not each detector processing is enabled, and information on commanding.

A critical second component of this data stream is the reject event and good event counters, which are used to determine the deadtime of the instrument and convert the primary science data into absolute fluxes during periods of intense SEP radiation. Event counters, detailed below, are all 16-bit counters that "freeze" at the maximum count until reset. Every second the current-count value is transferred to a holding register (which is inserted into the secondary science telemetry) and the counter is cleared.

- The *reject event counter* is a 16-bit counter that is clocked every time an event is rejected by the discriminator accept mask test.
- The *good event counter* is a 16-bit counter that is clocked every time a qualifying event is processed. This counter is clocked for all events, even if they are not written into static random access memory (SRAM), due to buffer overflow, or read-out of SRAM the following second (telemetry overflow).
- The *total event counter* is simply the sum of the reject and good event counters; it is provided for data processing convenience.
- The *singles counters* for each detector is a 16-bit counter that is clocked by the "Singles Count" signals received from the analog electronics subsystem. Since these inputs are asynchronous, they are sampled and filtered. Singles "incidents" with pulse widths $<\sim 200$ ns may not be recognized. Threshold energies for the singles counters are approximately 110 keV for the thin detectors and approximately 35 keV for the thick detectors.

3.3.4 Instrument Housekeeping

Instrument housekeeping packets are read every 16 seconds whenever 28 VDC power is supplied to the instrument. The following describes the variables monitored in the housekeeping data stream.

Variables Monitored

Bias Voltages: The thin and thick detector bias voltages are recorded.

Temperatures: Temperatures recorded include the telescope electronics board, instrument chassis bulkhead, the APB, the DPB, and the DC/DC converters.

Total Dose: The APB board will contain a device for measuring the total radiation dose experienced by the electronics on the APB board. The device provides three analog outputs which are converted by the DPB into a 24-bit counter, the minimum step size of which is 0.020 mRad, and full scale reading of 335 Rad. The counter wraps after reaching full scale.

Currents: The detector leakage currents will be monitored for each individual thin and thick detector. For the thin detectors the monitor will cover the current range from 0–0.24 μA. For the thick detectors the monitor will cover the range from 0–0.12 μA.

3.3.5 Commands

The spacecraft provides a time-of-next interval command every second which is used to uniquely time-tag each telemetry packet. Commands available via ground command include:

- LLD levels (described above)
- Internal pulser calibration signal amplitude and rate
- Detector bias enable
- Individual detector processing enable/disable inputs
- Coincidence mask settings to define "good" events

3.3.6 Observation Strategy

Due to CRaTER's immovable mounting to the spacecraft (see Figs. 14 and 15), telescope orientation is determined by the pointing of the LRO spacecraft. CRaTER is aligned, along with most other instruments on the spacecraft, with its telescope axis along the nadir/zenith line. As such, one of CRaTER's apertures (D6 end) will nominally be pointing towards the Moon, and the other towards deep space (D1 end). Some excursions from this pointing, including limb sounding yaws in the $+/-y$-axis directions, are expected for brief periods for calibration and secondary science activities.

CRaTER's nominal measurement configuration will be to accept any detector coincidences as valid events, and, both the thin and thick detector LLD settings will be placed just above the noise. With the usual expected flux of particles that are measurable by CRaTER, this will allow us to analyze every valid detected event. In periods of high particle flux, the LLD settings will be adjusted to suppress the more dominant proton population in order to devote a higher percentage of our telemetry to the higher LET heavy ion population.

Section 3.1.2 examined the maximum fluxes expected in the D2 detector (Fig. 6) due to the largest SEP events of the previous solar cycles. During these large events, the effects of the enhanced heavy ion spectrum may be more important scientifically than just measuring primarily the proton spectrum. Several features have been incorporated into the CRaTER signal processing design to allow for a reduction in the sensitivity of the instrument to SEP

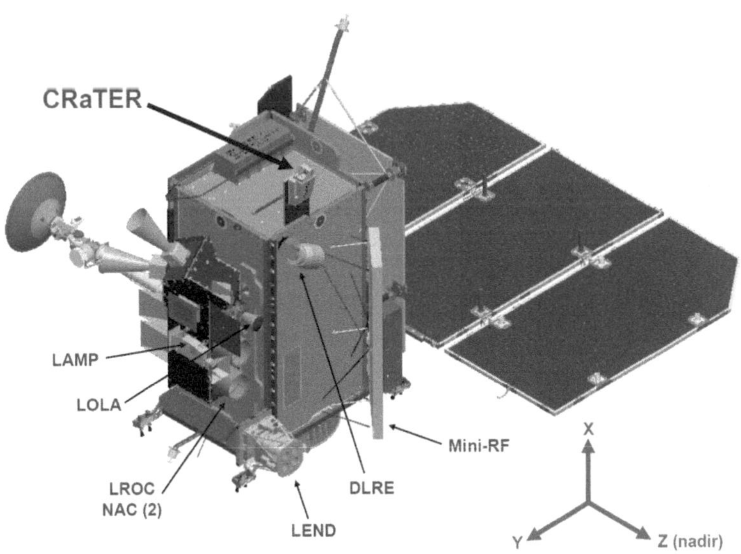

Fig. 14 CRaTER mounting location on the LRO spacecraft. The positive z-axis points toward lunar nadir during normal flight operations; the positive x-axis points nominally along LRO's orbital velocity vector

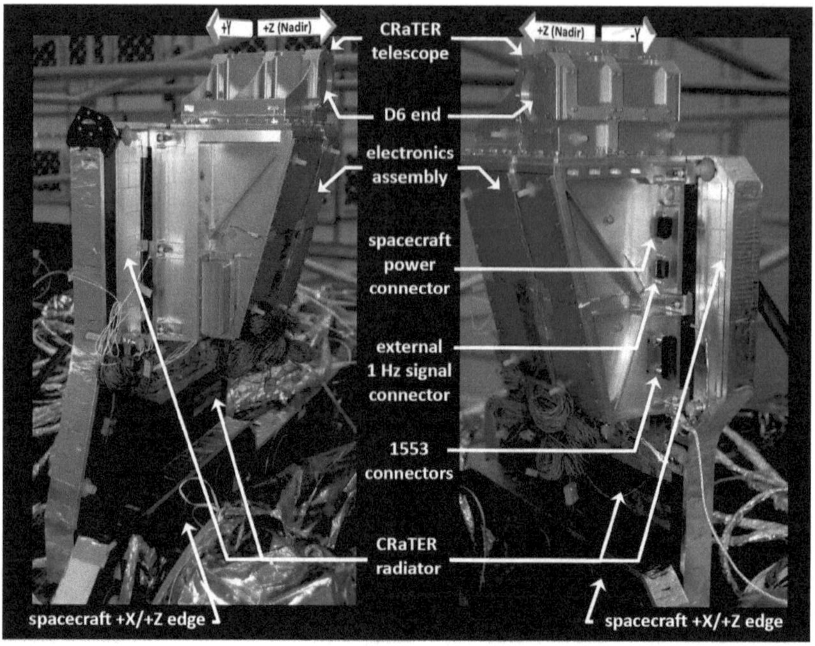

Fig. 15 Two labeled views of the CRaTER flight unit as mounted to the LRO spacecraft during final integration and before thermal blanket attachment. The nadir direction $(+z)$ is labeled in each view for perspective

protons in favor of the rarer heavy ions. First, the thickness and LET ranges of the thin and thick detectors have been optimized such that the thin detectors that measure high LET events do not see many of the protons. Furthermore, the FPGA is designed to be able to deactivate one or more of the detector signals for triggering a PHA by exceeding the LLD value. The LRO spacecraft monitors CRaTER secondary science event rates in real-time and is designed to reconfigure instrument settings accordingly within seconds.

The following describes the automated process that configures the instrument into three different operating modes depending on detection rates: low, elevated, and extreme. In the default "low rate" operating mode all events are sent to the ground; all detectors are in operation and the LLD thresholds for the thin and thick detectors are set just sufficiently high to avoid detector noise levels but otherwise are low. Once the event rate exceeds a certain value, the instrument switches automatically into an "elevated rate" mode, requiring at least two detectors to have detected a signal to record and telemeter a valid event. In the "extreme rate" case, the zenith-facing thick detector is disabled, the thin detector thresholds are raised above the signal level produced by protons, and detection is required by at least two of the enabled detectors. Mode switching can occur at a 60-second cadence and the mode is recorded in the telemetry stream.

This process will help ensure continued measurement of higher energy protons and all heavy-ion particles during the infrequent periods when the instrument response would otherwise become saturated by lower energy protons in its nominal configuration. During these times, the lowest LET portion of the spectrum (less than a few keV/micron) dominated by the protons is sacrificed in favor of the higher LET portions contributed by the heavy ions. The threshold event rates for mode transitions are reconfigurable through ground command.

4 Instrument Testing, Characterization, and Calibration

The purpose of this section is to describe the testing, characterization, and calibration of the CRaTER Flight Models (FM), based on lessons learned from experiences with the extremely high-fidelity Engineering Model (EM). In brief, considerable bench testing demonstrates that there is a desirable linear relationship between the digital value returned by the PHA and the original energy deposited in the detector. Gains and offsets for each detector are determined with high precision by calibrating the instrument with a beam of high energy protons produced by the Northeast Proton Therapy Center (NPTC) of Massachusetts General Hospital (MGH). A 230 MeV proton beam at the MGH NPTC (Cascio et al. 2003) is degraded in energy using sheets of plastic (Cascio et al. 2004) until a beam is produced with large energy dispersion and a peak energy tuned to the response of a pair of thin and thick detectors. The dispersed beam produces a characteristic track in energy deposition in the pair of detectors. The gains and offsets for each of the detectors is then determined by iteratively varying the free parameters of the instrument response until the measurements match the predictions of GEANT numerical simulations of the energy loss.

4.1 Instrument Characterization

4.1.1 Linearity of the Pulse Shaping

The most significant concern in instrument characterization is quantitatively understanding the relationship between the value of the energy deposited in the i-th detector and the resulting value produced by the ADC of the PHA. This relationship is demonstrated to be linear with an RMS residual of less than 0.1% using a stable external pulse generator. The external pulse generator is capable of generating stable, linear, and repeatable test pulses. A coaxial cable connects the pulse generator to any of the detector measurement chains before preamplification through a set of six connectors mounted directly on the telescope electronics board. An absolute scale for the external pulse generator is set using a series of attenuators and a coarse gain. Once the absolute scale is set a dial is used to generate pulse amplitudes with fractional amplitude variability of 0.01%.

We measured the height of the pulses generated by the external pulser and the best fit of a line to these data has an offset of 0.001 V, or approximately 0.03% of the peak value. The RMS of the residual to the linear fit is on the order 0.1%. It is important to note that both the small offset of 0.03% and the RMS residual from a linear relationship of 0.1% are upper limits imposed by the resolution of the sampling technique, and it is likely that the external pulse generator has even better performance. Nonetheless, these extremely low values are sufficient for our calibration needs.

Once we verified that the external pulse generator was linear, we connected it to each of the analog measurement circuits of the FM. The absolute amplitude of the external pulser was then set so that when the pulser amplitude dial was set to 1000 the resulting pulses just saturated the PHA. We then collected two minutes of data running at 100 events/second at each of nine input pulse amplitudes spaced between 100 and 850. The returned events were then analyzed and the PHAs at each pulser amplitude were characterized as a Gaussian distribution with a width and a center. Figure 16 is a plot of the center channel of the measured peak, in analog-to-digital units (ADU), as a function of pulser amplitude setting for all detectors in the FM. (The peak values for D1 are as measured; a successive offset of 100 ADU has been added so that the individual traces of D2 through D6 can be identified more clearly.) The relationship between pulser amplitude and measured peak is remarkably linear in each

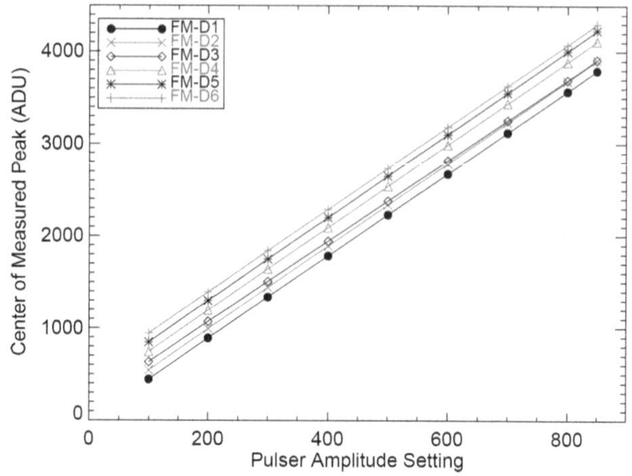

Fig. 16 Demonstration of the linearity of the pulse height analysis circuitry for all detectors of the FM using the external pulse generator, performed at a typical in-flight temperature of 10°C. The center of the measured peak value is offset by 100 ADU successively between detectors (starting from D1) in order to better demonstrate linearity of each chain

of the six detectors. Additional analysis, including finer pulser setting increments (not shown here), was used to quantify the linearity of the relationship, with the quadratic term only contributing ~0.3% at the highest values. The RMS residual from a simple linear model was significantly less that 0.1%. Since it is highly unlikely that the external pulse generator and the internal analog pulse shaping and PHA circuits would conspire to produce this linear agreement, we conclude that the relationship between E_i and C_i is also described with a linear relationship. Therefore, for the purpose of calibrating CRaTER at the accuracy required for the success of the mission, we may write,

$$E_i = G_i C_i + O_i \text{ [keV]}$$

where G_i [keV ADU^{-1}] is the gain of the i-th detector chain and O_i [keV] is its zero offset.

4.1.2 Electronics Noise Level

The number of samples collected at each pulser setting permits examination and determination of an upper limit on the electronics noise level. Figure 17 is a plot of the standard deviations of the pulse height distribution as a function of the mean of each distribution for a range of pulse amplitudes. These measurements were taken with the instrument operating in a refrigerator held at 10°C, a temperature representative of typical in-flight values. Results are plotted for all six detectors of the FM. The width of the measured peak (in ADU) is clearly a linear function of the center of the measured peak (in ADU), or a fixed fraction of the peak amplitude. For these measurements the noise is approximately 0.15% of the amplitude, well below the maximum noise level requirement. However, it is important to note that the measured noise is a combination of the noise in the pulse shaping and PHA circuitry within the instrument and any noise produced by the external pulse generator itself. These measurements therefore are an upper limit on the true noise level of the CRaTER analog electronics of 1 ADU or 0.02% of the maximum energy, far below levels needed to achieve our measurement requirements.

4.1.3 Stability

While the internal pulse generation system is mainly used for verifying instrument functionality and not for precise calibration purposes, it is useful for tracking CRaTER's stability

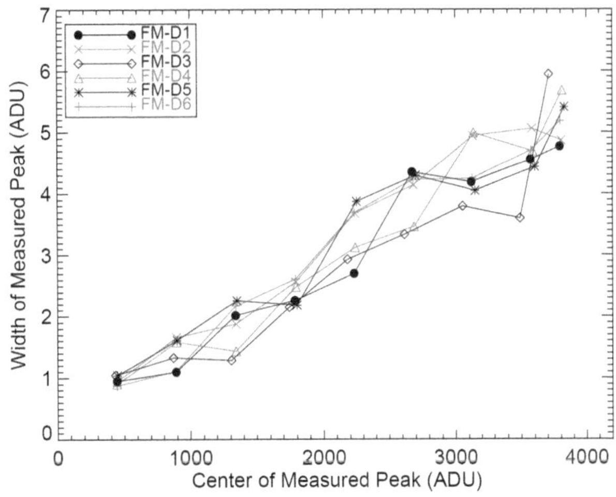

Fig. 17 Standard deviation of measured PHA peak values (in ADU) returned as a function of the center of the measured peak (in ADU) for nine externally generated pulse amplitudes in all six detectors of the FM with the telescope operating at 10°C. This shows that the noise level of the electronics is a constant fraction of the input pulse amplitude, and is less than about 0.15%

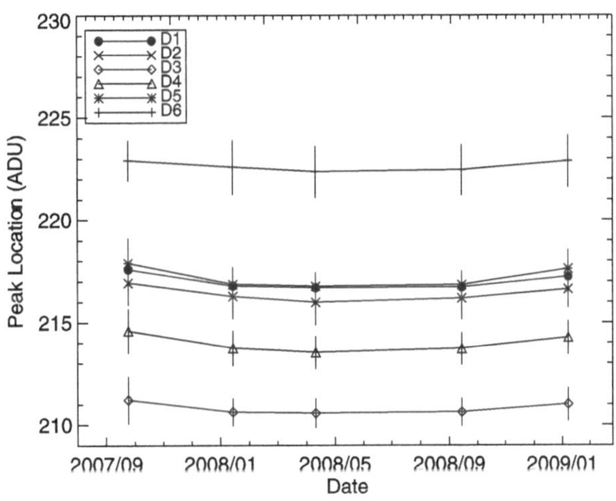

Fig. 18 Measured center of signal peak (in ADU) in the FM generated by the internal pulse generator at a fixed setting, during five internal calibration runs spaced over nearly 18 months. The FM response remained steady and reproducible at the ~0.06% level

over time. Each time the FM went into testing it was customary to conduct a full sweep through all of the possible amplitudes of the internal calibration system. Figure 18 is a plot of the measured center peak in each of the six FM detectors to a fixed internal pulse as a function of time. Five internal calibration runs span a period of approximately 18 months. Variation in the response over this period, taken under different operating situations and at different locations, remains stable at the ~0.06% level.

4.1.4 Temperature Dependence

The temperature dependence of the CRaTER electronics was determined by placing the instrument in a thermal chamber while feeding pulses through the chamber wall from the external pulse generator, which is maintained at room temperature. Figure 19 is a plot of the response of each detector chain to a fixed external amplitude pulse over the entire expected operating temperature range of the instrument. The gain has a very weak and non-linear dependence on temperature. However, note that over the entire temperature range, the total

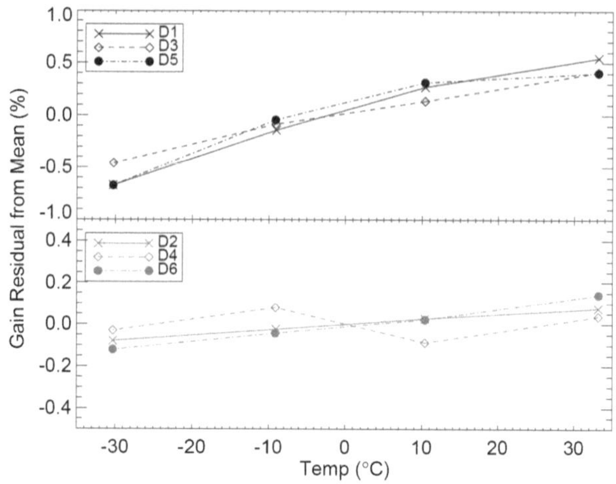

Fig. 19 Response of measured gain of a fixed external pulse generator amplitude versus temperature, ranging from $-30°$ to $+35°C$. Total percentage gain change across the full temperature range is only $\pm 0.1\%$ for the thick detectors (D2, D4, D6) and is only $\pm 0.5\%$ for the thin detectors (D1, D3, D5)

variation is only approximately $\pm 0.1\%$ for the thick detectors and $\pm 0.5\%$ for the thin detectors. We conclude that the thermal dependence of the instrument response is very small, on the order of $<0.5\%$ level effect, and that a simple temperature-dependent function describes the thick and thin detector gains sufficiently well.

4.2 Theory and Modeling

This section provides the theoretical basis for our ability to characterize CRaTER using a spread beam of protons. Figure 20 is a plot of the typical energy deposited into the components of CRaTER as a function of initial energy for a beam of protons. This calculation was performed by creating a table of typical energy loss rate as a function of energy for protons in silicon using the SRIM program. Note that the maximum energy deposited is by a particle that just stops at the back of the detector. The thicker detector can therefore produce a larger energy deposit of slightly more than 10 MeV, compared to the ~4 MeV of the thin detector. Once the particle has enough energy to pass through the detector, the total energy deposited decreases as the particle's energy loss rate decreases.

In reality, the situation is more complex than the relations shown in Fig. 20. This is because the SRIM code only describes proton energy loss in a statistical sense. It also does not include physics such as nuclear interactions, deep inelastic scattering, and other interactions such as pion production. GEANT4 is a more sophisticated toolkit for the simulation of the passage of particles through matter with an abundant set of physics processes to handle diverse interactions of particles with matter over a wide energy range. For many physics processes a choice of different models is available, and overall the results are much more realistic than the simple SRIM calculations. Figure 21 is similar to Fig. 20, but it has been generated using GEANT4 and shows energy depositions in the first pair of detectors. GEANT4 is used extensively in the community, and is especially well-validated at proton energies relevant to CRaTER measurements for calibration.

Figure 22 shows a two dimensional histogram of simulated energy deposits measured in, for example, detectors D3 (thin) and D4 (thick) expected from a normally incident beam of protons with a uniform distribution of energy, using GEANT4 and the actual CRaTER detector thicknesses and materials. For this figure, only particles that produced a simulated signal in both virtual detectors were included in the histogram. The vertical dashed line indicates the simulated D3 LLD threshold. Five key portions of the histogram are labeled,

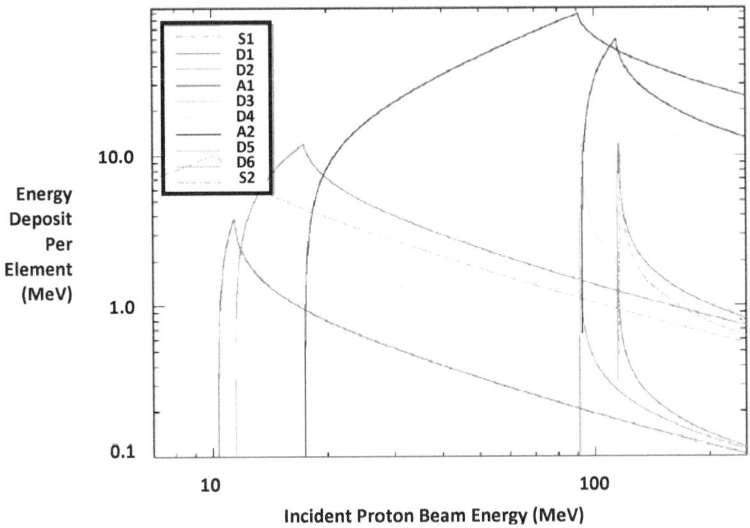

Fig. 20 Calculations of energy loss in each component of the CRaTER telescope stack as a function of initial energy of incident protons using the SRIM program, showing the regions of peak energy response for each detector in the telescope

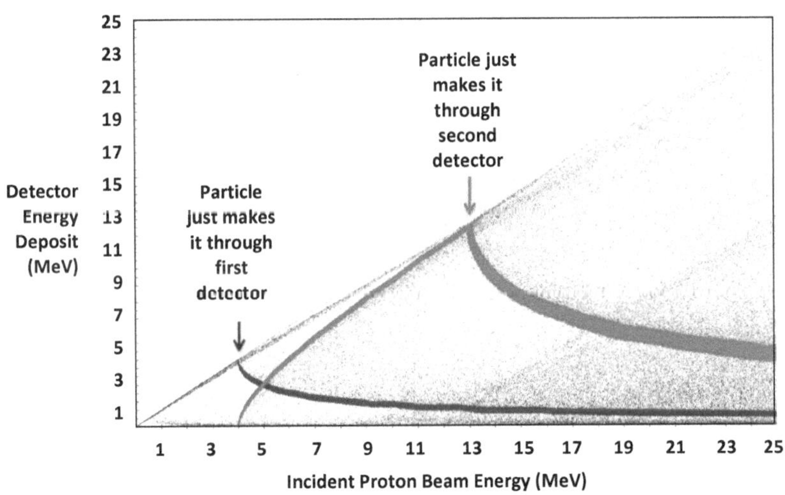

Fig. 21 A scatter plot of the energy deposited in a pair of thin and thick detectors as a function of incident energy. Unlike the previous two calculations shown in this section, this simulation makes use of the Geant4 code, which includes more accurate physics including scattering and nuclear interactions

and represent the track that proton energy deposition follows as a function of increasing incident proton energy: (1) protons pass into and stop in D3; (2) protons just stop in D3; (3) protons make it farther into D4 with increasing energy; (4) protons just stop in D4; (5) protons pass through D4. This strongly characteristic pattern is produced in each thin-thick detector pair at energies at and near the point where protons are just stopping in a particular detector. The encircled locations are these key locations in energy space. We use

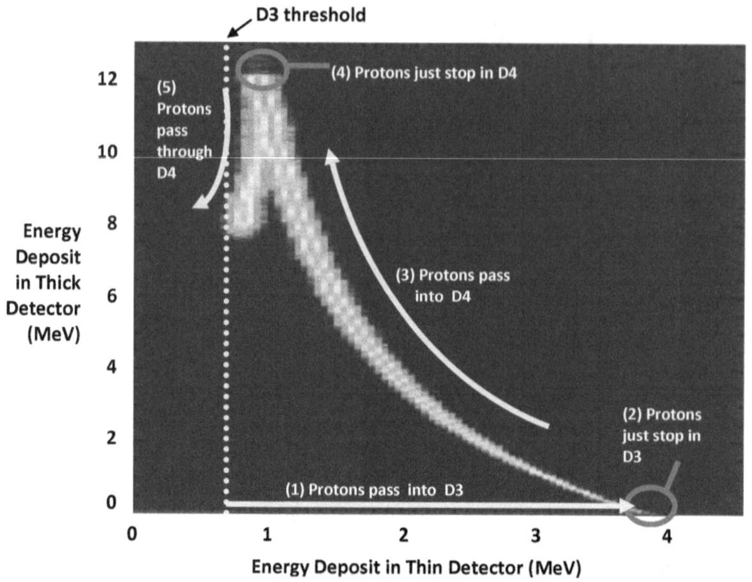

Fig. 22 Two-dimensional histogram of simulated protons depositing energy in D3 (*horizontal axis*), a thin detector, and D4 (*vertical axis*), a thick detector. The arrows illustrate the track that proton energy deposition follows in the histogram as a function of increasing initial proton energy. The encircled locations are the unique locations in energy space where protons are just stopping in either D3 (location #2) or D4 (location #4)

this characteristic signature as the theoretical basis in the end-to-end calibration procedure described next.

4.3 Calibration

Final calibration of the CRaTER Flight Models (FM) was based on lessons learned from extensive experiences developed with the high-fidelity, flight-like Engineering Model (EM). Techniques developed with the EM were refined and applied to the FM before instrument delivery. In brief and as designed, there exists a straightforward linear relationship between the digitized value returned by the PHA system (ADU) and the original energy deposited in the detector (keV).

Calibration consists of two major activities. In addition to an end-to-end calibration conducted on the fully integrated flight model (described below), laboratory testing of individual detector chains provided an independent, first-level of calibration. Before and during final assembly of the telescope, radionuclides such as Cobalt-60 and Americium-241 stimulate each detector chain while in vacuum. These sources produce well-defined particle and/or photon energies, which along with an external, high-precision pulse generator, provide a means to establish quantitatively the ultimate association between detected energy and ADU value. Once fully integrated, individual detectors can no longer be stimulated by particles from radionuclides. Therefore, a second stage of calibration for the fully-integrated CRaTER instrument complements the sub-system calibrations.

Final system gains and offsets for each detector are determined with high precision by calibrating the instrument with a beam of high energy protons produced by the Northeast Proton Therapy Center (NPTC) of Massachusetts General Hospital. A well-calibrated 230 MeV proton beam at MGH is degraded in energy using sheets of plastic until a beam is

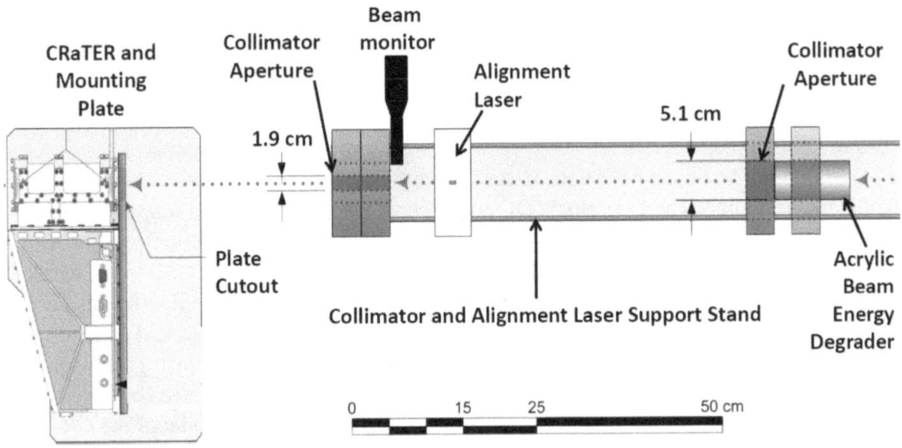

Fig. 23 Beam line geometry and configuration for typical testing of CRaTER EM and FM units at the MGH proton facility

produced with large energy dispersion and a peak energy tuned to the characteristic energy deposition response of a pair of thin and thick detectors (shown above in Fig. 22). Since we have previously shown that the relationship between the energy deposited in a detector and the channel number determined by our measurement chain is a simple linear function, the absolute response of each detector in the instrument can be determined. The gains and offsets for each of the detectors are determined by iteratively varying free parameters of the instrument response until the measurements match the predictions of GEANT4 numerical simulations of the energy loss.

The experimenter's beamline at the MGH proton facility has its own dedicated target room designed not for medical patients but rather for physics experiments. The geometry of the beam line for a typical CRaTER beam test is shown in Fig. 23. The beam is transported under vacuum and exits through a Kapton window (from the far right). There is a collimator just downstream of this window to allow precise positioning of apertures, beam degraders, and any sensing devices, including CRaTER. Alignment is provided with a laser system, and the distance between the window and the collimator is typically 90 cm. The equipment area is shielded, and experience has shown the shielding to be adequate to keep neutron doses low enough to avoid harming electrical equipment in the vicinity of the test device.

A primary population of 230 MeV protons is generated by an upstream cyclotron. After emerging from this cyclotron, the protons are degraded in energy by a rotating carbon wedge. Magnetic selection allows the user to define momentum and emittance. In this way the user can obtain monoenergetic beams from 70 through 230 MeV.

Dosimetry is provided by the insertion of a thin-foil ion chamber upstream of the instrument under test. This chamber is calibrated to fluence or dose by a Faraday cup, a thimble ion chamber, or a radiation sensitive diode placed at the target location. This calibration is normally done just prior to measurements. The ion chamber is set up to send information about flux and total dosage in real time to monitoring devices located in the control room.

The most commonly employed energies are 160 and 230 MeV. The 160 MeV configuration allows currents to 15 nA, while 230 MeV allows currents of 80 nA. In all cases these currents can be stepped down with energy degraders. Small diameter beams—less than 1 cm—or targets where uniformity is not crucial use the unscattered beam spot directly from the cyclotron. Medium diameter beams—from 1 to 5 cm—use a single scatterer to get a Gaussian spatial distribution of the right size and then employ a collimator to produce a central flat region. Medium diameter beams—from 1 to 5 cm—use a double scattering arrangement. The second scatterer is contoured to use incident protons efficiently, and allows beams to 20 cm. All beams make a circular spot, though this can be changed with custom collimators.

During beam testing at MGH, the CRaTER unit is mounted on a tripod so the telescope is within the beam line. The 1 Hz clock, DC power supply, and 1553 ground support equipment (GSE) interface are placed behind a wall to shield them from the radiation present within the room when the beam is operating. A router in the office connects the GSE to the wireless local area network, permitting computers in the office to command the instrument, record returned telemetry to the permanent archive, and display scientific data in real time.

The calibration procedure is as follows. First, a simulated PHA 2D histogram is generated by using the best guess calibration free parameters to convert simulated deposited energies into ADC channels. GEANT4 is used along with a high-fidelity model of the CRaTER detectors for this purpose. A full simulated response is created by generating synthetic histograms over a range of incident energies and then integrating over the histograms with weighting determined by the initial best guess for the beam center energy and width. The simulation is compared with the observations and the free parameters (gain and offset) are adjusted until the model converges with the observations. We begin by selecting all of the beam data that produced a signal in both of the detectors for a given detector pair. A 2D histogram is then generated, with histogram intervals varied appropriately to achieve good resolution of the proton trajectory traced out in the histogram. An initial guess is then made for the beam parameters along with the free parameters of the instrument response model. The important free parameters fixed by this calibration technique are the detector gains and the energy offsets.

An example of an iteration step is shown in Fig. 24. The top panel displays a histogram of energy deposits (expressed in ADU) in D3 (horizontal axis) and D4 (vertical axis). The color coding indicates the number of energy deposits at each point in the two-dimensional histogram (purple is the zero level background, ranging to red at the highest level). The bottom panel displays a histogram of the comparable two-dimensional histogram as predicted by the GEANT4 model incorporating all the key geometries and materials of the detectors under calibration. At each step in the iterative process, detector gains and energy offsets needed to convert from modeled absolute energy to ADU are recorded. The best fit solution yields final calibration values.

As-delivered calibration values determined from the sub-system and full calibration process on the flight model are shown in Table 6. These values provide the means to convert digital values of pulse-height in each detector to the energy deposited in each detector, a key step needed to estimate LET. Ever more definitive calibration coefficients continue to be refined through ongoing analysis of extant data; definitive values will be updated as part of the formal delivery of CRaTER data to NASA's Planetary Data System archive.

In addition to this final calibration process, the CRaTER EM instrument response was characterized further at the NASA Space Radiation Laboratory (NSRL) at Brookhaven National Lab (BNL). While these beam runs did not contribute directly to establishing final calibration coefficients for the FM, they did provide another critically important data source for establishing instrument response over the full range of LET we can expect in space. The BNL NSRL facility is ideal for this purpose as it is designed for simulating the heavy ion component of GCR which produces inherently high LET response, in addition to protons. Instrument characterization runs at BNL included: protons, silicon, iron, and chlorine ions at various energies between <300 and 1000 MeV/nucleon. The results from these CRaTER

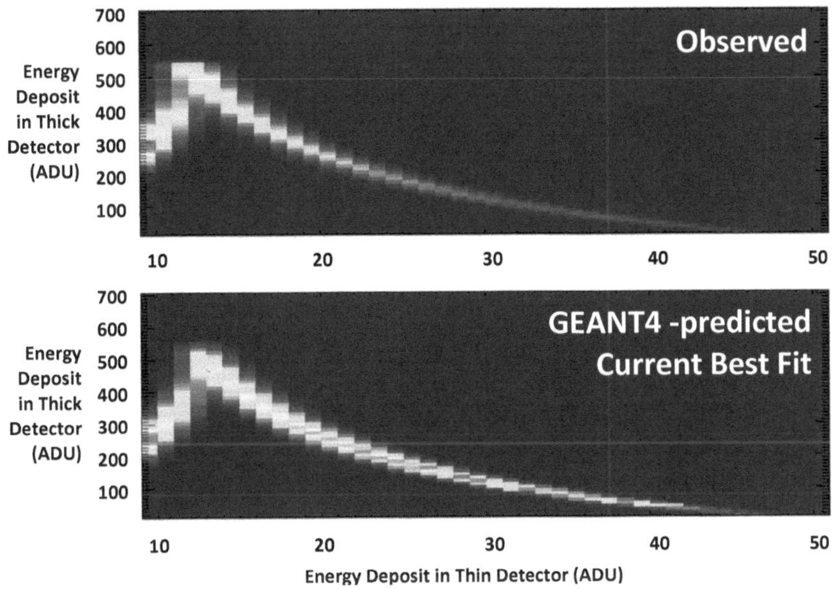

Fig. 24 (Color online) *The top panel* displays a histogram of energy deposits (expressed in ADU) in D3 (*horizontal axis*) and D4 (*vertical axis*) obtained when the CRaTER FM was being irradiated with protons. Color coding indicates relative number of energy deposits at each point in the two-dimensional histogram (purple is the zero level, ranging to red at the highest level). *The bottom panel* displays a comparable histogram as predicted by the GEANT4 model, demonstrating one iterative step in the end-to-end calibration process

Table 6 As-delivered calibration values for the CRaTER flight model

Parameter	Units	D1	D2	D3	D4	D5	D6
Thickness	Microns	148	1000	149	1000	149	1000
Gain, G_i	keV/ADU	76.3	21.8	78.6	21.6	76.3	21.9
Offset, O_i	keV	105.1	50.0	152.8	74.7	119.1	46.6

runs and comparison with models have already been published (Charara 2008) and are the focus of future publications.

5 Data Analysis, Interpretation, and Modeling

The CRaTER Science Operations Center (SOC) is located in Boston University's Astronomy Department, part of the University's Charles River Campus. The SOC's main function is to receive process, display, archive and distribute data from the CRaTER instrument and the ancillary spacecraft engineering and positional information necessary for a more detailed analysis of the data. The basic architecture of the CRaTER SOC is shown in Fig. 25. The SOC's core consists of two independent sets of secure Linux-based workstations and a RAID-based data archive. Within each set are two identical machines—a prime and backup—to provide a mechanism for rapid recovery in the case of hardware failure. Both sets are part of the Boston University network domain (bu.edu).

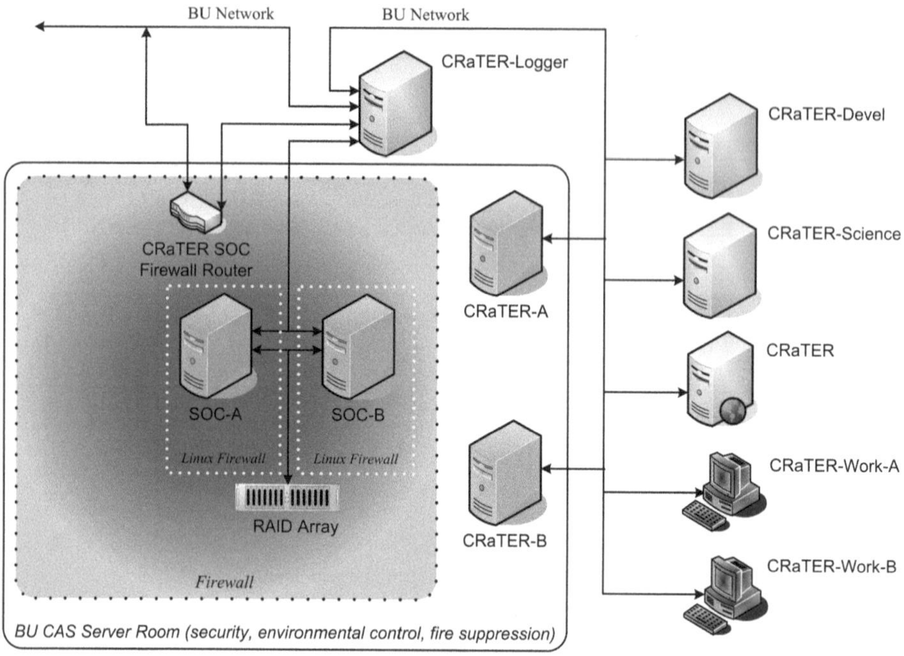

Fig. 25 The basic architecture of the CRaTER science operations center

The first set, designated SOC-A/SOC-B, receives the real-time telemetry stream, instrument data files, and ancillary flight data files and other products produced by the LRO Mission Operations Center (MOC). The SOC-A/SOC-B computers host the CRaTER data pipeline process—a set of scripts and software that convert the data in the instrument data and housekeeping files from binary to ASCII format, apply calibration and engineering conversion factors, perform data quality checks, and merge the data with corresponding spacecraft ephemeris and pointing data. Copies of all the received files and pipeline process products are stored in a local archive on the RAID array. Every three months, beginning six months following the end of the initial mission commissioning period, the most recent three months of flight data files, MOC products, and pipeline process products, are assembled into an archive and submitted to NASA's Planetary Data System's (PDS's) Planetary Plasma Interactions (PPI) data archive node.

Starting from the Level 0 data, the CRaTER SOC will develop and deliver higher level data products to the UCLA node of the Planetary Data System. The Level 0 and higher order data products are all directly related to the primary CRaTER measurement: the LET in six silicon detectors embedded within the TEP telescope. The pipeline flow from raw data to Level 2 data delivered to the PDS, including the description of each data product, is shown in Fig. 26. In addition, NASA/JSC's Space Radiation Analysis Group and NOAA's Space Weather Prediction Center contributed significantly to defining tailored data products to support astronaut operations and space weather customers. For example, higher-level, user-motivated data products will include estimates of "surface", "tissue", and "deep tissue" dose and dose rates in near real-time over the course of the LRO mission.

The Monte Carlo transport code HETC-HEDS (Townsend et al. 2005) is used to model the anticipated radiation environment at the Moon during the LRO mission (Charara 2008). Initially, the code was used to perform benchmarking studies of the simulations by compar-

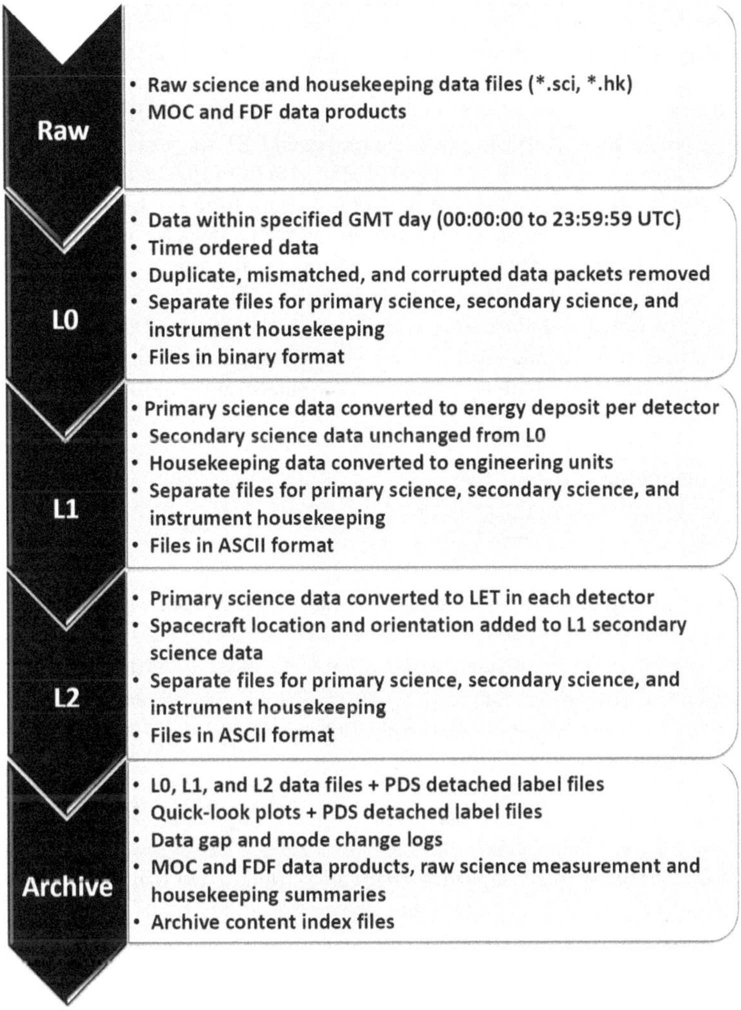

Fig. 26 The pipeline processing flow diagram outlining up to the Level 2 CRaTER data delivered to the planetary data system archive

ing them to measurements from CRaTER calibrations performed with proton beams from the cyclotron at Massachusetts General Hospital, and various heavy ion beams at the NASA Space Radiation Laboratory at Brookhaven National Laboratory. Next, an LET database that includes simulations of all the relevant components of the GCR spectrum with their corresponding energies was generated for the CRaTER flight model configuration. Particles with charges between $1 \leq Z \leq 26$ (protons through iron) and kinetic energies between 20 MeV per nucleon and 3 GeV/nucleon were used for these simulations (Particles heavier than iron have fluences that are too low to be of practical concern for radiation protection purposes). Then estimates of total energy lost per path length of material by any particle produced, whether it is a primary beam particle or a secondary particle generated from a fragmentation event from the projectile or the target are used to obtain an average LET for each particle type, energy and detector component. The complete LET data base for these GCR elements (protons through iron) was developed for use by the CRaTER team for future analyses of, and comparisons with, real time data obtained with CRaTER during the actual mission.

Specific HETC-HEDS model predictions of the expected LET spectrum from the anticipated GCR environment during the mission time frame (2009–2010) have been made and are available for prompt comparison with the measured LET spectrum for the mission. The response of the detector to various historical SEP spectra from 1989 and 2003 have also been simulated with HETC-HEDS using the measured spectra from those events. For instance, a large event (November 1997) that occurred during the comparable phase of the previous solar cycle, was modeled as an example of a solar energetic particle event that might occur during the prime or extended mission phases. Finally, theoretical calculations of the level of nuclear interactions that will occur within the TEP and other elements of the telescope, along with observations of fragmentation collected both on the ground at heavy ion accelerators and ultimately at the Moon, will be the subject of separate publications.

6 Discussion and Summary

6.1 Other Representative Science Goals

In addition to its exploration-enabling measurements of the lunar radiation environment, CRaTER also provides high-quality observations which in turn enables many other scientific investigations. While not the primary goal of the LRO mission during its primary phase, these areas serve as secondary science goals during the first year of operations and targeted science topics for any extended mission scenarios. Several example science studies using CRaTER measurements follow next.

6.1.1 Lunar-Plasma Environment Interactions

The Moon is electrostatically charged by the surrounding space plasma environment, including the spatially-varying regions of hot plasma within the Earth's distant magnetotail (Spence and Kivelson 1990) and time-varying solar wind plasma in and near the lunar wake (Clack et al. 2004), as well as by solar ultraviolet (UV) radiation. Observations from Lunar Prospector indicate that during SEP events the shadowed lunar surface can charge up to a few kilovolts negative when the Moon is in the solar wind or the plasma sheet region of the magnetotail (Halekas et al. 2007). Even during less extreme conditions, it is likely that strong local electric fields can form on the lunar surface, particularly near the terminator and in the polar regions, which could drive the transport of charged lunar dust (e.g., see Colwell et al. 2007). There is evidence to suggest that a submicron dust population exists in the lunar exosphere at altitudes in excess of 100 km (Zook and McCoy 1991), and it has been proposed that this population could have been electrostatically lofted by surface electric fields (Stubbs et al. 2006). However, lunar surface charging and dust transport processes remain poorly characterized and understood. In addition to the intrinsic scientific interest of such processes, these environmental factors and their dynamics will likely pose a hazard to human and robotic explorers, thereby further motivating and justifying their scientific study (Stubbs et al. 2007).

By using CRaTER and other LRO instruments and spacecraft capabilities (including off-nadir pointing) we will perform the following studies of the exospheric dust environment around the Moon. We will analyze and interpret any CRaTER signals suspected to have been generated by microphonics produced by direct meteoritic or exospheric dust impacts. When pointing off-nadir, we will analyze and interpret secondaries that were likely created by GCR and SEP spallation from exosphere dust. If exospheric dust is detected, then CRaTER data could help place important limits on its spatial and temporal mass distribution.

6.1.2 Sources of Galactic Cosmic Ray Variability

Galactic cosmic rays represent a low-level but continuous source of very energetic charged particles which bathe the solar system. Earth-based measurements reveal variability of GCR that range from the well-known solar cycle timescales (11 and 22-year), to partly understood intermediate timescales associated with solar rotation, corotating interaction regions and Forbush decreases (days), to nearly completely unexplored short timescales (hours and less).

Recent high-time-resolution measurements of GCR taken beyond Earth's shielding magnetic field and protective atmosphere have demonstrated that new physical understanding of the intermediate and short timescale variations are possible (cf., Quenby et al. 2008; Mulligan et al. 2009). These recent studies have been accomplished with instruments designed principally for other measurements, rather than instruments dedicated to cosmic rays. CRaTER therefore represents an important and highly capable scientific instrument that will provide important new insights on GCR variability. Specific topics that we expect to address with CRaTER observations include: comparison of present solar minimum GCR with past cycles and its variability toward solar maximum; exploration of substructure observed in Forbush decreases and the possible role that planar magnetic structures might play in modulating GCR flux (Jordan et al. 2009); and characterizing GCR "scintillation" (Owens and Jokipii 1973) and assessing the source of such variability. Finally, we will quantify the degree to which any measurable shielding of >10 MeV ions is provided by the terrestrial magnetosphere as a function of lunar orbit phase and compare with recent contrasting predictions suggested by Winglee and Harnett (2007) and challenged by Huang et al. (2009).

6.1.3 Solar Proton Studies

Intense solar proton events are episodic and approximately follow the solar cycle. In addition to being a second source of ionizing radiation risks, the structure and dynamics of SEPs have intrinsic scientific interest. SEPs represent a population of particles with typically power-law spectra up to 100's of MeV/nucleon that occasionally are accelerated to near the speed of light in explosive solar phenomena, including shock waves in the solar corona associated with flares, and shock waves in the interplanetary medium associated with coronal mass ejections as they propagate into the heliosphere (e.g., Cane and Lario 2006; Mewaldt 2006). A major challenge with SEPs is predicting their arrival time, which is often tied intimately to predicting the arrival time of interplanetary shocks (Case et al. 2008). Solar protons thereby are tracers of particle acceleration as well as their transport mechanisms (field-aligned transport versus cross-field diffusion) that allow them to arrive at any given location in the heliosphere from their source location.

CRaTER will provide high time-resolution measurements of SEPs which, combined with other simultaneous heliospheric measurements at multiple locations, can be used to probe these processes and mechanisms in new ways. SEP topics that we will study include: temporal (spatial) structure of SEP onset, dispersionless high-time resolution SEP drop-outs, and time-of-flight energy dependence of SEPs for particle acceleration and transport identification (cf., Mazur et al. 2000).

6.1.4 Exploring Surface Interactions with Secondaries

Though never a primary science driver, we nevertheless hope to exploit the as-designed capabilities of CRaTER in order to perform off-nadir measurements to explore possible lunar surface interactions with GCR. GCR with sufficiently high energies and incident with the lunar surface particularly at near grazing incidence can produce secondary by-products through spallation which may be detected at the LRO orbit. Radiation transport codes show that the relative population of secondaries depend upon the properties of the surface material being spallated. This process could be used to remotely sense some aspects of the surface using the lunar albedo, with the charged secondaries or neutrons (cf., Adams et al. 2007). This is a purely exploratory measurement we will do during the commissioning phase and periodically during the main mission. Though this is a speculative measurement, we would look for any potentially small signal against the GCR and SEP foregrounds by looking not only at rate effects, but also tell-tale features in the LET spectra.

6.1.5 Comparative Studies of Lunar and Martian Radiation Environment

Comparison of simultaneous measurements of the radiation environment at the Moon and at Mars can be used to study the processes which modify GCR and SEP fluxes throughout the inner heliosphere. SEPS in particular are of interest for such studies; their arrival and time history depends critically on magnetic connection to particle acceleration regions and the detailed propagation of shocks through the heliosphere. Data from CRaTER will be combined with data from the ongoing Mars Odyssey and upcoming Mars Science Lander missions, as available. On Odyssey (Saunders et al. 2004), even though MARIE is no longer functional, other instruments are working and provide integral charged particle fluxes, including the GRS instrument. Though designed primarily to measure gamma rays, the GRS instrument has provided a near-continuous record of GCR and SEP flux in the Martian orbit from 2002 to present. The Odyssey HEND instrument measures high-energy neutrons, a proxy for solar protons during SEPs. In addition to currently available Odyssey observations, we anticipate similar comparison opportunities with radiation measurements from the RAD instrument planned for the Mars Science Lander. During the prime mission, Mars and Earth will not be particularly well-aligned to study solar proton transport and connectivity. However, comparison of LRO measurements to Mars data may be more interesting during an extended mission period, especially if SEP's are observed when the Moon and Mars are approximately connected magnetically by the large-scale interplanetary magnetic field.

6.2 Summary

The Cosmic Ray Telescope for the Effects of Radiation (CRaTER) on the Lunar Reconnaissance Orbiter (LRO) characterizes the lunar radiation environment by quantifying the evolving LET spectrum through different amounts of TEP owing to GCR and SEP particle populations. By characterizing energy loss from the interactions of biologically-relevant ions with energies above ~10 MeV, CRaTER LET measurements serve as in-flight constraints needed to validate state-of-the-art radiation transport codes. In addition to these human exploration goals, CRaTER measurements also provide new windows on the science of SEP and GCR phenomena and their interactions with the lunar surface.

Acknowledgements We greatly acknowledge the many talents, supreme efforts, and tireless contributions brought to this science investigation by the dedicated and professional staffs at CRaTER institutions. We therefore express our sincere gratitude to Alex Boyd, David Bradford, Jennifer Brown, Christopher Claysmith, Nick Gross, Chia-Lin Huang, Andrew Jordan, Cynthia Kowal, Brian Larsen, Eric O'Dea, Maureen Rodgers, Kristin Sacca, and Huade Tan of Boston University; John Richardson and Demitrios Athens of MIT; and Eddie Semones of the NASA JSC Space Radiation Analysis Group. We are especially grateful for the incredibly valuable support and help in instrument characterization and calibration provided by Ethan Casio of the MGH Proton Facility and by Adam Rusek and Michael Sivertz of the BNL National Space Radiation Laboratory. Next, we are ever thankful for the incredible efforts in making the CRaTER investigation a reality by the LRO Project Team at NASA GSFC, including JoAnne Baker, Gordon Chin, John Keller, Ron Kolecki, Craig Tooley, and Richard Vondrak. A most special thank you to our NASA GSFC Instrument Manager Arlin Bartels; we are grateful to have had such a professional and dedicated instrument manager who made the entire process minimally troublesome and nearly exclusively joyous from instrument concept to launch. Finally, we gratefully acknowledge the unique contributions of Ann Vasaly, of the Classics Department at Boston University, who helped us develop the Latin motto for CRaTER, which is: "Luna Ut Nos Animalia Tueri Experiri Possimus" ("In order that we might be able to protect and make trial of living things on the Moon").

Support for the CRaTER project was funded by NASA under Contract number NNG05EB92C.

References

J.H. Adams, M. Bhattacharya, Z.W. Lin, G. Pendleton, J.W. Watts, The ionizing radiation environment on the moon. Adv. Space Res. **40**, 338–341 (2007)

G.D. Badhwar, M.J. Golightly, A. Konradi et al., In-flight radiation measurements on STS-60. Radiat. Meas. **26**(1), 17–34 (1996)

J.B. Blake, J.F. Fennell, L.M. Friesen, B.M. Johnson, W.A. Kolasinski, D.J. Mabry, J.V. Osborn, S.H. Penzin, E.R. Schnauss, H.E. Spence, D.N. Baker, R. Belian, T.A. Fritz, W. Ford, B. Laubscher, R. Stiglich, R.A. Baraze, M.F. Hilsenrath, W.L. Imhof, J.R. Kilner, J. Mobilia, H.D. Voss, A. Korth, M. Güll, K. Fischer, M. Grande, D. Hall, CEPPAD: comprehensive energetic particle and pitch angle distribution experiment on POLAR. Space Sci. Rev. **71**, 531 (1995)

H.V. Cane, D. Lario, An introduction to CMEs and energetic particles. Space Sci. Rev. **123**, 45–56 (2006)

A.W. Case, H.E. Spence, M.J. Owens, P. Riley, D. Odstrcil, Ambient solar wind's effect on ICME transit times. Geophys. Res. Lett. **35**, L15105 (2008). doi:10.1029/2008GL034493

E.W. Cascio, J.M. Sisterson, J.B. Flanz, M.S. Wagner, The proton irradiation program at the Northeast Proton Therapy Center, in *NSREC Proceedings* (2003)

E.W. Cascio, J.M. Sisterson, B. Gottschalk, S. Sarkar, Measurements of the energy spectrum of degraded proton beams at NPTC, in *NSREC Proceedings* (2004)

Y.M. Charara, Characterization of the cosmic ray telescope for the effects of radiation (CRaTER) detector. PhD Dissertation, The University of Tennessee, Knoxville, Tennessee, December 2008

D. Clack, J.C. Kasper, A.J. Lazarus, J.T. Steinberg, W.M. Farrell, Wind observations of extreme ion temperature anisotropies in the lunar wake. Geophys. Res. Lett. **31**, 6 (2004)

J.E. Colwell, S. Batiste, M. Horányi, S. Robertson, S. Sture, Lunar surface: dust dynamics and regolith mechanics. Rev. Geophys. **45**, RG2006 (2007). doi:10.1029/2005RG000184

M.J. Golightly, K. Hardy, W. Quam, Radiation-dosimetry measurements during US space-shuttle missions with the RME-III. Radiat. Meas. **23**(1), 25–42 (1994)

J.S. Halekas, G.T. Delory, D.A. Brain, R.P. Lin, M.O. Fillingim, C.O. Lee, R.A. Mewaldt, T.J. Stubbs, W.M. Farrell, M.K. Hudson, Extreme lunar surface charging during solar energetic particle events. Geophys. Res. Lett. **34**, L02111 (2007). doi:10.1029/2006GL028517

C.-L. Huang, H.E. Spence, B.T. Kress, Assessing access of galactic cosmic rays at Moon's orbit. Geophys. Res. Lett. **36**, L09109 (2009). doi:10.1029/2009GL037916

A.J. Jordan, H.E. Spence, J.B. Blake, T. Mulligan, D. Shaul, M. Galametz, Multipoint, high time resolution galactic cosmic ray observations associated with two interplanetary coronal mass ejections. J. Geophys. Res. **114**, A07107 (2009). doi:10.1029/2008JA013891

Y. Kim, J.W. Wilson, S.A. Thibeault, J.E. Nealy, F.F. Badavi, R.L. Kiefer, Performance study of galactic cosmic ray shield materials, NASA Technical Paper 3473, November 1994

J.G. Luhmann et al., STEREO IMPACT investigation goals, measurements, and data products overview. Space Sci. Rev. **136**, 117–184 (2007)

J.E. Mazur, G.M. Mason, J.R. Dwyer, J. Giacalone, J.R. Jokipii, E.C. Stone, Interplanetary magnetic field line mixing deduced from impulsive solar flare particles. Astrophys. J. **532**, L79 (2000)

R.A. Mewaldt, Solar energetic particle composition, energy spectra, and space weather. Space Sci. Rev. **124**, 303–316 (2006)

I.G. Mitrofanov et al., Lunar exploration neutron detector for NASA'S lunar reconnaissance orbiter project. Space Sci. Rev. (2009). doi:10.1007/s11214-009-9608-4

T. Mulligan, J.B. Blake, D. Shaul, J.J. Quenby, R.A. Leske, R.A. Mewaldt, M. Galametz, Short-period variability in the galactic cosmic ray intensity: high statistical resolution observations and interpretation around the time of a Forbush decrease in August 2006. J. Geophys. Res. **114**, A07105 (2009). doi:10.1029/2008JA013783

A.J. Owens, J.R. Jokipii, Interplanetary scintillations of cosmic rays, 502. Astrophys. J. **181**, L147–L150 (1973)

J.J. Quenby, T. Mulligan, J.B. Blake, J.E. Mazur, D. Shaul, Local and nonlocal geometry of interplanetary coronal mass ejections: Galactic cosmic ray (GCR) short-period variations and magnetic field modeling. J. Geophys. Res. **113**, A10102 (2008). doi:10.1029/2007JA012849

R. Saunders, R. Arvidson, G. Badhwar et al., Mars Odyssey mission summary. Space Sci. Rev. **110**, 1–36 (2004)

H.E. Spence, M.G. Kivelson, The variation of the plasma sheet polytropic index along the midnight meridian in a finite width tail. Geophys. Res. Lett. **17**(5), 591–594 (1990)

E.C. Stone, C.M.S. Cohen, W.R. Cook, A.C. Cummings, B. Gauld et al., The cosmic-ray isotope spectrometer for the advanced composition explorer. Space Sci. Rev. **86**, 285–356 (1998)

T.J. Stubbs, R.R. Vondrak, W.M. Farrell, A dynamic fountain model for lunar dust. Adv. Space Res. **37**(1), 59–66 (2006)

T.J. Stubbs, R.R. Vondrak, W.M. Farrell, Impact of dust on lunar exploration, in *Proceedings of Dust in Planetary Systems 2005*, ed. by H. Kruger, A. Graps (Eur. Space Agency Spec. Publ., 2007)

L.W. Townsend, T.M. Miller, T.A. Gabriel, HETC radiation transport code development for cosmic ray shielding applications in space. Radiat. Protect. Dosimetry **116**(1–4), 135–139 (2005)

W.R. Webber, J.A. Lezniak, The comparative spectra of cosmic-ray protons and helium nuclei. Astrophys. Space Sci. **30**, 361–380 (1974)

J.W. Wilson, F.F. Badavi, F.A. Cucinotta, J.L. Shinn, G.D. Badhwar, R. Silberberg, C.H. Tsao, L.W. Townsend, R.K. Tripathi, HZETRN: Description of a free-space ion and nucleon transport and shielding computer program. NASA Technical Paper 3495, US Government Printing Office, Washington, DC, 1995

R.M. Winglee, E.M. Harnett, Radiation mitigation at the moon by the terrestrial magnetosphere, Geophys. Res. Lett. **34** (2007). doi:10.1029/2007GL030507

W.-M. Yao et al., Review of particle physics. J. Phys. G: Nucl. Part. Phys. **33**, 1–1232 (2006). doi:10.1088/0954-3899/33/1/001

J.F. Ziegler, J.P. Biersack, U. Littmark, *The Stopping and Range of Ions in Solids*, vol. 1 (Pergamon Press, New York, 1984)

H.A. Zook, J.E. McCoy, Large scale lunar horizon glow and a high altitude lunar dust exosphere. Geophys. Res. Lett. **18**(11), 2117–2120 (1991)

Space Sci Rev (2010) 150: 285–302
DOI 10.1007/s11214-009-9607-5

The Lunar Reconnaissance Orbiter Miniature Radio Frequency (Mini-RF) Technology Demonstration

Stewart Nozette · Paul Spudis · Ben Bussey · Robert Jensen · Keith Raney · Helene Winters · Christopher L. Lichtenberg · William Marinelli · Jason Crusan · Michele Gates · Mark Robinson

Received: 17 November 2008 / Accepted: 6 November 2009 / Published online: 13 January 2010

Abstract The Miniature Radio Frequency (Mini-RF) system is manifested on the Lunar Reconnaissance Orbiter (LRO) as a technology demonstration and an extended mission science instrument. Mini-RF represents a significant step forward in spaceborne RF technology and architecture. It combines synthetic aperture radar (SAR) at two wavelengths (S-band and X-band) and two resolutions (150 m and 30 m) with interferometric and communications functionality in one lightweight (16 kg) package. Previous radar observations (Earth-based, and one bistatic data set from Clementine) of the permanently shadowed regions of the lunar poles seem to indicate areas of high circular polarization ratio (CPR) consistent with volume scattering from volatile deposits (e.g. water ice) buried at shallow (0.1–1 m) depth, but only at unfavorable viewing geometries, and with inconclusive results. The LRO Mini-RF utilizes new wideband hybrid polarization architecture to measure the Stokes parameters of the reflected signal. These data will help to differentiate "true" volumetric ice reflections from "false" returns due to angular surface regolith. Additional lunar science investigations (e.g. pyroclastic deposit characterization) will also be attempted during the LRO extended mission. LRO's lunar operations will be contemporaneous with India's Chandrayaan-1, which carries the Forerunner Mini-SAR (S-band wavelength and 150-m resolution), and bistatic radar (S-Band) measurements may be possible. On orbit calibration, procedures for LRO

S. Nozette · P. Spudis
Lunar and Planetary Institute, 3600 Bay Area Blvd, Houston, TX 77058, USA

B. Bussey (✉) · R. Jensen · K. Raney · H. Winters
Applied Physics Laboratory, Laurel, MD 20723, USA
e-mail: Ben.Bussey@jhuapl.edu

C.L. Lichtenberg
Naval Air Warfare Center, China Lake, CA 93555, USA

W. Marinelli · J. Crusan · M. Gates
National Aeronautics and Space Administration, Washington, DC 20546, USA

M. Robinson
Arizona State University, Tempe, AZ, USA

Mini-RF have been validated using Chandrayaan 1 and ground-based facilities (Arecibo and Greenbank Radio Observatories).

Keywords Lunar Reconnaissance Orbiter · Mini RF · Lunar poles

1 Introduction

The Lunar Reconnaissance Orbiter (LRO) Mini-RF technology demonstration is the product of over a decade of development. Its objectives are: (1) Flight verification of an advanced lightweight RF technology for future NASA and DoD communications applications; (2) Demonstration of a hybrid-polarity Synthetic Aperture Radar (SAR) architecture; (3) Obtaining measurements of the lunar surface as a function of radar band (S and X) and resolution (150 m, 30 m) which could identify water ice deposits in the permanently shadowed polar regions; (4) Production of topographic data using interferometry (S-band) and SAR stereo techniques; and (5) Mapping of areas of interest identified by the Chandrayaan-1 Forerunner Mini-SAR experiment and other lunar instruments. Because Mini-RF provides its own illumination and can penetrate the near subsurface at meter scales, it will acquire data not obtained by any other LRO payload.

Over the previous decade, the Department of Defense (DoD) and commercial industry made significant strides in developing advanced lightweight RF technology for wireless communication, Unmanned Airborne Vehicles (UAVs), and tactical missiles. The Mini-RF hardware is based on DoD communications technology and methodology. Precursor Mini-RF technology was flight-tested by the Naval Research Laboratory (NRL) in the low Earth orbit on the USAF MightySat-2 and XSS-10 missions as a Space Ground Link System (SGLS). Other technologies developed for commercial wireless systems, UAVs, manned aircraft such as the F-18, and tactical missiles were also incorporated into the payload.

In 2004, the DoD and NASA initiated the Mini-RF program to develop and flight-test advanced lightweight radar and communication systems and NASA elected to apply the technology to lunar exploration by building two payloads. The first, "Forerunner" Miniature SAR (Mini-SAR), was developed and integrated into the Indian Space Research Organization (ISRO) Chandrayaan-1 as a NASA guest payload and the second, on the LRO spacecraft as a technology demonstration. Chandrayaan-1 was launched on October 22, 2008 and is conducting a two-year systematic lunar mapping investigation. The Forerunner Mini-SAR is currently mapping the lunar poles at S band with a resolution of 150 m and is providing heritage and experience to the LRO Mini-RF system. The Forerunner Mini-SAR had to operate in the lunar thermal and radiation environment, yet was simpler in design and operation, providing significant experience and reduction of risk for the more advanced LRO Mini-RF system (Spudis et al. 2009). The LRO Mini-RF affords NASA and the DoD an opportunity to flight-qualify lightweight technology for a range of applications, including deep space communications. The flexibility, reconfigurability, and capability of Mini-RF will be demonstrated by a communications and radar mode utilizing the same hardware. The constraints of a lunar mission (range, limited duty cycle over the poles) and the low mass of advanced lightweight RF technology allows a technology demonstration which met the payload constraints of both the Chandrayaan and LRO spacecraft, and provided an opportunity to collect unique and valuable lunar science data. The new technologies being qualified on LRO Mini-RF include: Microwave Power Module (MPM) based multi-frequency transmitter, wideband dual-frequency panel antenna, all digital receiver and waveform synthesizer incorporating

Table 1 Mini-RF technology comparison

Radar System Comparisons with Mini RF Technology

	Radar Mass	DC Power Input	RF Power (average)	Band	Spacecraft Dry Mass	Orbit Altitude
DoD TACSAT (notional)	40 kg	1000 W	200 W	X	350 kg	500 km
Magellan	154 kg	1000 W	325 W	S	1035 kg	249 × 8543 km
SEASAT			1000 W	L	2300 kg	800 km
SIR-C	11,000 kg	3000–9000 W		L, C, X		Shuttle
Chandrayaan Mini-SAR	8.1 kg	<100 W	15 W	S	525 kg (includes 12 Instruments)	100 km
LRO Mini-RF	~13 kg	~150 W	25 W	S, X	~1000 kg (includes 6 Instruments, 1 Tech Demo)	50 km

Communications System Comparison with Mini RF Technology

System	Mass	DC Power Input	RF Power	Band	Data Rate	Spacecraft Dry Mass	Orbit Altitude
Standard DoD SGLS	5 kg (electronics)	30 W	5 W	L, S	1–2 Mbs	Various	LEO
NRL Mini RF SGLS	1.5 kg (electronics)	30 W	5 W	L, S	1–2 Mbs	~100 kg	LEO
Clementine	4.0 kg (xponder) (14 kg system total w 2nd xponder)	30 W	5 W	S	128 kbs (Lunar)	224 kg	Lunar 5000 × 400 km
Commercial GEO/TTC	17 kg total (dual redundant)	40 W	8 W (single)	Ku	8 kbs (GEO)	2,600 kg	GEO
NASA Discovery	21 kg	170 W	50 W (average)	S, Ka	100 Mbs (Lunar)	1000 kg	Super GEO
Mini-RF Optimized for Communications	~14 kg (includes Antenna)	~175 W	50 W	S, Ka	100 Mbs	450 kg	Lunar

field programmable gate array (FPGA) and analog-to-digital conversion at 1 GHz sampling. The Mini-RF parts qualification program, which included commercial technology, allowed innovative components to gain space qualification. A comparison of the Mini-RF radar and communications performance with existing and previously flown technology, illustrating mass and performance improvements, is shown in Table 1.

1.1 Background of Investigation

The LRO Mini-RF payload will address key science questions during the LRO primary and extended missions. These include exploring the permanently shadowed polar regions

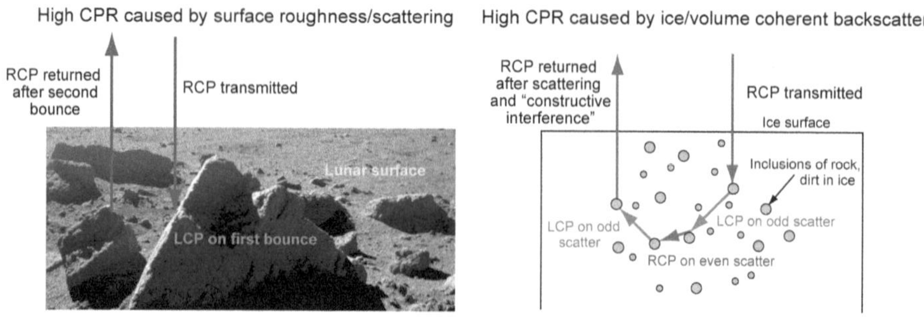

Fig. 1 Dual scattering caused by ice or roughness (from Spudis 2006)

and probing the lunar regolith in other areas of scientific interest (e.g. pyroclastic deposits). The nature and distribution of the permanently shadowed polar terrain of the Moon (e.g., Arnold 1979) has been the subject of considerable controversy. Initially, Arecibo monostatic radar measured the circular polarization ratio (CPR) in the region of the lunar South Pole; these data were interpreted to indicate the possible presence of ice on the lower wall of Shackleton crater (Stacy 1993). The presence of ice in the regolith causes a distinct, but not unique, CPR signature caused by the low-loss properties of ice and the Coherent Backscatter Opposition Effect (CBOE) (Fig. 1). Frozen volatiles have much lower transmission loss than silicate rocks, producing a higher average radar reflectivity. Total internal reflection also preserves the transmitted circular polarization sense in the scattered signal, which is also enhanced by constructive interference. The CBOE requires scattering centers (cracks or inhomogeneities) imbedded in a low-loss matrix such as ice (Mishchenko 1992). High CPR has been observed by radar in shadowed polar craters of Mercury. These results are generally attributed to total internal reflection and/or CBOE produced by low-loss frozen volatiles (e.g. ice) although other mechanisms have been proposed (Butler 1997). Data collected by the Clementine bistatic radar experiment (Nozette et al. 1996, 1997, 2001) also measured anomalous CPR in Shackleton as a function of bistatic angle, suggesting the presence of patchy, "dirty," ice deposits on the lower Earth facing wall. The high CPR region observed deepest within Shackleton crater also has a local angle of incidence of 50°, estimated by Stacy (1993). This wall has a monostatic CPR of 1.19 ± 0.12 (Stacy 1993) comparable to the radar bright feature at the North Pole of Mercury measured by Harmon and Slade (1992) that has generally been attributed to an ice deposit. Stacy (1993) noted that because this area may be permanently shadowed, the coherent backscatter opposition effect (CBOE) from ice deposits could be responsible for the enhanced CPR signature.

Subsequently, higher resolution Arecibo data were reported to be inconsistent with the previous measurements (Campbell et al. 2006), and it was suggested that all anomalous high CPR areas observed by Arecibo, near the lunar South Pole, were caused by rough surfaces as only some portions of these areas were believed to be permanently shadowed. It was also postulated (Simpson and Tyler 1999) that the Clementine polarization anomaly is only due to roughness and/or random noise in the data and not by the presence of ice. Campbell et al. (2006) suggested that the high CPR area within Shackleton crater could be due to diffuse backscatter from wavelength scale structures since other non-shadowed structures exhibit similar behavior.

Craters with diameters similar to Shackleton in regions that are fully illuminated (e.g., Shumberger G; Campbell et al. 2006) also show enhanced CPR signatures. The upper parts of the radar-facing inner rims of illuminated craters typically show greater backscatter cross sections, most likely due to increased blockiness. However, the high CPR in Shackleton is deeper (further down the wall), isolated, discontinuous, and not associated with other geological units (roughness or layering), but its true depth is not observable from the Earth (Spudis 2006). Additionally, a high degree of correlation exists between the decreasing epithermal neutron flux and the shadowed terrain at the lunar South Pole, including Shackleton, which has been attributed to water ice deposits (Feldman et al. 1998, 2000; Elphic et al. 2007). New results from the Kaguya mission, which include low light imaging of the floor and lower walls of Shackleton, show these areas to be smooth and block free (Noda et al. 2008).

1.2 Mini-RF Measurement Objectives

The lunar ice controversy cannot be resolved with existing data. New measurements are required. Previous studies of mechanisms of deposition and preservation of lunar polar ice suggest that it may be buried at a depth of 0.1–2 m (Feldman et al. 2000). There is no precedent for a spaceborne imaging radar instrument that could probe the regolith at this depth for ice and meet the stringent mass and power constraints of the LRO spacecraft.

The Mini-RF architecture is new. The hybrid-polarity design (transmitting circular polarization, and receiving coherently two orthogonal linear polarizations) provides data sufficient to measure the 2×2 covariance matrix of the backscattered field, which in turn lead to the four Stokes parameters (Raney 2007). Analysis of these data, either by standard radar astronomy methods or by applying matrix decomposition techniques, extracts all information available in the radar reflections, thus providing a sharper tool than CPR alone to help differentiate between "true" (ice) and "false" (regolith blockiness) lunar returns.

2 Mini-RF Investigation Description

Mini-RF can operate as a dual-frequency, hybrid-polarity imaging radar designed to collect information about the scattering properties of the permanently dark areas near the lunar poles at optimum viewing geometry. As the LRO Mini-RF system probes the lunar regolith at two frequencies (S-band and X-band) it will provide additional information on the physical properties of the upper meter or two of lunar surface. Under the proposed observational constraints, Mini-RF can identify areas of high CPR ($\sim$1), which could be caused by ice deposits. Areas that do show high CPR can be analyzed with greater sensitivity through their backscattering features. It is hypothesized that "ice" and "regolith" will have differentiable characteristics as seen through their respective Stokes parameters at two wavelengths. When supported by Chandrayaan-1 and other LRO data (e.g. neutron spectroscopy, shadow and lighting, roughness and surface texture, thermal environment), the LRO Mini-RF measurements should provide more conclusive evidence as to the likelihood that ice deposits occur in permanently shadowed areas. Specific Mini-RF instrument requirements, performance, and data collection modes are shown in Table 2.

2.1 Technology Demonstrations

The Mini-RF observations are made possible within the mass and power constraints imposed by LRO via application of a number of technologies. Two key technologies are a wideband Microwave Power Module (MPM) based transmitter and a lightweight broadband antenna and polarization design. The LRO Mini-RF also has an S-band-only interferometric

Table 2 Mini-RF requirements and performance

	Requirements	Performance
Frequency	S and X Band	2380 Mhz (±10 Mhz) and 7140 Mhz (±10 MHz)
Mass	≤ 16 kg	13.9 kg total 8.5 kg electronics 5.4 kg antenna
Radiation Tolerance	20 KRad, SEU/SEL 75 MeV	Selective Wavers
Thermal Requirement	−20°C–30°C	
DC Current	< 4.8 A Transmitter < 4.7 A Remainder	5.2 A Transmitter 1.7 A at 27 V ±0.25 dBm ±0.3 dBm
RF Transmitter Power		42.6 dBm S Band 41.1 dBm X Band
Operating Time (Duty Cycle)	3 min on, 50 min off, 3 min on	10 min on, 20 min off, 10 min on (limit of test)
Transmit Polarization	Circular Polarization	Circular Polarization
Polarization Isolation	1.7 dB axial ratio	≤ 2.0 dB axial ratio (to be verified in flight)
Receive Polarization	H and V	≥ 40 dB Isolation
Channel to Channel Power Calibration	±1 dB "knowledge", −0.1 dB X band	±0.2 dB S Band, ±0.2 dB X Band
S Band Baseline Resolution	150 m azimuth × range	150 m×150 m
S Band Noise Equivalent Radar Cross Section	−30 dB at 50 km altitude	−33.6 dB
Number of Looks	16	16
Range Swath	6 km S band, 4 km X band	6 km S band, 4 km X band
S Band Zoom Capability	15 m × 30 m	15 m × 30 m
X Band Baseline Resolution	150 m	150 m × 150 m
X Band Noise Equivalent Radar Cross Section	−24 dB at 50 km altitude	−26.3 dB
Number of Looks	16	16
Pulse Rep Frequency (2300–5000)/Ambiguity	−18 dB	−18 dB
X Band Zoom Capability	30 m	15 × 30 m
Radiometric Stability	±0.5 dB	±0.5 dB
Communications Demonstration	S Band	2380 Mhz
Mode	Half Duplex	Half Duplex
Transmit	Modulated Signal	BPSK, Manchester Variable Data Rate and Modulation Index
Received Signal	Digitize Signal	Continuous 500 kbs

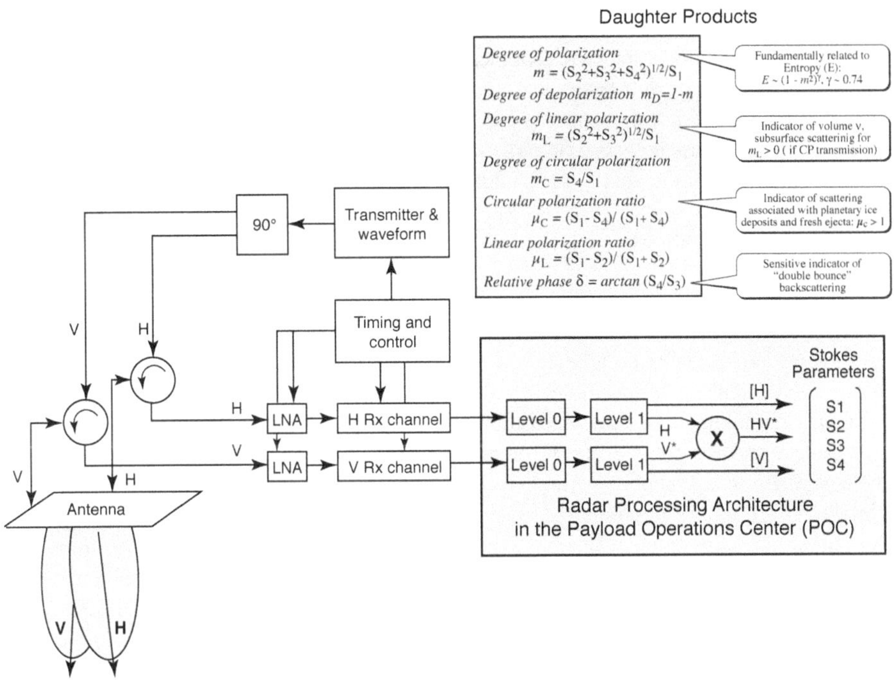

Fig. 2 Hybrid-polarity architecture

mode with 3.5-km wide strips with ±15-m mapping accuracy to measure lunar topography. This will be the first demonstration of interferometric SAR techniques in a planetary mission. The Mini-RF antenna architecture is shown in Fig. 2. This architecture is unique in planetary radar; it transmits right circular polarization radiation and receives the horizontal (H) and vertical (V) polarization components coherently, which are then reconstructed as Stokes' parameters during the data processing step. Both the communications and the radar astronomical objectives impose a requirement for circular polarization on the transmitted field. Conventional radar that would measure CPR then would have to be dual-circularly-polarized on receiver. The hybrid-polarity approach provides weight savings by eliminating circulator elements in the receiver paths, which reduces mass, increases RF efficiency, and minimizes cross-talk and other self-noise aspects of the received data. The H and V signals are passed directly to the ground-based processor. It is well known that the Stokes parameters comprise a full characterization of the backscattered field. The values of the four Stokes parameters do not depend on the choice of receiver polarization, so this architecture minimizes hardware while maintaining full science value. The result provides significant advantages over the conventional "CPR-measuring" dual-circular-polarized approach, yet the radar is simpler. The use of possible Stokes parameter-based products (e.g. CPR, degree-of-depolarization, degree-of-linear-polarization, phase "double bounce") have a number of significant advantages over traditional radar systems: less hardware is needed, resulting in fewer losses and a "cleaner," simpler flight instrument. The signal levels are comparable (within ∼2 dB) in both channels allowing relatively relaxed specifications on channel-to-channel cross-talk and more robust phase and amplitude calibration. The processor has a direct view through the entire receiver chain; including the antenna receives patterns and other radar parameters (e.g., gain and phase). These parameters are applied in processing

"Levels" (Level 0, 1) which correspond to successive data processing stages, as shown in Fig. 2. The design allows selective Doppler weighting to maximize channel–channel coherence (e.g., reduce the H & V beam mismatch). As CPR is less sensitive to channel imbalance by at least a factor of 2 with respect to explicit RCP/LCP, Stokes parameter-based backscatter decomposition strategies can help distinguish "false" from "true" high CPR areas (e.g., analysis of "m-δ" feature space; Raney 2007).

2.2 Concept of Operations

The Mini-RF concept of operations is shown in Fig. 3. Mini-RF has been allocated limited observation time so as not to interfere with collection of mapping data from other spacecraft instruments, particularly LROC imager, which like Mini-RF produces large amounts of data. During the normal mapping phase of LRO, Mini-RF is allocated two 10-minute collects to conduct communications demonstrations. Additionally, Mini-RF is allocated a 4-minute SAR collect and a 4-minute interferometry collect every month of the nominal mission. The designations "L" and "R" refer to the radar looking either "left" or "right" of the LRO orbital track. It may be possible to collect additional observations when the LROC imaging illumination is unfavorable. An imaging "season" refers to this period. The current plan is to identify high-value geological targets from the Forerunner Mini-SAR (Chandrayaan-1) systematic mapping and observe these areas using both the dual frequency of the LRO Mini-RF and the high-resolution zoom mode for specific areas. These data collects should allow better characterizations areas of high CPR, their geological settings, and the likelihood of a water ice composition.

By the end of the nominal LRO mission, Mini-RF will have collected regional S-band SAR maps from Forerunner Mini-SAR on Chandrayaan-1, 20 targeted SAR strips from Mini-RF on LRO, and correlation with higher resolution neutron and other data from LRO. Mini-RF could use observation opportunities from an extended LRO mission to acquire more data of both polar and non-polar regions. These data collection opportunities are currently defined as supplemental measurement goals, including additional SAR and interferometry data, additional communications experiment data, and bistatic options with the

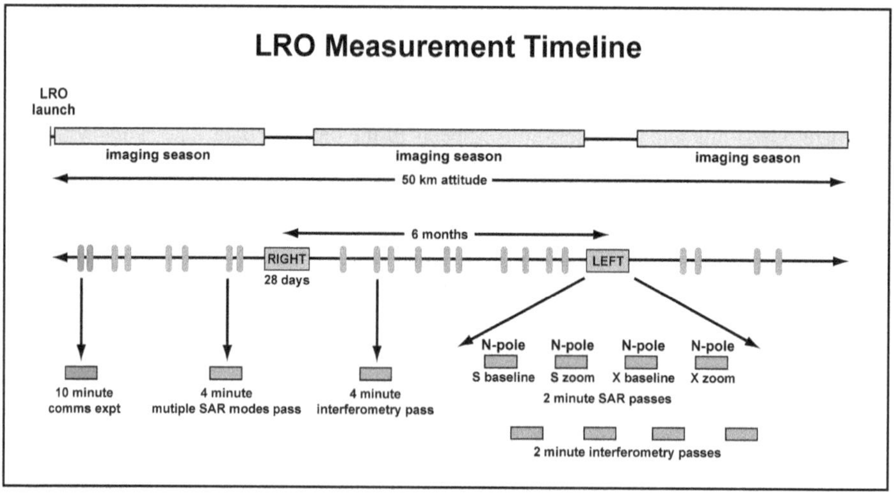

Fig. 3 An illustrative example of a possible LRO measurement sequence

Chandrayaan-1 instrument. High resolution "zoom" SAR imaging is also planned in support of the LCROSS experiment. Because both Chandrayaan-1 and LRO will be in lunar orbit at the same time, the Chandrayaan-1 and LRO Mini-RF units were designed to operate in a bistatic mode, with either Chandrayaan-1 or LRO transmitting and receiving in S-band. Both orbits cross at the poles and it may be possible to obtain bistatic measurements of selected polar targets to determine if high CPR is correlated with the bistatic (phase) angle. These observations, if possible, would be the most definitive in separating the effects of surface roughness from those of ice owing to the directional response of CBOE scattering from ice. The bistatic opportunities will require maneuvers to place both spacecraft near the same altitude so as to minimize the bistatic angle ($\leq 5°$).

3 Communication Technology Demonstration

The demonstration of Mini-RF in communications mode is a major objective of the LRO technology. By exercising the Mini-RF technology in this manner it may be space-qualified for future communications applications. While the data rates achieved are modest (85 kbs), these could be improved in future systems optimized for communications as shown in Table 1. Existing systems while capable of much higher data rates are heavier and not capable of multiple frequency multimode operation within the mass budget of Mini-RF (Table 1). Some of the communications objectives can be achieved with ground-based tests, but end-to-end far-field measurements under space radiation and thermal conditions, and space qualification of high voltage MPM technology, require flight testing. The Mini-RF frequency allocation (2380 MHz transmit and receive) is specified so as not to create interference with the LRO S-band communications system and is an approved planetary radar band. This frequency selection is compatible with Arecibo Observatory, which allows LRO to the Earth communications demonstrations. The LRO Mini-RF system will demonstrate half-duplex communication operation with ground assets during two 10-minute opportunities on Days 19–37 with additional times available during Days 38–57 of the mission. The demonstration opportunities are separated by minimum of 24 hours, followed by a 5-minute downlink demonstration, followed by a 5-minute uplink demonstration/calibration. In S-band (2380 MHz) uplink mode, the ground transmits a phase-modulated S-band signal; Mini-RF digitizes and records the raw data. The telemetered data is post-processed to achieve carrier lock and demodulation from Arecibo 2380 MHz transmissions by using a predefined waveform at 5 minutes per demonstration. In the downlink mode the Mini-RF phase modulates the transmitted signal while ground receivers perform carrier lock, bit sync, and data detection. The bit error rate is evaluated as a function of Eb/No to establish overall communication system performance. In the communications mode, Mini-RF provides 5–10 W peak RF power from its transmitter. The duty cycle is a function of input voltage and power consumption.

The transmit demonstration is limited to approximately 5-minute duration due to spacecraft average power requirements. S-Band direct modulation, including modulation direct on carrier, of BPSK at 85 kbs on the radar's transmit frequency will allow measurements of the Bit Error Rate (BER) performance. The link can support 220 kbps to the APL ground station. Mini-RF can gather statistically significant data and operate the payload within the allocated 5 minutes. Additional opportunities enabled by using antenna side lobes in view of the APL ground station will also be identified. These tests will modulate at a much lower rate to close the link using the side-lobe antenna pattern while the spacecraft is nominally pointed toward the APL ground station.

During the communication experiments, Mini-RF digitizes and records raw data with 100% duty cycle with a range decimation factor of 16, compatible with an LRO Space Wire data rate <30 Mbps. The limit to uplink power is due to large-aperture antenna (Arecibo) not overdriving the receiver. A compatibility test is not required and the communications test will be performed with commercial test equipment. In addition, a demonstration modulation using pseudo-random sequence at S-Band with the APL Ground System is planned with a BPSK modulation at 85 kbps modulation directly on carrier based on a Generator Polynomial within the radar with a sequence length of 32,767 bits. This strategy allows many repetitions of the sequence during the demonstration. The polynomials and modulation formats are compatible with CORTEX-XL ground receiver used at the APL ground station and with most conventional bit-error rate equipment. Compatibility testing with the APL ground station has been conducted.

4 Mini-RF Instrument Description

The Mini-RF Instrument is comprised of the following elements: (1) antenna, (2) transmitter, (3) digital receiver/quadrature detector waveform synthesizer, (4) analog RF receiver, (5) Control Processor, (6) interconnection module, and (7) supporting harness, RF cabling, and structures. The LRO Mini-RF system functional block diagram is shown in Fig. 4. The Mini-RF layout is shown in Fig. 5 and its installation on the LRO spacecraft is shown in Fig. 6.

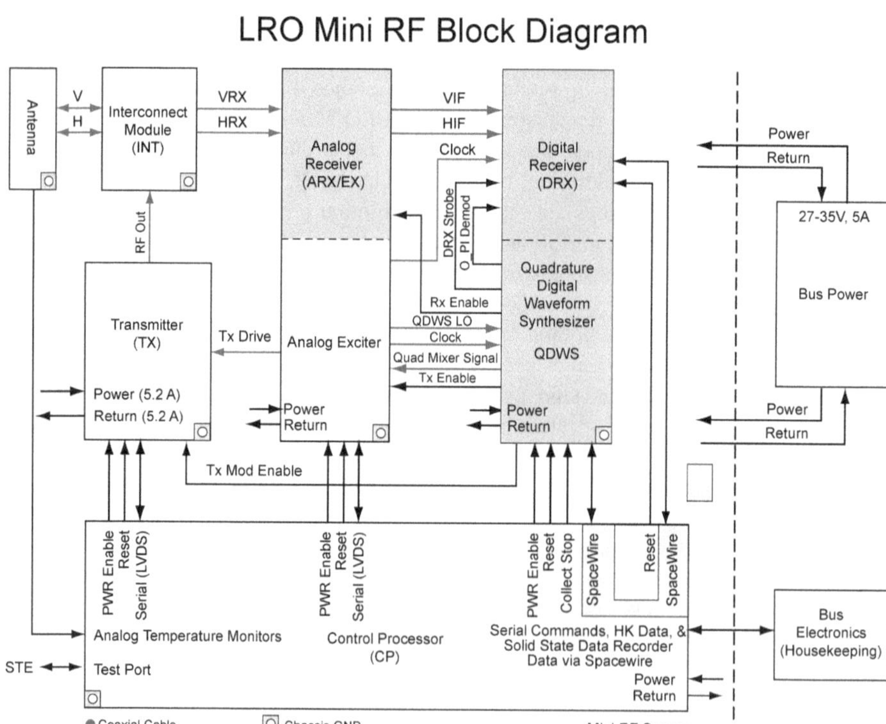

Fig. 4 LRO Mini-RF block diagram

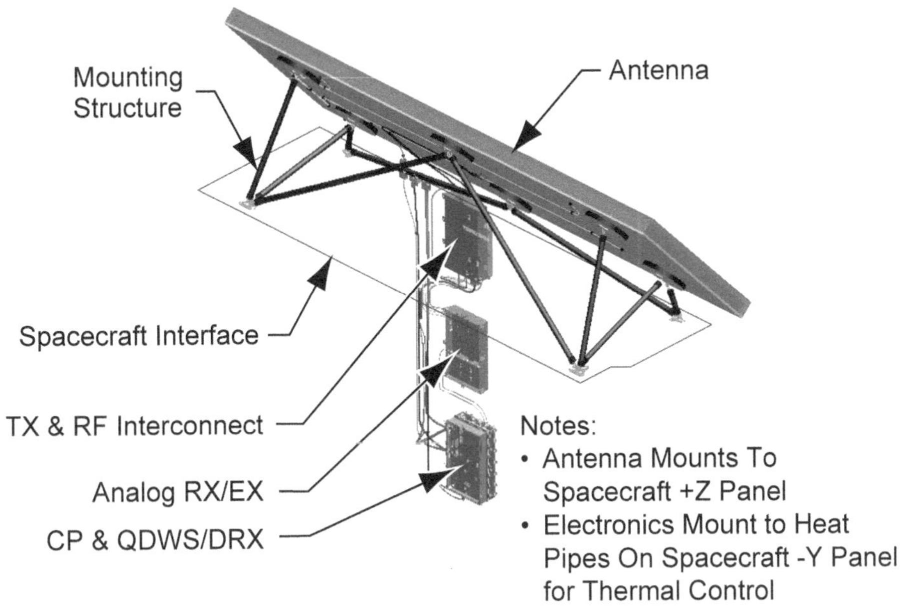

Fig. 5 LRO Mini-RF mechanical configuration

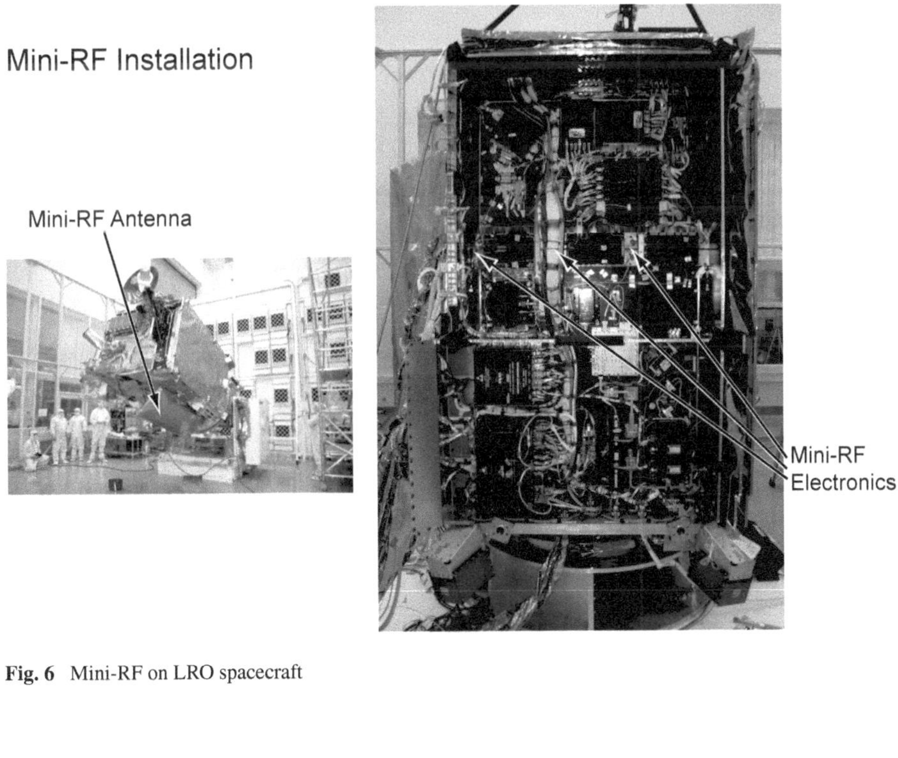

Fig. 6 Mini-RF on LRO spacecraft

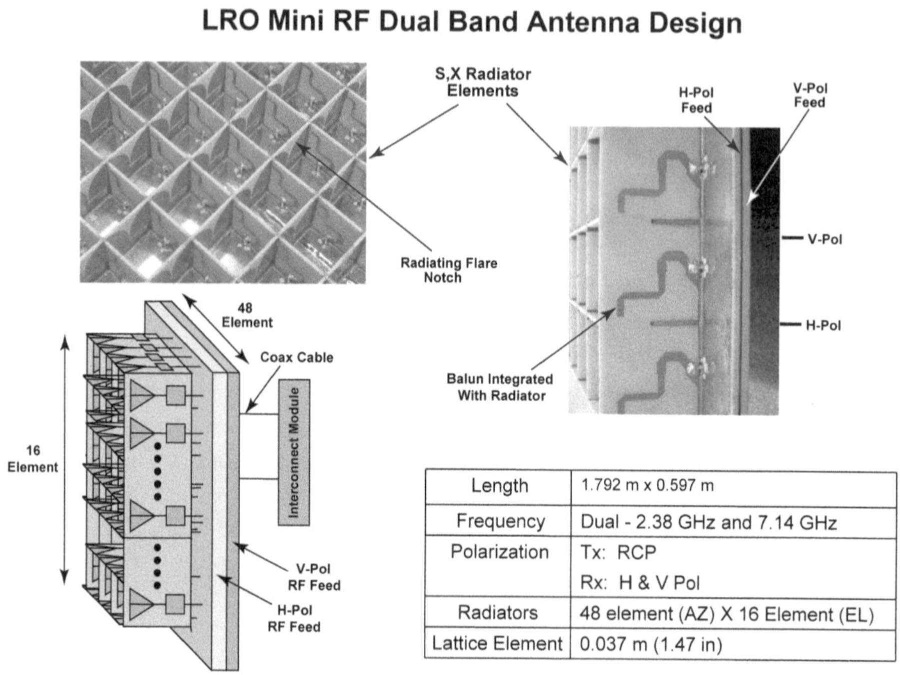

Length	1.792 m x 0.597 m
Frequency	Dual - 2.38 GHz and 7.14 GHz
Polarization	Tx: RCP Rx: H & V Pol
Radiators	48 element (AZ) X 16 Element (EL)
Lattice Element	0.037 m (1.47 in)

Fig. 7 Mini-RF antenna design

4.1 Antenna

An "egg crate" antenna (Fig. 7) allows a broadband, dual-frequency design with a single antenna panel, without any deployable mechanisms (e.g. feeds) while also meeting stringent weight and volume constraints. The elements are sized to allow a 3:1 frequency range. Each element incorporates radiators and physical phasing combines their power. The thermal design, materials selection, manufacturing, and test qualification heritage of single-frequency Chandrayaan Mini-SAR antenna was applied to the dual frequency LRO Mini-RF unit. Because of this heritage, the Mini-RF antenna is robust and lightweight (4 kg) while satisfying all technical requirements.

4.2 Transmitter

The LRO Mini-RF transmitter (Figs. 8, 9) takes full advantage of the capabilities of the wideband antenna. The transmitter is the first implementation of Microwave Power Module (MPM) technology on a long-duration spaceflight, which affords a significant breakthrough in available bandwidth and power efficiency with reduced mass as compared to previous traveling wave tube (TWT) systems (Fig. 10). The MPM combines a solid state RF driver/preamplifier with a traveling wave tube amplifier, a hybrid approach combining the advantages of both solid state and vacuum electronic technology. Flight-testing the MPM technology is a major goal of the Mini-RF demonstration. The MPM is enabling in giving Mini-RF its dual band capability within the challenging mass, power, and volume constraints of the LRO spacecraft. MPM technology has extensive heritage in airborne and other tactical systems but the Mini-RF development program had to make significant efforts to qualify

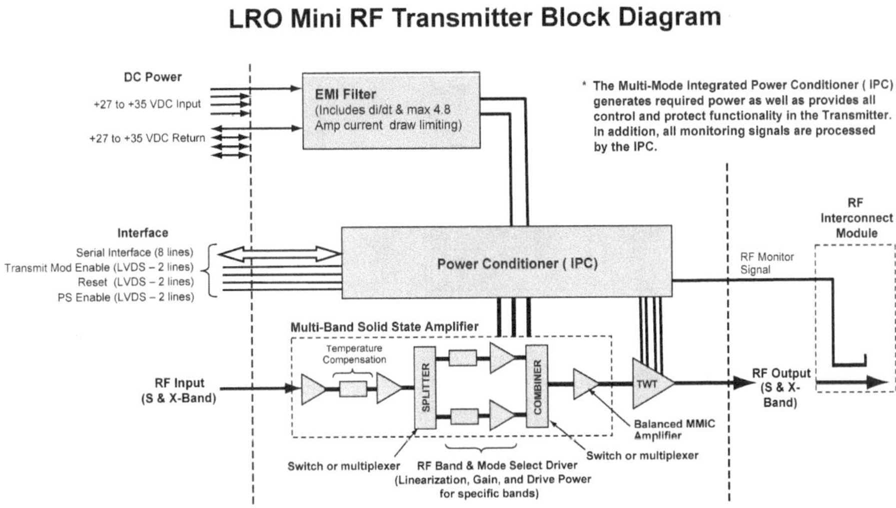

Fig. 8 Mini-RF transmitter block diagram

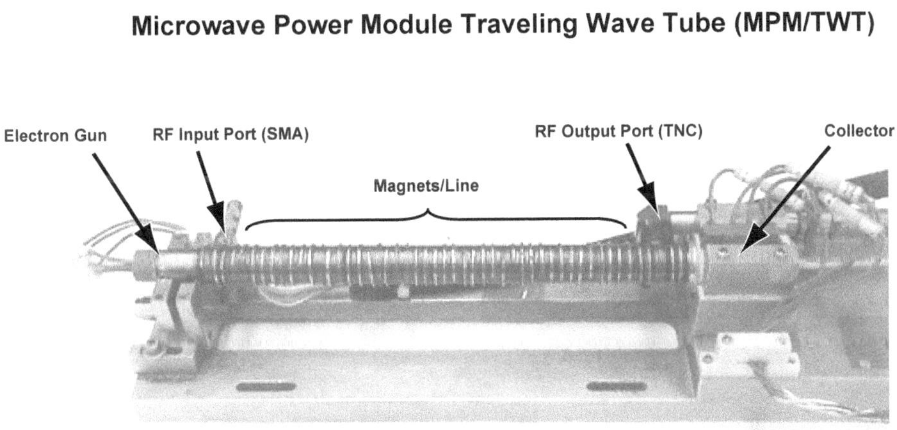

Fig. 9 Mini-RF Microwave Power Module Traveling Wave Tube (MPM/TWT)

it for spaceflight. The Mini-RF transmitter is based upon designs and processes developed over the last 15 years, with over 400 airborne units/year produced over the last two years. The primary Mini-RF transmitter challenge is adapting these proven airborne designs for space application. These include materials and part selection, outgassing, reliability, and radiation tolerance. The technical challenges overcome in the transmitter included developing low outgassing, high voltage insulators and space qualification parts screening for miniature high-voltage power supplies. The transmitter complied with the overall Mini-RF parts screening program which screened parts to a total dose of 20 kRad, no destructive latch-up, and tolerance of non-destructive latch up at 75 MeV. Meeting the stringent mass and power limitations required some parts to carry wavers but the overall parts' program was compatible with the Class S and LRO Class B requirement with de-rating criteria in accordance

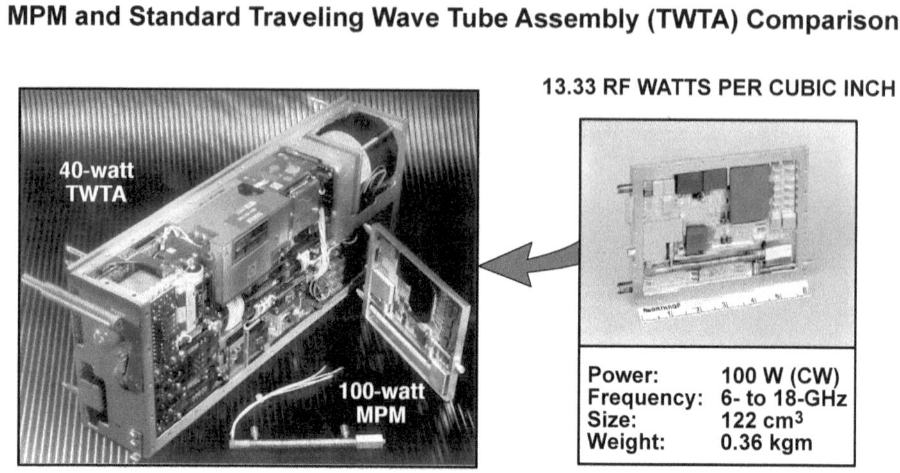

Fig. 10 Mini-RF transmitter module comparison

with established procedures. Mini-RF uses PEMs (Plastic Encapsulated Microcircuits) with the screening operable over the temperature range of −55 to +125°C. The MPM thermal design necessitated integration with the LRO heat-pipe system, which allowed for effective dissipation of heat generated by the transmitter.

Sandia National Laboratory (SNL) developed the Digital Receiver (DRX) and Quadrature Digital Wave Form Synthesizer (QDWS). These systems afford Mini-RF a degree of flexibility and reprogrammability not previously available in space-based systems. Over 1000 waveforms may be programmed. A processor module based on a heritage RAD 750 processor and associated firmware and software controls the Mini-RF system.

4.3 Interconnect Module

The Interconnect Module (IM) combines and splits the RF energy and serves as the interface between the transmitter, receiver, and antenna. Its design specifically handles issues such as multipaction using selected materials and geometry.

5 Calibration

Laboratory calibration data was acquired during spacecraft integration and test. The overarching goal of these activities was to insure production of a calibrated instrument. All waveforms in the waveform table were tested on brassboard hardware while selected waveforms were tested on flight hardware. This waveform testing is inherent in the overall Mini-RF integration and test program. Additional waveform testing was conducted on the flight instrument during thermal vacuum temperature ramp cycles. Internal calibration data are acquired every time that Mini-RF takes a data collect; a chirp, noise, and tone calibration is done both immediately before and immediately after a data collect. No end-to-end range tests were possible during integration and test, which necessitated the use of in-flight external calibration.

External calibration is planned in-flight by in conjunction with ground-based assets at the Greenbank and the Arecibo Radio Telescopes. These measurements will include polarization purity or axial ratio and antenna pattern. A transmitted signal from the LRO mini-RF is received by Greenbank while the antenna pattern is scanned over a range of angles. Specifically, the scan will be $\pm 12°$ from boresight in both elevation and azimuth, sampling at 0.5° increments. At each orientation, mini-RF will transmit for approximately 40 ms. Subsequently, each axis (azimuth or elevation) of the antenna will be parallel to the Earth's equator, with the boresight pointed towards Greenbank. The antenna will then be scanned parallel to the Earth's equator, at 0.4° per second 12° in one direction, then back to boresight, then 12° in the other direction, then back to boresight. During scanning, Mini-RF will transmit for 40 ms every 1.25 seconds, corresponding to an angle change between transmits of 0.5°. The scan should take approximately two minutes to complete. Next, the spacecraft will be rolled 90° so the other antenna plane is parallel to the Earth's equator, and the procedure will be repeated. The total time for the calibration should be approximately five minutes. An S-band received calibration will also be conducted using signals transmitted from Arecibo following the same geometry as the transmit calibration. An Arecibo X-band transmit calibration is under consideration. During commissioning, Mini-RF can acquire calibration data before the nominal science-mapping phase begins. At least two lunar equatorial areas will be imaged for calibration purposes most likely covering latitudes from 10° to −10° at 33° west longitude and 33° east longitude. Since each of these data acquisitions requires approximately six minutes and LRO Mini-RF may only transmit for four minutes, each equatorial target will be subdivided into two separate sequences. First, Mini-RF will image from 10°N to 2°S at 33° longitude on one orbit and then on the next orbit from 2°N to 10°S. These areas were chosen to have similar incidence angles as ground-based radar images collected over the same terrain to provide independent verification of Mini-RF measurements. The longitude of this second data take would be either 32° or 34°. Every six months there will be a window, centered around the dawn–dusk orbit, where calibration activities can be conducted, particularly calibration activities that require a spacecraft maneuver. During these times, Mini-RF can conduct the following calibration activities, in priority order; transmit calibration to Greenbank, two 1-minute nadir mode observations (ideally in a polar region), and a receive calibration from Arecibo using the same scan pattern used for the transmit calibrations. These calibration procedures have been tested successfully during the Chandrayaan-1 mission and experience gained will be applied to the LRO Mini-RF instrument. An example equatorial calibration image obtained by Chandrayaan-1's Forerunner Mini-SAR is shown in Fig. 11. An additional calibration is currently planned at X-band using the Goldstone Planetary Radar System. Radar data will be collected by Goldstone from the same lunar equatorial areas imaged by Mini-RF from lunar orbit at similar incidence angles.

6 Data Analysis, Interpretation, and Modeling

Following in-flight calibration, a set of coefficients will be derived and applied to the radar data as part of the standard processing stream (Fig. 12) prior to computing the Stokes parameters that in turn are used to calculate daughter products (e.g. CPR, degree of linear and circular polarization). A joint Chandrayaan/LRO Mini-RF Payload Operations Center (POC) is located at APL to support the Mini-RF experiments. The POC will provide the following functions: forward data acquisition sequences to GSFC, receive raw telemetry from GSFC, process raw telemetry, produce mosaics, and catalog data for the PDS and other repositories. The POC team is comprised of the POC lead engineers, Science Team representatives and the POC engineers. Figure 13 shows a high level view of the Mini-RF ground system and POC subsystems.

The POC will submit, to Goddard Space Flight Center (GSFC) for LRO, the command sequences that control the execution of the data acquisition and storage to the onboard flight

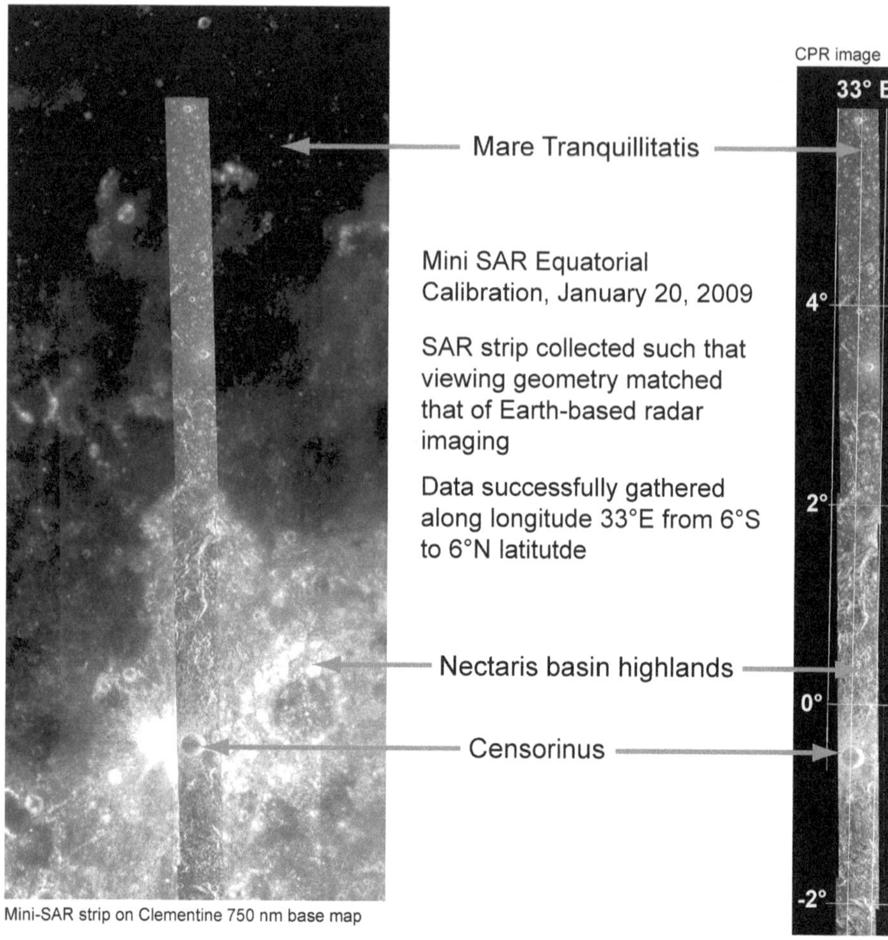

Fig. 11 Chandrayaan-1 flight calibration image

recorder. The Calibration and Collect commands have embedded argument lists that configure the radar with respect to waveform, pulse repetition frequency (PRF), pulse width, burst time, burst duty factor, number of bursts, position of receiving gate, bandwidth, start frequency, and other supporting parameters.

Once the acquired data and the required ancillary data have been received by the POC, the data will be processed according to the data type. This processing will use the ephemeris data supplied to the POC to create products to be checked for completeness and quality. All processed science products, science and housekeeping telemetry, ancillary products and command files are archived within the POC. Science products are also delivered to the Planetary Data System (PDS) for public release. The data processing architecture and calibration procedures to be used for LRO have been tested and validated using data from Forerunner Mini-SAR on Chandrayaan-1.

Acknowledgements We would like to acknowledge Susan Selkirk of Arizona State University School of Earth and Space Exploration for her support of graphics and related efforts in preparation of this paper. This paper is an LPI contribution #.

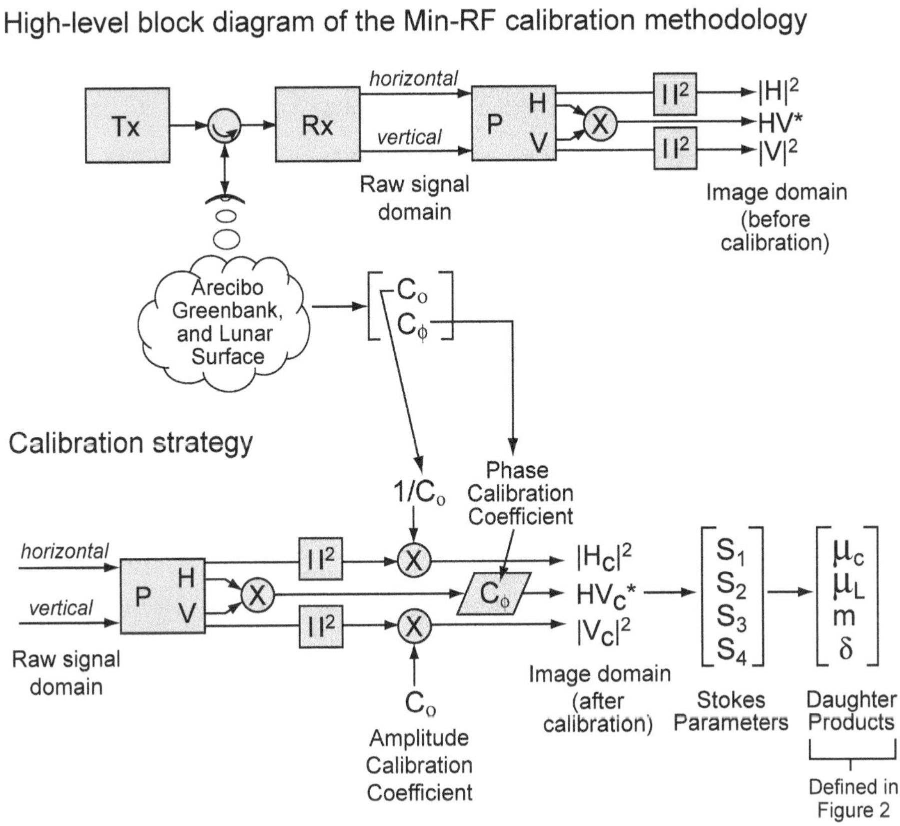

Fig. 12 Mini-RF calibration strategy

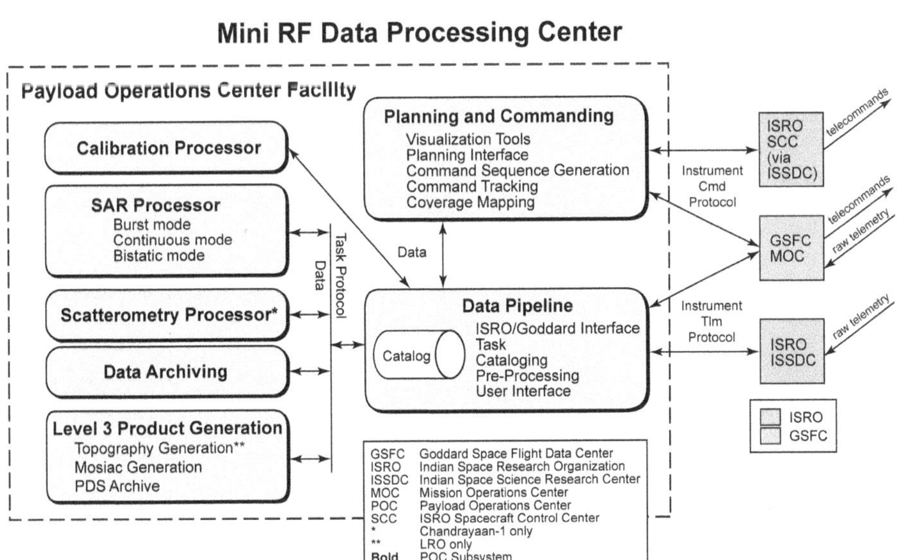

Fig. 13 POC overview

References

J.R. Arnold, Ice in the lunar polar regions. J. Geophys. Res. **84**, 5659–5668 (1979)

B. Butler, The migration of volatiles on the surfaces of Mercury and the Moon. J. Geophys. Res. **102**, 19,283–19,291 (1997)

D.B. Campbell, B.A. Campbell, L.M. Carter, J.-L. Margot, N.J.S. Stacy, No evidence for thick deposits of ice at the lunar South Pole. Nature **443**, 835–837 (2006)

R.C. Elphic, V.R. Eke, L. Teodoro, D.J. Lawrence, D.B.J. Bussey, Models of the distribution and abundance of hydrogen at the lunar South Pole. Geophys. Res. Lett. **34**, L13204 (2007). doi:10.1029/2007GL029954

W.C. Feldman, S. Maurice, A.B. Binder, B.L. Barraclough, R.C. Elphic, D.J. Lawrence, Fluxes of fast and epithermal neutrons from lunar prospector: evidence for water ice at the lunar poles. Science **281**, 1496–1500 (1998)

W.C. Feldman, D.J. Lawrence, R.C. Elphic, B.L. Barraclough, S. Maurice, I. Genetay, A.B. Binder, Polar hydrogen deposits on the Moon. J. Geophys. Res. **105**(E2), 4175–4195 (2000)

J.K. Harmon, M.A. Slade, Radar mapping of Mercury: Full-disk Doppler delay images. Science **258**, 640–643 (1992)

M.I. Mishchenko, Polarization characteristics of the coherent backscatter opposition effect. Earth Moon Planets **58**, 127–144 (1992)

H. Noda, H. Araki, S. Goossens, Y. Isihara, K. Matsumoto, S. Tazawa, S. Sasaki, N. Kawano, S. Sasaki, Illumination conditions at the lunar polar regions by Kaguya (SELENE) laser altimeter. Geophys. Res. Lett. **35**, L24203 (2008). doi:10.1029/2008GL035692

S. Nozette, C. Lichtenberg, P.D. Spudis, R. Bonner, W. Ort, E. Malaret, M. Robinson, E.M. Shoemaker, The Clementine bistatic radar experiment. Science **274**, 1495–1498 (1996)

S. Nozette, E.M. Shoemaker, P.D. Spudis, C.L. Lichtenberg, The possibility of ice on the Moon. Science **278**, 144–145 (1997)

S. Nozette, P.D. Spudis, M. Robinson, D.B.J. Bussey, C. Lichtenberg, R. Bonner, Integration of lunar polar remote-sensing data sets: Evidence for ice at the lunar South Pole. J. Geophys. Res. **106**(E19), 23253–23266 (2001)

R.K. Raney, Hybrid-polarity SAR architecture. IEEE Trans. Geosci. Remote Sens. **45**, 3397–3404 (2007)

R.A. Simpson, G.L. Tyler, Reanalysis of Clementine bistatic radar data for the lunar South Pole. J. Geophys. Res. **104**, 3845–3862 (1999)

P.D. Spudis, Ice on the Moon. Space Rev. (2006). http://www.thespacereview.com/article/740/1

P. Spudis, S. Nozette, B. Bussey, K. Raney, H. Winters, C.L. Lichtenberg, W.M. Marinelli, J.C. Crusan, M.M. Gates, Mini-SAR: An imaging radar experiment for the Chandrayaan-1 mission to the Moon. Curr. Sci. (India) **96**, 533–539 (2009)

N.J.S. Stacy, High-resolution synthetic aperture radar observations of the moon. Ph.D. dissertation, Cornell University, Ithaca, NY (1993)

N.J.S. Stacy, D.B. Campbell, P.G. Ford, Arecibo radar mapping of the lunar poles: A search for ice deposits. Science **276**, 1527–1530 (1997)

SPRINGER NATURE

GPSR Compliance

The European Union's (EU) General Product Safety Regulation (GPSR) is a set of rules that requires consumer products to be safe and our obligations to ensure this.

If you have any concerns about our products, you can contact us on ProductSafety@springernature.com

In case Publisher is established outside the EU, the EU authorized representative is:

Springer Nature Customer Service Center GmbH
Europaplatz 3
69115 Heidelberg, Germany